$17.50

# VENTURE SIMULATION IN WAR, BUSINESS, AND POLITICS

## ALFRED H. HAUSRATH

Formerly Chief, Military Gaming Division,
Research Analysis Corporation

The concept of SIMULATION, as a device for analyzing a wide variety of problems, is expanding into many fields today, becoming a particularly significant new tool in the military, political, and business worlds. The author describes the entire concept from its evolution to its processes and applications, all based on actual examples.

The book provides a "look behind the scenes," showing how simulation techniques have been developed and used to contribute to solutions of important problems in many fields.

Simulation can be used to explore the consequences of alternate plans in many ways and many fields without risking the costs of actual ventures. It offers a laboratory for experiment, under semi-controlled conditions, to study problems involving complex systems, patterns and consequences of cooperation, competition, conflict, and human decision. It also permits the tryout of new policies, strategies, and courses of action that would otherwise be prohibitively costly or even disastrous, if possible to study at all.

Until now, there has been no book remotely similar in scope. It is comprehensive, covering gaming of land, sea, and air warfare (foreign and American) as well as problems in management science, business enterprises, and political and strategic

*(continued on back flap)*

# ABOUT THE AUTHOR

Alfred H. Hausrath, former Chief, Military Gaming Division, Research Analysis Corporation, is the author of textbooks and many military studies and reports, and has directed team research on many vital military projects. He has conducted or directed studies in the Far East, Southeast Asia, the Middle East, Europe, and the United States on inter- and intra-national problems, including political-military affairs, strategic studies, civic action, cold war, war gaming, personnel management, and utilization, training, and staff development. He has been active in the fields of natural sciences, public administration, and personnel development, and has served as advisor, consultant, and staff member to the President's Advisory Commission on Education, the U.S. Armed Forces Institute, the Research and Development Board, the Military Assistance Program of the Department of State, and The President's Scientific Research Board. He is a Fellow of the American Association for the Advancement of Science, a charter member of the Operations Research Society of America, and a former staff member of Iowa State University and of The Johns Hopkins University. Now an independent consultant, he spent the last five years before his "retirement" as head of one of the largest sophisticated senior-level war gaming activities in the nation.

# Venture Simulation in War, Business, and Politics

**Alfred H. Hausrath**
Formerly Chief, Military Gaming Division
Research Analysis Corporation
McLean, Virginia

McGRAW-HILL BOOK COMPANY

New York   St. Louis   San Francisco   Düsseldorf   Johannesburg
Kuala Lumpur   London   Mexico   Montreal   New Delhi
Panama   Rio de Janeiro   Singapore
Sydney   Toronto

VENTURE SIMULATION IN WAR, BUSINESS, AND POLITICS

Copyright © 1971 by the Research Analysis Corporation. All Rights Reserved. Printed in the United States of America. No part of this publication may be reproduced, stored in a retrieval system, or transmitted, in any form or by any means, electronic, mechanical, photocopying, recording, or otherwise, without the prior written permission of the publisher. *Library of Congress Catalog Card Number* 72-136178

07-027230-1

1234567890 MAMM 754321

This book was set in Linofilm Caledonia by Progressive Typographers, and printed and bound by The Maple Press Company. The designer was Ernest W. Blau. The editors were Tyler G. Hicks and Stanley E. Redka. Stephen J. Boldish supervised production.

*To* GENERAL THOMAS T. HANDY, BRIGADIER GENERAL JOHN G. HILL, GENERAL JAMES E. MOORE, GENERAL CARTER B. MAGRUDER, GENERAL CHARLES D. PALMER, *all of whom are retired officers of the United States Army, and the many other military officers who, as counselors and friends, contributed to my understanding of the Art of War; and to the many professional colleagues who have advanced the arts of simulation and gaming, thereby helping to solve problems confronting the modern world.*

# Contents

Preface    xv

## 1. New Uses for an Old Technique .................    1

    The Study of War    1
        Games of War Started Early    3
        Birth of the True War Game: KRIEGSSPIEL    5
    What Is a War Game?    8
        The Function of a War Game    10
        Gaming, Innovation, and Creativeness    10
        War Games Are of Many Types    11
        New Emphasis on War Gaming    15
    Chapter Notes    16

## 2. The Role of Gaming .......................    18

    Gaming to Train    19
        The Early Period    19
        Gaming for Training Today    21
    Gaming to Test Plans    22
        Russian War Gaming    22
            April Games; August Guns    23
            Climax—Tannenberg    25
        German War Gaming    25
            Prussian Military Leadership    25
            The Ardennes Offensive: The "Battle of the Bulge"    28
            The German Navy Also Gamed    29
        Japanese War Gaming    30

United States War Gaming   32
British Gaming—Map Maneuvers, World War II   32
Allied Gaming and Operations   33
   Biggest Military Operation of All Time   33
   Planning and Testing of Combined Operations   34
The Arab-Israeli Wars   35
   Masters of Lightning War   35
   A New Classic in Military Operations   36
   Israeli War Gaming   37
Gaming for Research   37
   First U.S. Operations Research Group   39
   Operations Research Case Study: Mine Warfare
      Against Japan   40
   A Classic Example   43
Military Participants and Computer Support   44
Gaming of New Concepts   44
   Nuclear Weapons   45
      Doctrine for Tactical Nuclear Weapons   45
      Doctrine for Strategic Use of Nuclear Weapons   45
   The Airmobile and Air-assault Concepts   47
   Vietnam and the Airmobile Division   47
Chapter Notes   49
The Japanese Attack on Pearl Harbor, War Gamed in
   Tokyo (Photo Folio)   53

## 3. Operations Research and Analytical Gaming ...... 61

Development of Gaming in Operations Research   61
Analytical War Gaming   63
   TIN SOLDIER   64
   The MAXIMUM COMPLEXITY COMPUTER BATTLE   67
The Rise and Spread of Analytical War Gaming   68
   Concept Crystallized   68
   Early Gaming at CORG, CONARC   69
   British Analytical Games   71
      The AORE Games   74
      The RARDE Games   75
      Comparative Gaming   78
   Extension of Analytical Gaming   78
   Proliferation of Games and Simulations   79

## 4. Essentials for Gaming ...................... 82

Rudiments of War Games   82
   Basic Requirements for War Games   83

Personnel    83
Facilities and Equipment    84
Standard Rules, Data, and Procedures    85
Description of the Situation    85
Periods in Game Sequence    86
Pre-play Period    86
The Playing Period    88
The Post-play Period    90
Costs of Wars, Maneuvers, and Games    91
Cost of Wars    91
Cost of Maneuvers and Exercises    92
Chapter Notes    96

## 5. Models: The Anatomy of Games .............. 98

Why Models Are Used for Simulations    98
Macro- and Micro-models    99
Modeling of Combat Operations    100
Mathematical Models    101
Mathematical Models of Battle    101
Available Data    102
Unavailable Data. Example: Determination of
    Firepower Scores    102
Other Input Data    103
Qualitative Data    104
Deterministic Models    105
Probabilistic Models    109
Probability and Random Numbers    110
Game Rules as Part of Game Models    112
Game Consistency    112
A Conceptual Model    113
Logistics and Mathematical Models    114
Terrain Analysis    115
Terrain Features and Military Operations    115
The Grid System    116
A Model of a Battle    117
Chapter Notes    119

## 6. Types and Characteristics of War Games ........ 123

Types of Games Classified by Techniques    123
Rigid Games    123
Free Games    124
Open and Closed Games    124
Hand-played and Computer-assisted Games    126

Map Maneuvers   127
Types of Games Classified by Purpose   130
  Training Games   131
  Analytical Games   132
  Analytical Games and Their Uses: INDIGO;
      TACSPIEL   132
  Games for Small-unit Operations   136
    CARMONETTE III   137
    MINIGAME   138
  Tactical-level Games   142
    The United Kingdom and Canadian War Games   143
    U.S. War Games   144
    SYNTAC as an Example   145
    The MARINE CORPS LANDING FORCE WAR GAME,
        Another Example   146
  Division to Theater-level Games   147
  Other Types of War Games   149
  Extent of War Gaming   150
  Chapter Notes   151

## 7. Short-cut and Special-purpose Games .......... 152

Time-saving Gaming Techniques   152
  Quick Gaming   153
  Super-Quick Gaming   154
  Computerized Quick Gaming   155
  SCHNELLSPIEL: A Seventh Army Planning Game
      in Europe   157
Gaming for Combat Readiness: An Eighth Army Operational
    Game in Korea   157
A Large-scale Coordinated Logistic Study   159
  Study Objectives   159
  Component Models and Relationships   160
  The QUICK GAME   160
  The SIGMALOG Model   162
  The Coordinated System   164
Theater Staff Games   165
TBM-68, A Family of Manual War Games   168
  The Theater War Game Model   168
  The Theater QUICK GAME   169
  The Division Operations Model   169
  The Amphibious Warfare Model   170
  The Counter Guerilla Warfare Model   171
  Summary   171

1. TACSPIEL   172
    A Division-level Computer-assisted Analytical War Game   172
2. THEATERSPIEL   179
    A Theater-level Computer-assisted Analytical War Game   179

## 8. Business Games and Management Simulations .... 188

Evolution of Management Science   188
   Professionalizing of Business Education   189
   Techniques in Business Education   190
Management Simulations Get Started   191
Business Games Begin   194
Business Games Flourish   197
Similarities with War Games   202
What Happens in a Business Game   204
   How Business Games Are Played   207
   Steps in Building a Business Game   209
General-management Games   210
   The UCLA Executive Decision Games   211
   The Carnegie Tech Management Game   212
   INTOP   212
   Small-business Games   213
Functional Games   214
   The In Basket Exercise   215
   UCLA's Inventory Game   215
   Inventrol   216
   Maintenance Management Game   216
   DISPATCH-O Game   216
   MIT Marketing Game   216
   The Airline Sales Game   217
   Operation Federal Reserve Game   217
   SOBIG   217
Some Concluding Thoughts   218
   Maximizing Learning   218
     How Specific?   220
   Other Considerations   220
   In Summary   221

## 9. Strategic and Political Games: An Instrument of National Policy ....................... 223

Strategic Matters and Political Affairs   224

Strategic Warfare    225
Strategic War Games    226
Political Games    229
Political Gaming in the Universities    230
Inter-Nation Simulation (INS)    230
The MIT POLEX Games    232
Political Gaming in Research Organizations    232
Industry-sponsored Research    234
The Douglas Threat Analysis Model    234
Political-Military and Crisis Games    242
FAME, POMEX Games    243
Gaming Guerrilla and Insurgency Operations    244
The Counterinsurgency Game    251
The THEATERSPIEL Cold War Model    253
The TACSPIEL Guerrilla Model    261
Strategic Games and Simulations    264
Free-play Strategic Games    264
Rigid Strategic Games    266
The TEMPER Simulation    266
Some Values of TEMPER    267
Constraints of TEMPER    268
Chapter Notes    269

## 10. Gaming in Retrospect: An Appraisal ............ 275

Limitations of War Games    276
Logic of Battle    277
The Fog of War    278
Doctrine and Obsolescence    279
Assumptions and Constraints    279
Assumptions    279
Constraints    280
Parity Assumptions    281
Simplifying Assumptions    282
Significant Assumptions    282
Probability and Statistical Assumptions    283
The Element of Chance    284
Human Decisions    286
Maneuvers versus Games—What Games Cannot Do    288
The Big Limitations    290
Computer Support    290
Impact of Limitations    290
An Overall Evaluation    291

## 11. Gaming in Prospect: A Look to the Future ........ 294

    Automation    295
        Data Banks    295
        Potential Data-bank Uses    296
    Improved Computer Models    296
    Simplified Terrain Analysis    297
    Simplification of Daily Input Data    298
    Programming Improvements    300
        Models Programmed for Data Extraction and Analysis    300
    New Technology    301
    New Fields for Gaming    302
    A New Era in Warfare    304
        New Challenges    305
        Peace Gaming    305
    The Search for an End to War    307

## A Gaming Glossary ............................ 309

## Selected Bibliography ........................ 321

    Nonmilitary Games and Simulations    322
    References on Business Games and Management Simulations    322
    References on Military and Political Games and Simulations
        and Related Source Materials    327

Subject Index    377

Index of Persons Active in Simulation and Gaming    390

Index of Games and Simulations    395

# Preface

The purpose of this book is to describe the nature and purpose of simulation designed for use in analyzing the probable results of decisions in risk-taking situations. Drawing upon extensive experience in war gaming and in applications to other fields, the book describes the general usefulness of simulations, and cites important decisions to which gaming has contributed. It tells how simulations and games evolved, how they are conducted, the input data they employ, the output data and results they generate, the kinds of problems to which they may be applied, the uniqueness of their capability, and their limitations.

Simulation has become a flexible and practical device which can be used to analyze a wide diversity of problems in many different fields. It can be employed to explore the consequences of alternative plans and actions without running the risks inherent in actual trial-and-error ventures. It offers a laboratory for experimentation under quasi-controlled conditions on problems involving complicated systems, cooperation, competition and conflict, and human decisions. It permits the tryout of policies, strategies, and courses of action which would be prohibitively costly, dangerous, or even disasterous, and in some cases impossible to study adequately by other means. It has been used as a pre-testing technique in war, business, politics, and diplomacy and is spreading rapidly into other fields such as education and staff development. In the military where it all began the process has evolved from classic board games like chess to analytical map maneuvers which are commonly called *war games*. In this book an effort is made to trace this evolution, indicate the processes and details involved in actual examples, and to suggest wider applications of this comprehensive exploratory and analytical tool. Games may bring out unanticipated responses, may indicate the broad probabilities of certain outcomes, and in some cases may yield quite precise quantitative measurements.

A simulation normally is built upon a model. The model is a simplified representation of a system or of an activity to be studied. A model provides for the essential elements and forces to interact and function in much the same way they do in the real-life system. A model may be a physical device such as a simple toy balloon or a scale model of a jet plane in a wind tunnel to study its behavior in wind currents. A model also may be a theoretical construct consisting of a set of data, rules, and mathematical formulas such as may be used to simulate a comprehensive business enterprise, or a diplomatic crisis or a military confrontation, or an economic model to enable a nation's economy to be studied.

Before a model can be designed it is necessary: first, to identify and collect the significantly relevant facts, elements, and principles and their limiting operational characteristics; second, to analyze and measure their interactions; third, to determine the spectrums of outcomes of such interactions; and fourth, to organize and synthesize all of these into a coherent interrelated system. Then the model can be built into the form of a working model, the actions of which match those in the real system. The model can be programmed for a computer if desired, but simpler models can be exercised by human operators.

Once the model is ready to use, any set of a wide range of circumstances can be specified as starting conditions, i.e., "givens" to be carried through in order to determine what outcomes will result. The model may provide for the injection of human decisions or not, as planned. Each set of "givens" can be formulated or considered as a set of hypotheses to be tested. Thus the simulation process provides a laboratory or testing ground to try out concepts before implementing them with the accompanying risk of greater expenditure of money, effort, lives, and destiny.

Employed in the form of human decision, man-managed simulations as in business and war games, the process has been alluded to as synthetic or forecasted history. In this form individual games vary widely in scope and purpose. A one-person, one-hour game may be used to provide a learning experience, or testing procedure, in school situations or in on-the-job training of workers. At the other extreme are several large games which may involve scores of high ranking executives, officials, or officers engaged full time for a week or two, or even for many months.

## ACKNOWLEDGMENTS

During the several years in which material for this book was being assembled the author drew upon a very large number of other peoples' work and experience. Each is credited where his contribution is specifically cited. There are a number of others whose contributions were so continuous or so pervading that it was impractical to mention them in the context. Therefore the author takes this opportunity to express his personal indebtedness and appreciation to all whose work or efforts contributed to this book in direct or indirect ways. The book was initiated at the suggestion of Mr. Frank A. Parker, President, Research Analysis Corporation, supported as a RAC Institutional Research project, and with Dr. George S. Pettee, Chairman, RAC's Research Council and RAC's Open Literature Committee, and Dr. Hugh M. Cole, Vice-president, RAC, and Generals Thomas T. Handy and James E. Moore as advisers.

The entire manuscript has been reviewed for military accuracy by Major General James G. Christiansen, USA(Ret.) and by General Charles D. Palmer, USA(Ret.), both of whom were long associated with war gaming after eminent military careers; and by Dr. Pettee, widely recognized as a political scientist. Portions of the manuscript were reviewed by Dr. Cole, a leading military historian, and by Dr. Charles A. H. Thomson, a political economist of note, who succeeded to Dr. Pettee's responsibilities after the latter's retirement. In addition, many other specialists reviewed those portions which refer to their own fields of activity, most of whom have been mentioned in the text or notes. Brig. General John G. Hill, who was closely associated with the author during most of his gaming experience, was a constant source of expert advice and was largely responsible for the development and leadership of one of RAC's most successful war games.

A large number of staff members of the Operations Research Office of The Johns Hopkins University (RAC's predecessor organization) and of RAC, particularly those in the Strategic, Conflict Analysis, and Military Gaming divisions while the author successively headed the divisions, were the real instrumental performers in developing the gaming art in those organizations. In like fashion those in other organizations, many of whom are named in the book, have been the architects of other important simulations and games. During the period of maximum gaming activity in ORO and RAC the unnamed active duty officers and civilian scientists of the sponsoring or cooperating units of the Department of the Army, the office of the Joint Chiefs of Staff, and of the Department of Defense were major contributors to the development of game models and their application to the urgent problems of defense and national security.

RAC staff units (particularly the Library, under Miss Margaret Emerson, the Art staff under Mr. Vaughn Jackson, and the Editorial staff under Mr. William J. Wood) facilitated and supported the author's efforts with their own expertise. Photographs and illustrations not otherwise credited were produced at RAC. Extensive manuscript assistance was provided by Mrs. Avonale L. Stephenson, and by Mrs. Pauline S. Matthews and Mrs. Margaret P. Whalen. The book was completed as a post-retirement activity of the author.

The whole manuscript was reviewed and approved for publication by the U.S. Department of Defense, the Departments of the Army, Navy, and Air Force, and by the State Department. Although the work of many others is cited in this book, the context in which their statements are used, the selection and organization of material, and the interpretations and opinions expressed, are those of the author alone and do not necessarily represent the views or opinions of any of the individuals or organizations mentioned above.

*Alfred H. Hausrath*

# 1 New Uses for an Old Technique

Man always has competed for food, for a mate, for domain, even for prestige. Life itself has been a continuing struggle to survive, to protect and provide for loved ones, to subdue natural forces, to overcome enemies, and to achieve cherished goals. Nothing has stimulated more effort, thought, ingenuity and resourcefulness than such ventures. And heretofore no collective or more meticulously organized efforts have been made than in preparation for and conduct of war.

## THE STUDY OF WAR

Like the sword over Damocles' head, war threatens the very existence of nations even in today's world. All advances of civilization, science, and communication have found no means of removing the peril that may engulf any nation in unsought and unwanted war. The State of Israel is a prime example. The only alternative man's genius has been able to devise is the development of ever more powerful military forces and weapons for self-defense in event of hostilities or for use as a deterrent to war. (1-1)[1]

How can any nation be secure against the possibility of being overwhelmed by an ambitious and powerful adversary? The lessons of history have taught that a poorly armed nation invites attack. As no reasonable precaution can be omitted in matters of national security,

---

[1] These numbers refer to Chapter Notes listed at the end of the chapter.

military budgets have become the major portion of the cost of governments. The practice has been to develop one's military power to such a degree that a prudent enemy will not initiate war against it. But with the power of modern weapons and the element of surprise, even a lesser power, by striking first, can paralyze another much greater than itself. To offset the various threats of possible military attack and be prepared for such contingencies, all major powers engage in acquiring intelligence on possible adversaries, develop advance plans to meet each potential threat, and use diplomatic channels to avoid conflict and to form and maintain strong alliances.

Even such efforts leave some uncomfortable uncertainties. Are the plans workable? Can one nation meet with success an attack by another at a certain time and place? Can an initial blow be struck in such strength that a nation could not recover in time to defeat the aggressor? These and a thousand other similar questions are the concern of national leaders and those who are entrusted with national security. The answers to such questions are too vital to be left to judgments alone. Are there more reliable means of getting answers? There are. War gaming is one such means.

In recent years war-gaming techniques have provided a way to tackle elusive problems in business, management, education, politics, international relations, diplomacy, arms control, disarmament, peace, and other deep concerns of our times. Gaming is a means of studying activities in which alternative courses of competitive or variable actions are possible.

Some of what man learned in his preoccupation with war has spread into his more peaceful pursuits. This book endeavors to present a composite of such learning and to indicate how it carries over into man's nobler activities: balancing belligerency with reason, destruction with construction, war with mutual welfare. No small part of this process has been related to man's learning techniques of analyzing and managing complex enterprises.

Effective management of collective enterprises started—or at least developed—with military leaders in the serious business of working out military problems. Much of the process of developing battlefield tactics and military campaigns centered about *simulation techniques* in such forms as field maneuvers and the portrayal of troop deployments and movements on maps or terrain models. These representations of military conflict came to be called *war games*. Although called "games," no frivolity is involved. The process of gaming is a means of designing and using *models, computers,* and *simulation* techniques for practical purposes.

Consideration of the methodology itself and of details encountered in applying the techniques in military situations should contribute to an understanding of the processes involved. This understanding also should point the way to possible applications in other human affairs in today's world.

Games of War Started Early

Games of war, like warfare itself, predate written history. Artifacts found in tombs and the remains of long-deceased civilizations indicate formalized efforts to represent and manipulate symbolized military forces in mock warfare on playing boards. Miniature figurines and graphic representations of soldiers, war equipment, and playing boards have been found in Egyptian tombs and in archeological excavations in Greece and Asia Minor, Persia, and India. Games of the Checkers and Chess types are believed to have originated as games of war and have been handed down from antiquity in India, Iraq, China, Japan, and other countries.[1]

Rameses II, who reigned in Egypt from 1292 to 1225 B.C., is pictured on the walls of his palace at Thebes as engaged in a game of Draughts or Checkers. See Fig. 1-1. Plato and Homer mention the game some seven to four centuries B.C.[2] Chess, which represents a single battle in each game, is believed to have originated in China some 4,000 years ago and is reported in Chinese writings of about 1000 B.C.[3] The "national game of Japan," Go, which represents a whole military campaign, resembles Chess as a board game and is also rooted in antiquity.[4]

The next significant change in games of war seems to have been stimulated by the Renaissance in the fourteenth through sixteenth centuries, when attention was focused on the development of innovations and advances in the art of warfare. Leonardo da Vinci

---

[1] Checkers is a modern name for the older game of Draughts; the latter term is still used in Great Britain and several other countries.
Frank Dunne, *The Draughts-Players Guide and Companion*, pp. 9–20, Warrington, England, Frank Dunne, 1890.
Sir J. Gardner Wilkinson, *Manners and Customs of the Ancient Egyptians*, vol. I, pp. 190–193, London, John Murray (Publishers), Ltd., 1837.
H. J. R. Murray, *A History of Chess*, Oxford, Clarendon Press, 1913.
[2] *Encyclopedia Britannica* (1963), vol. 5, pp. 358, 457–461.
[3] Alice Howard Cady, *Go-Bang*, pp. 3, 4, New York, American Sports Publishing Company, 1896.
[4] Arthur Smith, *The Game of GO, The National Game of Japan*, p. vii, New York, Moffat, Yard & Co., 1908.

FIG. 1-1. King Rameses of Egypt, circa 1250 B.C., playing Draughts with a lady of his household. Carvings on wall of his palace at Thebes. Photo of wall as it stands today. (Courtesy of The Oriental Institute, University of Chicago. Upper left: drawing of carving published in Sir J. Gardner Wilkinson's book, A Popular Account of the Ancient Egyptians, vol. 1, p. 193, 1854.)

(1452–1519), "the supreme example of Renaissance genius," offered his services as a "military engineer" to the Renaissance prince Cesare Borgia[1] and to the court of Louis XII, King of France. In this capacity Leonardo, among his many other talents, spawned a remarkable series of astonishing military inventions and concepts never matched by any other man.[2]

This surge of attention to advances in the art of warfare induced similar efforts to make war games more realistic and to reflect the newer techniques of war. One milestone in this movement was the

---

[1] Son of Pope Alexander VI, Cesare was appointed a cardinal at the age of seventeen and was made Duke of Romagna at twenty-six.
[2] Among Leonardo's military inventions, portrayed in his notebooks, were improved designs of fortifications and weapons, including the concepts and practical designs for a flying machine, tanks, and submarines.

development of the King's Game by Christopher Weikhmann at Ulm in 1664. The King's Game was a modification of Chess which Weikhmann claimed embodied "a compendium of the most useful military and political principles." The game was very popular with the German nobility and was played for many years.[1] Another milestone was War Chess, developed by Helwig in 1780 at the Court of Brunswick in Germany, where he was master of pages. Helwig developed the game to train his young royal charges for their future careers as officers in the army. The game was played on a board of 1,666 squares, each of which could be made to represent a particular type of terrain as well as lakes, villages, and similar realities. The game quickly spread to France, Austria, and Italy and was played with enthusiasm. These games seemed to stimulate a plethora of other games of the War Chess type and led to the next great advance.

Birth of the True War Game: KRIEGSSPIEL

The war game, as we recognize it today, originated in Prussia in 1811 and was designed for training purposes. Students of the history of war gaming generally credit Baron von Reisswitz and his son Lieutenant von Reisswitz with the origin of the modern war game.[2] Although other forms of esoteric and rudimentary war games had been in use, work on them was interrupted by the Napoleonic Wars (1805-1815) and was not resumed with any degree of professionalism until Baron von Reisswitz, the civilian War Counselor (Herr Kriegs- und Dömanenrath) at the Court of Breslau in Prussia, made some major adaptations. Herr von Reisswitz transferred the earlier type of war game from the chess board to a sand table on which the terrain was modeled to a scale of 1:2,373.[3] This game was called to the attention of Prussia's King Frederick William III, who encouraged its development and use. This led Herr von Reisswitz to develop a finished plaster model that displayed terrain features in relief and

---

[1] Farrand Sayre, *Map Maneuvers and Tactical Rides*, 3d ed., pp. 5, 6, Fort Leavenworth, Kans., The Army Service Schools Press, 1908-1910.
John P. Young, *A Survey of Historical Developments in War Games*, pp. 7, 9-11, Bethesda, Md., The Johns Hopkins University, Operations Research Office, ORO-SP-98 (AD 210865), March 1959.
[2] *Militair-Wochenblatt*, no. 402, 1824; nos. 56 and 73, 1874; *Allgemeine Deutsche Biographie*, vol. 28, pp. 153-154, Leipzig, 1889.
[3] *Ibid*. Also, H. O. Heistand's translation, "Foreign War Games," in *Sources of Information on Professional Subjects*, p. 244, [U.S.] War Department Bulletin, 1898. Also referred to by Totten, p. viii; Young, ORO-SP-98, p. 15; and Anderson (all listed in the Selected Bibliography).

represented water courses, roads, villages, and woods in color. Troops and weapons were represented by little porcelain cubes. Pleased with the game, the Prussian King arranged for its introduction at Potsdam where matches were organized and played. The game became something of a showpiece to senior officers and visiting foreign dignitaries. In 1816, when Grand Duke Nicholas visited Potsdam, the King demonstrated the game to him. The Duke became enthusiastic and introduced the game into the Russian court the following year. (1-2) Interest in this gaming concept soon spread to other European countries.[1]

Herr von Reisswitz's introduction of realism represented by three-dimensional terrain gave war gaming the impetus it needed and nudged it in the direction of realistic modernization even though the rules of the game were carried over from War Chess. It remained for his son, George Heinrich Rudolph Johann von Reisswitz, to become the real innovator of the modern war game (Fig. 1-2) The younger von Reisswitz, a First Lieutenant of Artillery in the Prussian Guard, was encouraged by the interest in his father's efforts and enthusiastically applied himself to the development of the game. His purpose was to bring into the game the best military experience and thinking of the time. He eagerly pursued this task, drew up rules based on actual experience, and endeavored to popularize the game among his companions. He conceived the idea of adapting the game to actual military operations and is credited with the development of the game in usable form.

Lieutenant von Reisswitz, while stationed with the Second Artillery Brigade at Stettin, developed the game to the point where, by 1816, several matches had been played. He demonstrated the suitability of the game for study of a battalion in action as well as its use for the observation of operations of several army corps.[2] The game met with considerable favor among officers in the German Army.

In 1824, at the urging of subordinates, General von Müffling, the Chief of the German General Staff, consented to witness an exhibition of the game although he did so with an initial attitude of disinterest, if not disdain. As the operations portrayed on the map during the demonstrations unfolded, it is reported that the old General's face lighted up, and at last he exclaimed with enthusiasm, "Why, this is not a game, it is a veritable war school! It is my duty to recommend it to the

---

[1] Heistand, *op. cit.*, pp. 244–245; *Allgemeine Deutsche Biographie*.
[2] *Militair-Wochenblatt*, nos. 56 and 73, 1874; Heistand, *op. cit.*, p. 246; *Allgemeine Deutsche Biographie*.

FIG. 1-2. Lieutenant Johann von Reisswitz, the younger, who developed KRIEGSSPIEL to reflect actual military operations, the first true war game. (Photo source: the author.)

whole Army." He lost no time in doing so, and the modern war game won a foothold in the military profession as a training and planning device. (1-3)

At Berlin in 1824 Lieutenant von Reisswitz published a manual containing a detailed set of rules for his game under the title "Instructions for Representing Military Maneuvers by the Apparatus of the Kriegsspiel." With these rules and directions, the game was to be conducted on map-like charts showing terrain features. It was claimed that any military situation could be represented. The manual was followed by supplements and revisions in 1825 and 1828,[1] and Lieutenant von Reisswitz was called upon to conduct demonstrations for the crowned heads of Europe. The prestigious German professional military journal, *Militair-Wochenblatt*, as early as March 6, 1824, publicized the game with flattering comments. Fifty years later the same publication, under dates of July 11 and September 9, 1874, reported the history of the origin and development of the game by the von Reisswitz father-son team.[2]

As so often happens, this success of young von Reisswitz aroused so much envy and resentment among his contemporaries that he was transferred to an isolated border fortress at Torgau. He became so

---

[1] *Militair-Wochenblatt*, nos. 56 and 73; Totten, pp. ix, x.
[2] *Militair-Wochenblatt*, nos. 56 and 73; *Allgemeine Deutsche Biographie*, p. 154; Heistand, *op. cit.*, pp. 244, 249, 256; and Sayre, *op. cit.*, p. 8.

disheartened by the injustices he suffered as a result of these jealousies that he committed suicide there in 1827.[1]

Between 1827 and 1869 various officers attempted to simplify and update the von Reisswitz game but it remained for successes of the Prussian campaigns of 1866 against Austria and in the War of 1870–1871 with France to create a full realization of the game potential. Starting about 1872, KRIEGSSPIEL, in a variety of forms, spread to practically every major nation of the world, along with other innovations associated with the Prussian successes. Among these innovations was the vaunted Prussian General Staff type of staff organization.

What Is a War Game?

Many important military problems are amenable to analytical study. In fact, in World War II an entirely new profession was developed to make such studies. The profession, called *operations research* (recently also termed *systems analysis*), applies the techniques of mathematics and other sciences to the solution of military problems. One of the techniques or branches of this new profession has been developed into a rather sophisticated specialty of war gaming. See Fig. 1-3. This is not meant to imply that in earlier times military problems were studied with less care. The difference results from the development and use of improved methods of making the studies. War gaming has become a recognized technique in relatively recent times, but its essence has been a part of the pre-planning that military commanders have done through the ages.

Broadly considered, a war game may be any of several military activities, including a maneuver, a field exercise, a command post exercise (CPX), a map maneuver, a map exercise, or a certain kind of simulation. In a true war game human decisions always are involved.

The military services, keenly aware of the need for uniform usage of terminology, define a war game as "a simulation, by whatever means, of a military operation involving two or more opposing forces, conducted, using rules, data and procedures designed to depict an actual or assumed real life situation."[2]

---

[1] *Militair-Wochenblatt*, nos. 56 and 73; *Allgemeine Deutsche Biographie*, p. 154; Heistand, *op. cit.*, p. 249.
[2] *Dictionary of U.S. Army Terms*, AR 320-5, Headquarters, Department of the Army (D/A), April 1965.
Later edition, _____, AR 310-25, Hq., D/A, March 1969, p. 478, adds note that Department of Defense (DOD) and D/A delete "conducted."

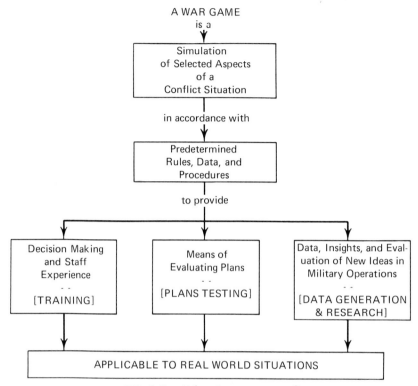

FIG. 1-3. What is a war game?

Simulated warfare provides a means of gaining experience, identifying errors or shortcomings, and improving skills without paying the penalties of the real world. Anything in the real world as costly as war impels a search for a substitute.

Terms used in war games are not fully standardized. The word "game" persists and, probably, will continue to be used. Accepting this fact, an organized group[1] of official war gamers of Canada, the United Kingdom, the United States, and Australia made a modest start toward a standardization of terms. The group set itself the task of formulating definitions for the more frequently used gaming terms (see Gaming Glossary) and concurred in the following: "Game.[2] A

---

[1] A four-country ad hoc Working Group on War Gaming established by the Quadripartite Conference on Army Operational Research.
[2] Definition officially approved by the English-speaking nations: the United States (including the U.S. Joint Services and the U.S. Army), the United Kingdom, and Canada.

simulation of a situation or conflict in which the opposing *players decide* which course(s) of action to follow on the basis of their knowledge about their own situation and intentions and their (often incomplete) information about their opponents." (Italics supplied.) This definition does not restrict the purpose(s) of the game and fits the various applications of games, whether used in connection with military, political, business, or management problems. It follows that a *war game* is a game representing military conflict.

War games as practiced in the operations research community are not regarded in any sense as an act of frivolity. Similarly, in the military forces of the principal nations of the world, war gaming is a serious matter. The training and insights that games provide, and the information they yield, are important to national security and even to national survival.

**The Function of a War Game:** War gaming is a systematic method of studying military problems. It provides a laboratory for the study of military operations. Actual warfare and actions available to an enemy and one's own forces are simulated. The game may duplicate existing conditions and capabilities or may focus on situations involving hypothetical forces, weapons systems, and equipment not yet developed. The locale of these postulated military operations may be any portion of the earth's surface for which adequate maps exist, or the games may be played on hypothetical terrain to simulate any set of conditions.

The terrain selected is not limited to areas controlled by the nation sponsoring the game, as must be the case in field tests and maneuvers that employ real military forces. Thus Red China could game a military takeover of Thailand, Malaysia, India, Pakistan, or any other area of the world without the cooperation or even the knowledge of the sovereign nations involved. This great flexibility and the adaptability of war games are an asset of extreme importance to military planners.

In like manner any nation may employ war gaming to assess or test military requirements and the contingencies with which military forces must be able to cope to ensure the security and survival of the nation.

**Gaming, Innovation, and Creativeness:** Possibly one of the greatest values in war gaming, or in the gaming technique when applied in other fields, is that it can provide an impelling stimulus to innovation, strong motivation, and a favorable climate for creativeness. Much of the progress of the world and of civilized man has come from creative minds. Yet there is precious little understanding of how creativeness may be stimulated and nurtured.

The gaming technique, first and most fully developed in war gaming, establishes an environment that challenges and motivates a responsible participant. He must bring all past learning and his most mature judgment to bear on analyzing the situation confronting him and then employ the best possible approach in meeting that situation. Moreover, he knows that he is matched against a competent and resourceful opponent, whose performance could make his own look somewhat less than favorable in the profession in which he wants to excel.

Games or map exercises have been used extensively to train officers in military forces throughout the world. Gaming challenges the competitive spirit and spurs the contenders to do their best in a given situation. It stimulates the search for new and more effective ways of meeting situations and encourages innovation. In these respects, motivations aroused in war gaming serve as incentives which, though less intense than those in real warfare, may have carryover values that will pay off in the ultimate test of actual combat.

Each opponent is impelled to seek out and recognize essential and critical elements of the situation and the limitations of resources. More importantly, each opposing participant is spurred, not only to employ the best of known past experience, but to originate, invent, and employ a still better or newer concept or innovation. Thus gaming is a seed bed for the germination of new ideas and a nursery in which to develop and nurture those ideas.

War Games Are of Many Types

War games, irrespective of type, have common features. They are two-sided; represent forces opposed to each other; provide for the employment and tactical movements of units in each force; include the clash of opposing forces in fire fights, battles, or even prolonged campaigns; and employ some graphic means of identifying units, weapons, and positions of forces engaged. Further, war games provide a system for taking into account the firepower and other capabilities of the forces and equipment involved and a means of assessing the effects of combat.

The conformation of war gaming may display great variance. In one guise, mock warfare is conducted with real troops using real equipment. Such field maneuvers can be quite large and expensive. Thousands of military personnel may participate in the battles and skirmishes, and the area of play may extend to hundreds of square miles. Army, Air Force, and Navy units ofttimes are employed in joint operations that cover great distances. A division and its equipment may be moved by air and sea from the continental United States to Hawaii or to Europe and then play through a mock war at the destination. A war game of this type is likely to cost millions of dollars. Or war games may be held as a training exercise or tactical field test and be conducted entirely within a military reservation. These smaller games usually involve a force of battalion size or less.

At the other extreme, war games are conducted by a few officers who use markers on a map or chart spread out on a table. Each of the opposing forces is assigned an officer as a Commander while another officer may serve as Umpire. In such cases troops and military equipment are not employed, and the costs are trivial. Other map maneuvers may be more complex and involve scores of people to operate the game even though troops and military equipment are not involved.

Then there is the automated game or *simulation*,[1] employing a mathematical model and, once started, not requiring players.

There is a characteristic physical setup for indoor war games. Each opposing Commander has his own operations center, i.e., his own "war room" where, with maps and other devices, he can lay out a model of the area over which he is fighting. In his Center he keeps a display of the disposition of his forces and such information as he can gain about his enemy's forces. Figure 1-4 shows a typical player-room setup.

Usually some form of Control Center is used. The Umpire or Controller may have a separate "war room" (a Control Center) where he keeps a composite display of all elements of the opposing forces, the positions, and the combat condition of each. The Center is presided over by an Umpire, Chief Controller, or a Game Director. Such a setup is shown in Fig. 1-5.

Usually, too, there are limits or restrictions imposed on the Commanders of the opposing forces. These restrictions are employed to impose such restraints as would exist in real warfare. For example, the distance a unit may move in a given time period must be equivalent to that possible in actual combat. In some games these restraints are left to military judgment and are employed as presented, or modified, at the discretion of the Umpire. In other games the guide-

---

[1] "Simulation" as used here is so defined by certain computer-oriented mathematicians who limit the term to applications completely represented by precise, mathematical quantities and interactions described by mathematical formulas that can be programmed and automatically processed in a computer without the intervention of human decision. An entire game or an input to a more comprehensive game may be so modeled.

New Uses for an Old Technique  13

FIG. 1-4. A typical player room. View of the Red game room in the TACSPIEL I series. Team of three players and Control technical advisor plan all operations, including artillery, engineer, and air support for one or more divisions normally played by the Red team. *Left to right:* Col. James B. Corbett, Marshall Andrews, Lt. Col. R. A. Montgomery, Jr., Lt. Col. R. B. Kadel. (1962)

lines, restraints, and means of assessing effects are spelled out in a set of predetermined, detailed rules.

In simple war games, employed to demonstrate a principle or test a concept, one map table or display may be used, with opposing Commanders and the Umpire all viewing the complete disposition of forces. One modification may be use of a curtain placed between the two sides of the map board to restrict each Commander's view of his opponent's layout. Figure 1-6 shows such a typical map-maneuver game board.

A universal feature of war games is the employment of a common timing system. First light, daylight, and nighttime are important factors in military operations, and the hour of the day is represented as it advances. Stop-action periods are another time factor given consideration. In this period Commanders consider "next moves," and the Umpire makes assessments and judgments. These stop-action

*14    Venture Simulation in War, Business, and Politics*

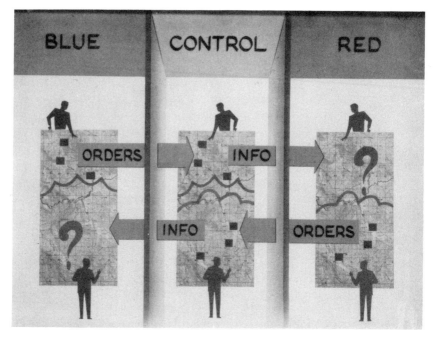

FIG. 1-5. A typical three-room game facility. Schematic diagram of game rooms and process of play. The Blue and Red players write mission orders to their units based on their analysis of the situation. The mission orders are transmitted to Control. Control reconciles the Blue and Red orders, determines what actions will result therefrom, makes assessments in terms of fixed rules and reference tables, and then passes back to the Red and Blue Commanders, respectively, the type of information they would get under actual combat situations. This setup of three rooms is characteristic of most war games in which human decisions are made, when two opposing forces are in conflict, and when a rigid, rule-governed type of game is conducted. (February 1966)

periods are, usually, for limited times, as they are in actual combat activities.[1]

Some variations in war games are not universal. For example, in some games a detailed record is made of each stage of play for later analytical studies; in other games such a record is not required. Another variation from real life is concerned with the question, Who won? Strange as it may seem, this question often is of little importance in the gaming of war. In this respect war games are unlike real warfare. War games are not "played for keeps" as wars are. They are often to

---

[1] Unlike Chess or Checkers, war games do not allow each opponent to alternate in making his moves. As in war, the operation is continuous.

teach or to train officers in some military principles or concepts or to ascertain what can or cannot be accomplished in an actual conflict. To "lose" can be an effective learning experience. When a new or proposed concept or a proposed weapon or weapon system is tested in a war game and proves unworkable or ineffective, serious and costly mistakes are avoided both in the cost of producing the weapons and the possible cost of lives in real war. More important than Who won? is the question What was learned? to lay a foundation for national security or even a bloodless path to peace without war.

New Emphasis on War Gaming

War gaming, however imperfect it may be in its present state of development, has become an important means of studying certain problems of modern warfare.

Military operations are too important to be planned and conducted without the most thorough consideration and use of all feasible means of pre-testing for weaknesses or more effective alternatives. Modern military powers employ war gaming as an extension of training, planning, and research activities. A nation seriously concerned

FIG. 1-6. A typical map-maneuver war-game board.

with military capabilities cannot afford to ignore a means that a potential enemy may use to advantage, nor to shun measures and studies that if neglected may prove disadvantageous, even disastrous, to its own security and national interests.

The McNamara policy in the Pentagon put great emphasis on achieving the greatest military effectiveness at the lowest practical cost.[1] The regime, although not different in objective from earlier administrations, approached the objective with vigorous methodology and increased aggressiveness in implementation. Searching and thorough studies and staff work were required, with firm deadlines, as inputs for consideration in the decision-making process. Characteristically, two questions were asked about studies submitted: (1) Has it been costed? and (2) Has it been gamed? In making important decisions much significance is attributed to the results of these two methodological approaches, in addition to the best of the other traditional techniques of rigorous analysis. (1-4)

With the potential of contributing to success and survival in shooting wars, war gaming can be an absorbing and worthwhile activity. To this end the remainder of this book goes into more detail on the various purposes for which war gaming and simulation have been and may be employed, how games have developed and are played, their features, values, and limitations.

## Chapter Notes

1-1. Among entities which have suffered disaster in recent times, without provocation on their part, are Mainland China, South Korea, Tibet, and South Vietnam. Arms control and disarmament efforts notwithstanding, the ever-present threat of war remains as long as no effective way exists to compel and assure compliance and to punish irrevocably such violations and aggressions as may occur. Both internation conflict and internal subversion need to be prevented and possibly policed. Internal subversion is the more difficult to counter; the ability to eliminate, or even suppress, it is perhaps more out of reach than elimination of internation conflict, at least in the near future.

1-2. *Militair-Wochenblatt*, the prestigious German professional military journal, appears to be a primary source of information on the origins and early development of KRIEGSSPIEL—The War Game. The

---

[1] In Pentagon colloquialism this policy is referred to as "more bang for the buck." In reality, dollar costs are only one element in considering the most effective security measures. The real criterion is military effectiveness.

journal's issue of March 6, 1824 is quoted in a note dated Feb. 25, 1824 and signed by Lieutenant General von Müffling, then Chief of the German General Staff, in which he attributes the origin of the game to the father, Counsellor von Reisswitz. He paid high tribute to the game; credited its further development to the son, Lieutenant von Reisswitz; and extolled the significant military contribution the game represented in the study of warfare. In the July 11, 1874 issue, pp. 527–532, the *Militair-Wochenblatt* carried an extensive review of the Reisswitz KRIEGSSPIEL over the signature of General Dannhauer.

1-3. Quoted from the articles in the *Militair-Wochenblatt* in July 1874 and *The Revue Militaire de L'Etranger* in August and October 1897; translated by Maj. H. O. Heistand, U.S. Army, and published under the heading, "Foreign War Games" in *Sources of Information on Professional Subjects*, p. 246, a War Department Bulletin, 1898.

1-4. Charles J. Hitch, *Decision Making for Defense*, pp. 27–28; 31–32, 43–58, Berkeley, University of California Press, 1965. This book reprints the first series of four lectures given by Charles Hitch, under the aegis of the H. Rowan Gaither Lectures in Systems Science, at the University of California. The third lecture (pp. 44–58) was on cost effectiveness and makes a strong point that dollar costs are not the final determining factor in security decisions.

 **The Role of Gaming**

Modern gaming originated with the Germans as a means of training military officers. The gaming of war proved worthwhile for this purpose. Next, war games were used to test tactical and strategic doctrine.[1] Then operational gaming was added as a means of developing and pre-testing operational plans. In these ways the Germans found war games contributed substantially to military success. Military gaming is now employed to serve three purposes:

1. Gaming to train military personnel
2. Gaming to test plans
3. Gaming for research, e.g., to explore new concepts

The use of games for each of these three specific purposes is, in many instances, clear-cut and well defined. In other cases there is a multiple purpose in that a game may be used for both training and the testing of plans. Although games for research usually are conducted to study matters associated with new concepts, they also include the evaluation of plans and/or the formulation of doctrine in the application of new concepts. Research games rarely are used for direct training for military purposes. The three major purposes of games not

---

[1] Although considerable initiative is left to the commander in *how* to accomplish his mission, it is presumed that he will do so in accordance with sound military practice. Such practice is represented by the "principles of war" and "doctrine" as taught in the nation's military schools.

only are interrelated but represent stages in the progressive development of war gaming.

There is evidence that war games of the field-maneuver type, using actual military forces were used in India as early as 500 B.C. Present-day uses of gaming for war show a remarkable correlation with these ancient applications. A case from India of unusual significance is cited.[1]

The Mahabharata, the great Indian epic, circa 200 B.C., recounts "a great mock battle fought which involved practically all the forces on the Indian subcontinent." The battle "lasted for 18 days and employed different types of battle arrays changed from day to day" to conform to the several types of battle array then in vogue. Among these were needle-shaped, heron-shaped, hawk array, crocodile-shaped, circular, and thunderbolt.[2] A close examination of the arrays "shows four kinds of movements were common: circular, crooked, separate and compact." (2-1)[3]

This case may be the earliest known example of a field-maneuver type of war game used for three implied purposes. The first, *training*, is not unique, as all maneuvers for thousands of years have served this purpose. The second, *field testing of tactical plans*, was not rediscovered until relatively recent times, i.e., within the last two centuries. The third purpose, *research*, was demonstrated by the comparative experimental trials of various tactics.

## GAMING TO TRAIN

The Early Period

From the earliest times, tribes, clans, city-states—and, later, sovereign nations—have organized fighting forces to overcome enemies and protect their own people and homelands. The profession of arms, one of the oldest specialized occupations of man, is unique in several ways.

First, the penalty for ineffectiveness may cost one's life, bring ruin and desolation—even annihilation—to one's family, community,

---

[1] The author is indebted to Dr. S. S. Srivastava, Director of the Scientific Evaluation Group, Defence Research and Development Organization, Ministry of Defence, Government of India, for the example. The implications of purposes are the interpretation of the author.
[2] These are the English-language equivalents of their Sanskrit/Pali terms, respectively: Suci, Kraunca, Syena, Makara, Mandala, and Varjra.
[3] These numbers refer to Chapter Notes listed at the end of the chapter.

country, and the civilization of which one is a part. It is probable that no other profession known to man exacts such a severe price and has such far-reaching consequences for lack of success. Moreover, few other professions require such risk of death and injury even when one performs effectively.[1]

Second, the profession calls for the development of skills which the individual may never be called upon to use. Even though few periods in history have been free of war, many professional soldiers have lived out full careers without being involved in actual combat. The number of wars in which nations have engaged over any extended period of time is a matter of wide interest and questionable statistics. Much depends on one's definition of war. One report lists forty conventional wars in the 20-year period immediately following 1945, as counted by the Pentagon.

One of the most widely quoted (but fictitious) summaries of statistics of war is attributed to the former president of the Norwegian Academy of Sciences, aided by historians from England, Egypt, Germany, and India, who reported "14,531 wars during 5,560 years of recorded history, an average of 2.6 wars per year, with only 10 of 185 generations of recorded human experience free of war." The military journals of a number of countries also published this report. The original report was fantasy, hypothetical in intent and plainly labeled as such when originally published—a fact which escaped attention in its requoting. See Chapter Note 2-2 for details of this "Great Hoax of War Statistics."[2] Although some members of every graduating class between 1805 and 1962 of the U.S. Military Academy (West Point) participated in shooting wars (if the Indian campaigns, Korea, and Vietnam are counted) (2-2), many career officers have missed personal participation in actual combat.

A unique characteristic of war is that the skills of the military profession must be learned vicariously. Most professional training provides for participation in or contact with the profession as it is conducted in

---

[1] As a hunter of predatory animals, early man, armed only with club, knife, or spear, ran similar risks to his own safety in his normal way of life. This type of personal risk carries over to persons engaged in only a few of today's hazardous vocations, such as that of astronauts, aircraft test pilots, wild-animal trainers, and high-wire acrobats.

[2] *Time*, "On War as a Permanent Condition," Sept. 24, 1965, pp. 30–31; *Military Review*, June 1960, p. 72; *U.S. Naval Institute Proceedings*, Nov. 1962, p. 51, quoting from the *Irish Defence Journal*, July 1960; F. L. Greaves, *Military Review*, Dec. 1962, p. 55; and Hanson Baldwin in *Army*, Aug. 1967, p. 38, to name a few. Greaves, at the U.S. Army Command and General Staff College in 1962, reported that in the 17-year period 1945–1962 "21 fair-sized wars" and "30 or more minor revolts and civil wars, and relatively small conflicts" occurred. (*Military Review*, Dec. 1962, p. 58.)

the real world, and practice of the profession involves continuing use of its skills in real-life situations. Not so in the art of war. The supreme and ultimate practice of the military art can be exercised only in actual warfare.

Military officers normally learn the skills of their profession in times of peace. Means of studying the art must involve substitutes for personal participation in actual war. Military schools can teach the principles and techniques of war, drawing upon historical accounts to provide examples. But the great battles of history are lessons of the past. Much of the weaponry, communications, equipment for transportation, and other products of technology become obsolete between wars. Moreover, historical records are typically incomplete and often may be misleading. Afteraction reports and reconstructed accounts usually suffer from limited observational opportunities, faulty memories, unavailability of key people, and lack of knowledge of details which influenced decisions and events. Records, though valuable, are an inadequate substitute for the realities of war, especially for wars as they will be fought tomorrow or 5 years hence.

Although war gaming was used sporadically in the nineteenth century, its role in planning and training was in the process of becoming firmly established. By the advent of the twentieth century, and particularly by World War I (1914–1918), war gaming as a training activity was well established and in some countries was conducted as a training procedure in the commands of active military units. More universally, war gaming was a part of the military school system.

Gaming for Training Today

The professional military man, and the nation he serves, must be concerned with the best possible training in order for soldiers to become proficient in the art of war. Means are needed to place trainees in situations similar to those faced in real war, and practice must be provided to increase those skills that will ensure effective tactical response to real war situations.

War games of the indoor type, used widely for training purposes, have many advantages and some disadvantages when compared with the outdoor type of field exercises. Advantages include the telescoping of time requirements, the opportunity to game situations for which a particular terrain is not available, precise control and management of the situation, greater exposure of students to a wide range of problems, and lesser costs. Disadvantages include lack of participation of lower echelons of command and enlisted personnel and the absence of the unique problems encountered when actual personnel and equipment are employed. Tin soldiers do not require food, sleep, medical attention, etc.

Other handicaps of gaming for training are game complexity, the slowly evolving application to practical aspects of some tactical problems, the development of a multiplicity of rules and tables of reference data, the lack of trained gaming directors, and the snail's pace at which games proceed. A search for ways to simplify the use of gaming continues. In one or more forms, war games are considered an important part of modern military training—not to replace but to augment other forms of study and experience; to provide a sense of reality, insights, and a depth of understanding of many problems of command and control in military operations.

War games supplement field exercises for training purposes. When the purpose is to study military problems such as tactical concepts, weapons mixes, and support requirements in the context of a present or future time period (or in an unavailable geographic area, or with force structures and/or weapons not yet in existence), and when these matters are to be studied in some semblance of an experiment or of scientific research, the war game offers opportunities unmatched by field exercises. (2-3)

## GAMING TO TEST PLANS

The use of war gaming to pre-test plans for military operations reached a significant level in World Wars I and II. Probably the most notable use of war games in modern military history was made by the Germans and Japanese who, in war games, developed and tested plans that led to unprecedented military successes. The following examples are drawn from the Russian, German, Japanese, United States, United Kingdom, and Israeli military forces to illustrate the use of war games and plans in several different situations.

### Russian War Gaming

Although references to war gaming in Russia are sparse in the open literature,[1] and few details are given, one significant example may be

---

[1] Several scholars in Russian political-military affairs were alerted to the author's search for references to war gaming. All read *Pravda*, *Izvestiya*, and more particularly the military journals including *Red Star* and *Voennaya Mysl* (Military Thought). They reported few cases in which war gaming has been mentioned and recalled such references as quite sporadic and inferential.

cited. That example comes from World War I. The gaming was to test Russian operational plans against German forces on the Russian Western Front in East Prussia.

**April Games; August Guns:** In April of 1914, prior to eruption of World War I,[1] the Russian General Staff had "played out the campaign in war games." The commanders participating in the games were the same commanders who led forces in executing the plans in August of 1914.

The Russian General Staff war games were presided over by General Sukhomlinov, who played the role of Commander-in-Chief (he subsequently became War Minister). General Rennenkampf, Commander of the First Army, and General Samsonov, Commander of the Second Army, also participated. The plan provided for the Russian First Army with some 200,000 men under Rennenkampf to engage the enemy while the Second Army under Samsonov delivered the decisive blow upon the German flank and rear. The two Russian armies then were to link in a common front in the area of Allenstein.

The war games, waged according to the plans, revealed the existence of a weakness that would prove fatal should the Russian Second Army start too late in an invasion of Eastern Germany. Under this condition of timing, Russia would meet the enemy with divided forces at an inopportune time and suffer decisive defeat. To link up on schedule at the Insterburg Gap, protected on the south by the Masurian Lakes, timing was critical. Samsonov's army would have to start some three days before the First Army, *an action not contained in the plans.* This change, so clearly indicated in the war games, was *never made in the plans or in their execution.*

The German General Staff also war-gamed this same general situation, as a means of formulating and testing another of the Schlieffen plans.[2] The German war games were predicated on Russian plans requiring a division of the attacking forces to move around natural obstacles for linkup beyond as the most probable Russian course of action. General Schlieffen's strategy was to attack the Russian armies

---

[1] Archduke Francis Ferdinand, heir to the Austrian throne, was assassinated by a Serb terrorist at Sarajevo, Bosnia, June 28, 1914. Austria declared war on Serbia July 28. This was followed by declarations of war by Germany against Russia, August 1, and against France, August 3, and Germany entered Belgium the same day; Britain declared war against Germany on August 4, and Japan against Germany August 23. Hence Barbara Tuchman's choice of *The Guns of August* as her book title.

[2] Count Alfred Schlieffen (1833–1913) was one of the great military thinkers and teachers of strategy. Chief of the German General Staff (1891–1906), General Schlieffen contributed much to German military ascendancy.

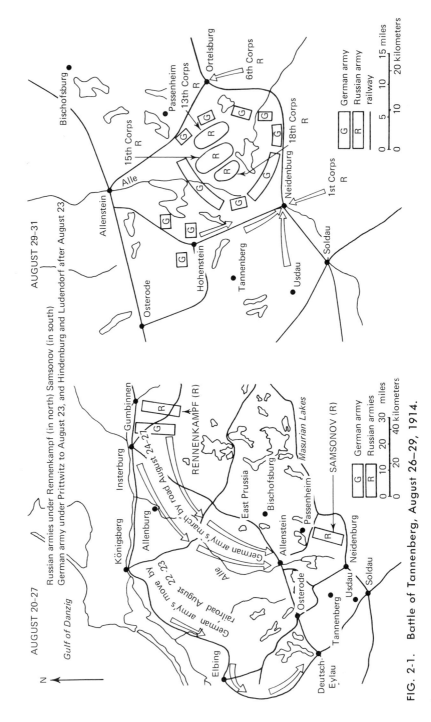

FIG. 2-1. Battle of Tannenberg, August 26–29, 1914.

while they were divided around the Masurian Lakes in Eastern Prussia, thereby enabling his numerically inferior army to achieve a victory. This victory was accomplished at Insterburg in August 1914 and became known as the Battle of Tannenberg.[1] See Fig. 2-1.

**Climax—Tannenberg:** Tannenberg is a classic battle of history and a classic example of war gaming to test plans. What might have been a decisive victory for the Russian forces at the Insterburg Gap became a Russian disaster, with the annihilation of Samsonov's army at Tannenberg August 26–31. So disastrous was this defeat that General Samsonov confided his intention of suicide to Potovsky, his Chief of Staff, and then shot himself during the early hours of August 30.[2]

Each of the contending nations had gamed the same situation, and with the same results. The generals of one nation meticulously applied the lesson of the games and won a decisive victory over the numerically superior forces of the opposing nation whose generals failed to apply the critical lesson the games had demonstrated.[3]

German War Gaming

**Prussian Military Leadership:** The Germans, long the leaders in war gaming, made effective use of war games prior to and during World War II. They used games for planning operations on a grand scale and took the game results seriously. A group of high German army officers in an official historical document, written under the heading "War Games Conducted to Test Plans for Future Operations," re-

---

[1] See Edward Mead Earle (ed.), *Makers of Modern Strategy*, chap. 8, "Moltke and Schlieffen; the Prussian-German School," by Hajo Holborn, passim and particularly pp. 189, 193, 201. Earle refers specifically to German war games of 1901 (p. 192) and again in 1914 in connection with "an oft-posed Schlieffen war game problem" (p. 189), which contributed to the "Schlieffen Plan of 1905" (p. 193) against France, employed in World War I.

[2] See Barbara W. Tuchman's *The Guns of August*, chap. 15, "The Cossacks are Coming," particularly pp. 269, 287–289, and chap. 16, "Tannenberg," and pp. 304–309. Mrs. Tuchman's meticulous account drew upon the authoritative report, Lt. Gen. Nicholas N. Golovine's *The Russian Army in the World War*, pp. 38–39 and 205, translated and published at Yale in 1931; and upon Gen. Max von Hoffmann's *War Diaries*, vol. 2, "The Truth About Tannenberg," pp. 273–275.

[3] Day-by-day accounts of the Tannenberg campaign are given in some detail in various Russian and German-language documents, in translations, and in many books. Among the latter are General Max von Hoffmann, *The War of Lost Opportunities;* Lt. Gen. Nicholas N. Golovine, *The Russian Army in the World War* (World War I); and British accounts by Maj. Gen. J. F. C. Fuller, *The Decisive Battles of the Western World*, vol. 3; Brig. Gen. Sir James E. Edmonds, *A Short History of World War I;* and Field Marshal Viscount Montgomery of Alamein, *A History of Warfare*, pp. 464–468.

ported that such games were generally directed by the Army High Command, or by an Army Group Headquarters, and explained: "Since it is the responsibility of every supreme military Commander to prepare the national defense . . . war games and training trips designed as a means of testing plans for the concentration and employment of forces were, *as a matter of course,* conducted constantly by the German Army."[1] (2-4) (Italics supplied.)

They went on to say: "In the German Army, impending operations were rehearsed in advance on a war game map or a general map if the available time at all permitted this. . . . It was not particularly difficult to play through the presumable course of the initial operations on this basis."[2] Furthermore, they pointed out that in many cases the enemy situation was known in great detail, and by using this knowledge it was possible to make preparatory theoretical exercises still more valuable.

During World War II German operational games were conducted as a regular procedure to work out and test plans prior to execution. War games preceded the German attack on France and the quick defeat of the French Army. The SEA LION operation plans for the invasion of England, never executed, were gamed, as were operation BARBAROSSA, the invasion of Russia; the Ardennes attack, the "Battle of the Bulge"; and German U-boat tactics against Allied shipping.[3]

Sometimes, as in the case of the sudden impulsive decision by Hitler to attack Poland in 1939, there was only a matter of hours, or at most a few days, depending on weather, before the attack was to be launched, and time was not available to re-game the operation under the current conditions.[4] But the *blitzkrieg* with its screaming Stuka dive bombers and fast armor penetrations, so effective as a new development in warfare, as the world well remembers, had been worked out and planned with the aid of war games.[5] These tested plans were ready for employment on the Eastern Front.

The initial campaigns against France were gamed and the lessons learned were put into effect, with the decisive and quick defeat of the French Army and its much vaunted "impregnable" Maginot Line.[6]

---

[1] *War Games,* p. 84, U.S. Army Historical Document MS P-094, Department of the Army, Office of the Chief of Military History, 1952. Referred to hereafter as OCMH, MS #94.
[2] *Ibid.,* p. 81.
[3] *Ibid.,* pp. 37–63, 76, 182.
[4] *Ibid.*
[5] *Ibid.,* p. 84.
[6] *Ibid.,* p. 182; Young, ORO-SP-98, pp. 83–86.

Games tested the plans, rehearsals trained the units in executing the plans, and the operations were conducted with coordination, speed, and telling success at a minimum cost in lives and material.

The projected invasion of England, Operation SEELOEWE (SEA LION), was gamed as a map exercise in the summer of 1940, with high officers of the German Army, Naval and Air Forces, harbor commandants, and antiaircraft officers participating. The exercise identified many of the problems to be overcome by special plans and rehearsals. Even then, the difficulties revealed made the whole operation questionable, and the operation, although rehearsed, was never carried out.[1]

In like manner the Germans gamed what they called the Otto Map Exercise, which became real in Operation BARBAROSSA,[2] particularly in the Ukraine. An evaluation of this exercise was recorded in the war diary of General Halder under the date of February 5, 1941. General Halder referred to the difficulty of effecting an envelopment west of the Dnieper from the northern wing alone, as proved out in the actual operations. As the games had indicated, the initial German successes in the Ukraine were later blunted and the whole operation bogged down, becoming as much a burden as a victory.[3] (2-5)

General Rudolph Hofmann cites an example to illustrate how closely war games "were sometimes connected with the actual events at the front in wartime." The episode he reported took place on November 2, 1944 before the Ardennes battle. The staff of the Fifth Panzer Army was conducting a war game, as directed by Field Marshall Model (of Army Group B) "to rehearse the defense measures against a possible American attack against the boundary between the Fifth and Seventh Armies." The leading commanders and their General Staff officers were assembled at the headquarters. The war game had hardly begun when a report was received that what appeared to be a fairly strong American attack had been launched in the Hurtgen-Gemeter area. Field Marshall Model immediately ordered that, except for commanders directly affected by the attack, all the same participants were to continue the game and use the current reports from the front as input information for the game.

During the next few hours the situation at the front—and as reflected in the game—became critical. To meet this situation, the 116th Panzer Division, being held in Army Group reserve, had to be

---

[1] OCMH, MS #94, pp. 53–57; Young; *op. cit.*, p. 86.
[2] Code name for the German invasion of Russia.
[3] OCMH, MS #94, pp. 37–53 and 53–57; Young, *op. cit.*

placed at the disposal of the threatened army. General von Waldenburg, the division commander, who was engaged in the game, received his operational orders one after another from the Army Group, the Army, and the Commanding General involved. Within a few minutes General von Waldenburg, instead of using "purely theoretical orders" at the game table, "was able to issue actual operational orders to his operations officer and his couriers. The alerted division was thereby set in movement in the shortest conceivable time. Chance had transformed a simple map exercise [war game] into stern reality."[1]

**The Ardennes Offensive: The "Battle of the Bulge"[2]:** On December 16, 1944, the Wehrmacht, as ordered by Hitler, in an attempt to make good on "Hitler's Last Bid,"[3] launched a great counteroffensive under Field Marshal Karl Rudolph Gerd von Rundstedt. Following the lessons of history and of war games[4] repeated over the years, the offensive was launched with a slashing drive in the Ardennes sector, which was least securely held by thinly spread American forces. By awaiting the prospect of a continuing spell of bad weather, to nullify overwhelming American air superiority, Hitler was determined to accomplish a breakthrough to Antwerp, thereby cutting off Allied forces in the north. Von Rundstedt, whom Hitler put in charge, foresaw grave weaknesses which he tried unsuccessfully to persuade Hitler to rectify. At best, Von Rundstedt hoped to capture some of the huge stockpiles of American supplies and thereby replenish the Wehrmacht's dwindling resources. He also strove to reach a strong natural defense position along the Meuse River to protect his southern flank before American forces could be moved in to block his advance. However, this possibility had been recognized by the American command. General Bradley counted on the great mobility of his own Twelfth Army Group, including General Patton's hard-hitting Third Army, to meet just such a drive. The intensity and strength of the attack by twenty-four divisions (including ten Panzer-type divi-

---

[1] OCMH, MS #94, pp. 19–20. Also reported in Young, *op. cit.*, p. 86.
[2] Hugh M. Cole, *The Ardennes: Battle of the Bulge*, one of the volumes in the authoritative series *United States Army in World War II*, published by the Office of the Chief of Military History, Department of the Army. Cole writes: "The plan for The Ardennes counteroffensive was born in the mind and will of Hitler the Feldherr." (p. 9.)
[3] Title of chapter in General Eisenhower's *Crusade in Europe*, describing the operations. Dwight D. Eisenhower, *Crusade in Europe*, Garden City, N.Y., Doubleday & Company, Inc., 1948.
[4] Reference is made to Von Rundstedt's participation and appreciation of war gaming of related Western Front situations in the book by General Erich von Manstein, *Lost Victories*, pp. 119, 120, Chicago, Henry Regnery Co., 1958.

sions) resulted in a penetration of the American lines to a depth of 50 miles before being stopped. With limited resources of men and materiel, and facing a losing war, the German Army had made an attack of astonishing effectiveness. (2-6)

"With the failure of the Ardennes offensive Germany had exhausted her last reserves," wrote General von Manteuffel, but for a short time Hitler and his Third Reich held on. After Berlin fell to the Russians and the advance guards of the Red Army and elements of the American forces met at the Elbe, on May 7, 1945, Germany—much too late—finally accepted unconditional surrender.[1]

**The German Navy Also Gamed:** Grand Admiral Karl Doenitz, during World War II, was successively Senior Officer of Submarines, Commander-in-Chief of the Navy, and finally Chief of State (Hitler's successor) of the Third Reich, the German government of that period. (2-7) In his memoirs he wrote of war gaming as the accepted and the usual thing in working out plans.

"In the winter of 1938–39, I held a war game to examine, with special reference to operations in the open Atlantic, the whole question of tactics—command and organization, location of enemy convoys and the massing of further U-boats for the final attack."[2]

He then reported the points that emerged from the war game and wrote: "I incorporated the conclusions which I had reached as a result of this war game in a memorandum which I submitted to the then Admiral commanding the Fleet, Admiral Boehm, and to the Commander-in-Chief of the Navy. The former came out in strong and unequivocal support of my contentions." Doenitz went on to explain "the basic principle on which the war game was conducted," namely, "that the enemy [Great Britain] would at once introduce the convoy system," a "belief [which] was not generally held." Upon this belief—or assumption—which proved true, Doenitz developed his highly effective, anti-convoy "wolf-pack" tactics.[3]

But gaming was not a guarantee of success, and the military defeat of Germany in World War II was complete and devastating. With it went the destruction of the military hierarchy, the vaunted German Staff, the proud military traditions. For more than 20 years Germany

---

[1] General Hasso von Manteuffel, in final chapter in H. A. Jacobsen and J. Rohwer (eds.), *Decisive Battles of World War II, The German View,* English translation, New York, G. P. Putnam's Sons, 1965. Quoted passage from p. 417.
[2] Admiral Karl Doenitz, *Memoirs, Ten Years and Twenty Days,* London, Weidenfeld and Nicholson, 1959.
[3] *Ibid.,* pp. 31–33.

has concentrated first on its economic rebirth and then on reconstruction. With its security guaranteed by mutual security pacts, and with its forces limited, West Germany has not reestablished its once preeminent position in war gaming.

Japanese War Gaming

The Japanese, like the Germans, used war gaming to train officers, to develop and test plans, and to rehearse the execution of the plans that were carried through in World War II. The Japanese attacks on Indo-China, Pearl Harbor, the Philippines, Hong Kong, Singapore, Malaya, Burma, Netherlands East Indies, the Solomon Islands, and other areas in the Pacific and Southeast Asia all were gamed in advance.

In 1940 the Japanese established the Total War Research Institute for the purpose of determining Japan's future courses of action, both military and diplomatic. Detailed military and economic plans were tested with the aid of war games and actually put into effect on December 8, 1941 (Japanese time). Much of the success of the Japanese sweep from the Philippines through Malaya in late 1941 and early 1942 was foretold by the Japanese war games.[1] (2-8)

The surprise attack on Pearl Harbor on December 7, 1941 was gamed by the Japanese from September 2 to September 13, 1941 at the Japanese Naval War College and, in part, at larger facilities at the Army War College in Tokyo. The games indicated devastation to that portion of the U.S. fleet and air support in the harbor area at the time of the attack, and a decisive victory for the Japanese. The assurance of success indicated by the games, augmented by continuous watch and intelligence reporting, was accepted, and the attack was carried out as planned. The greatest defeat ever suffered by the U.S. Navy was inflicted. But despite the high degree of success shown in the games, the Japanese war lords, aware of competing requirements in their plan of conquest, chose not to exploit this victory by amphibious landings or other appropriate augmentation or follow-up, including a possible invasion and take-over of Hawaii. (2-9) The belief was held that the Pearl Harbor attack would serve Japan's purpose by crippling the U.S. Navy's capacity to block Japanese actions in the far Pacific until further success was achieved and the newer conquests secured.[2] See Figs. 2-4 to 2-13 on pages 49 to 53, actual photographs of the attack.

---

[1] See accounts by Samuel Eliot Morison, Ellis M. Zacharias, Roberta Wohlstetter, and others referenced in Chapter Note 2-8.
[2] Roberta Wohlstetter, *Pearl Harbor, Warning and Decision*, p. 371.

As important as was the Pearl Harbor attack, the Japanese attached even more significance to the war games used to work out the campaign for taking and occupying almost all the East and South Asian area. As a result of these games the schedule for operations throughout this vast expanse was determined, along with the decision to make a simultaneous strike at the Philippines and Malaya. (2-10) The Japanese success was phenomenal. The only setback was the delay in schedule caused by the unexpectedly stubborn resistance of the U.S. and indigenous forces in the Philippines.

The Japanese continued to test plans by war games throughout World War II. A series of games initiated in September 1942 will serve as an example. These games were conducted to test plans for annihilating United States forces in the South Pacific and defeating the United States' "island hopping" approach to Japan. The Battles of the Coral Sea and Midway had been fought. Guadalcanal had already been assaulted by U.S. forces and fighting was in progress.[1]

The games were conducted in Tokyo by the Japanese Naval General Staff. These officers, familiar with the current status and resources of the Japanese military forces, played the Blue (Japanese) side. To obtain the best possible players for the Red (United States) side, the General Staff arranged for participation by the most thoroughly informed Japanese officers with the most up-to-date contacts with the United States. They found them among some outstanding Japanese Naval Intelligence Officers who had been assigned to duty in the Japanese Embassy in Washington, and who had been interned with all Japanese nationals in the United States when war broke out. In August 1942 arrangements were completed to repatriate internees, with a mutual exchange of Embassy personnel—the swap which brought the U.S. Ambassador, Joseph C. Grew, and his staff back to the United States. (2-11) The exchange was effected at Lourenço Marques, Mozambique, in neutral territory. The Swedish liner *Gripsholm* carried Japanese internees from New York and American repatriates to New York; the Japanese liner *Asama Maru* carried American internees from Japan and Japanese repatriates back. But before the *Asama Maru* docked at home port, the Japanese Naval Intelligence Officers were picked up in a Japanese Navy launch and whisked to Naval Headquarters in Tokyo where they were held incommunicado to seal them off from all news and contacts which might affect the "pristine information they carried concerning the U.S."

---

[1] Battle of the Coral Sea, May 7, 8; Battle of Midway, June 3–6; Guadalcanal assaulted Aug. 7, 1942; organized Japanese resistance ended Feb. 9, 1943.

The officers were told they were to play the Red (U.S.) force in war games, a task they performed thoroughly and exceedingly well. This Red team of Japanese intelligence experts demonstrated in the game that Japan's only hope was to achieve its conquests and consolidate them as early as possible, because the greater resources of the United States, once converted to full war-making capacity, would surely deprive Japan of its war potential and force her surrender. (2-12) Pineau,[1] who interrogated the Japanese officers, expressed his opinion that "the Japanese indulged in some sort of war game for every major operation of World War II." (2-13)

Like Germany, Japan s military machine was not only crushed in World War II, its faith in its military leadership came under serious challenge. First, under United States occupation and encouragement, Japan wrote its new constitution to repudiate any role as a military power. Only recently has Japan recognized its need to develop "self-defense forces" commensurate with its zooming economic development and the growth of other Far East Powers.

United States War Gaming

All U.S. services tested operational plans with war games. The late Fleet Admiral Chester W. Nimitz, lecturing at the U.S. Naval War College at Newport, Rhode Island, on October 10, 1960 is quoted: "The war with Japan had been re-enacted in the game rooms here (Naval War College) by so many people and in so many different ways that nothing that happened during the war was a surprise—absolutely nothing except the Kamikaze tactics toward the end of the war; we had not visualized those."[2] Army, Air Force, and Marine Corps examples will be discussed in later sections.

British Gaming—Map Maneuvers, World War II

The British also used gaming in World War II. Lieutenant General (later Field Marshal) Bernard L. Montgomery, in his North Africa campaign against Rommel's Afrika Korps, applied a technique for which he was well known. He would steal away from his headquarters to a distant hideout, where he could think through and plan

---

[1] Report to the author by Roger Pineau, Smithsonian Institution, Washington, D.C. (Lt. Roger Pineau, USNR, former U.S. Naval Intelligence Officer).
[2] F. J. McHugh, "Gaming at the Naval War College," *United States Naval Institute Proceedings*, March 1960, p. 52. Also quoted in Sidney F. Giffin, *The Crisis Game. Simulating International Conflict*, p. 23, Garden City, N.Y., Doubleday & Company, Inc., 1965.

his operations with a minimum of interruptions. His Chief of Staff made a daily visit to keep Montgomery informed about ongoing operations and to carry back new orders as needed.

At his hideout Montgomery pored over his maps, devised plans, and then tested them in a kind of gaming situation with his Staff officers and unit commanders. Each protagonist was quizzed on specific details of capabilities, requirements, reactions to enemy moves, and what he could deliver in relation to the situation and plans before him. Montgomery required his Intelligence Officers to play through the forthcoming battle, with the depositions of the enemy and his own troops spread out on a map. The Staff was required to imagine themselves the enemy, to react as the enemy would, and "to play against Montgomery [this] strange and fascinating parlor game, making move-for-move against the British." Most of this activity was accepted practice in planning, but Montgomery played with intensity and tried always to put himself in Rommel's place. He anticipated enemy actions by close study of his adversary and always asked himself, Now what would Rommel do about it? He kept pictures of Rommel and the other enemy commanders displayed before him in his van, believing this display would keep him ever mindful of his enemies and help him to anticipate their moves and reactions.[1]

This was Montgomery's special gaming-like, map-maneuver technique that helped turn the unprecedented success of the Desert Fox to a final defeat. The British Eighth Army campaign, started on October 23 at El Alamein, weakened the Afrika Korps and began the elimination of this German threat. The November 8, 1942 TORCH landings of the U.S. forces under Gen. Dwight D. Eisenhower's command (which included the II Corps under Maj. Gen. Omar N. Bradley) began the final Allied offensive against the remaining Axis forces (Italian forces and Rommel's Afrika Korps). Thus, at Tunisia on May 13, 1943 the German threat was completely removed from North Africa.

Allied Gaming and Operations

**Biggest Military Operation of All Time:** General Montgomery, engaged in the combined planning of OVERLORD, the Allied invasion of France, again applied his map-maneuver or gaming technique. On April 7, 1944 Montgomery made a presentation attended by General Eisenhower, England's Prime Minister Churchill, and the

---

[1] Field Marshal Montgomery, *op. cit.*, p. 17; and Alan Moorehead, *Montgomery—A Biography*, pp. 132-133, New York, Coward-McCann, Inc., 1946.

British King, and "with rare skill, traced his plan of maneuver as he tramped about like a giant through Lilliputian France" on a huge map of Normandy—the width of a city street—spread out on the floor of St. Paul's School in London.[1] The school then was being used by SHAEF (Supreme Headquarters Allied Expeditionary Force). This presentation, although a product of planning and pre-testing with full coordination of Allied land, sea, and air forces, was in itself a review or evaluation of the whole plan—not unlike a war game in nature and intent.[2] Once more, Montgomery was to face his old enemy, Rommel, now charged with the defense of Fortress Europe. Montgomery's knowledge of Rommel's tactics and personality indicated Rommel ordinarily committed his reserves just as quickly as he could drag them into the line, a factor to be reckoned with.[3]

**Planning and Testing of Combined Operations:** As a final example of World War I and II gaming to test and plan, a case may be cited from the U.S. Ninth Army's history of its participation in the post-Ardennes drive to defeat the German Army in Europe. This case illustrates the tie-in of an estimate of the situation with map maneuvers in a vital planning exercise. The Ninth Army was transferred from Bradley's Twelfth Army Group to become a part of the Twenty-first Army Group commanded by Field Marshal Montgomery. The historical account reports that a complete formal *estimate of the situation* was prepared. Each corps then was required to present its plan in the presence of all corps commanders and key Army and Corps staff officers. By this means each corps knew precisely what each other corps was to do and why. The several plans "were then 'war-gamed'—played out on the map—so that the action could be thoroughly previewed and every possible contingency discussed in detail."[4]

Considering the accumulation of war-gaming experience in relation to war plans, one researcher, writing in 1954, summarized it as follows: "War gaming is the traditional final step after the preparation of a war plan; it is universally regarded as the best peacetime test of a plan."[5]

---

[1] General Omar N. Bradley, *A Soldier's Story*, p. 239, New York, Henry Holt and Company, Inc., 1951.
[2] Bradley, *op. cit.*, pp. 238, 243–244; Eisenhower, *op. cit.*, pp. 243–245.
[3] Bradley, *op. cit.*, pp. 243–244.
[4] Ninth United States Army Staff, *Conquer, The Story of the Ninth Army*, p. 141, prepared under the direction of Maj. Gen. James A. Moore, assisted by Cols. Theodore W. Parker and William J. Thompson, Washington, D.C., The Infantry Journal Press, 1947.
[5] OCMH, MS #94, p. 121.

Gaming for planning purposes is not unique to World Wars I and II. The less formalized type of gaming, usually as a map exercise, or maneuver, has been employed by many of the great military leaders of the past.

## The Arab-Israeli Wars

The Palestinian War (1948–1949)[1] ended with the establishment of the sovereign state of Israel, a fact which the adjoining Arab countries were unwilling to accept. Confronted with avowed Arab hostility, continued sporadic terrorist incursions, and little prospect of lessened threat on all its borders, the small state of Israel chose to develop its military forces with the highest possible capability against its numerically superior antagonists.[2] That Israel succeeded in this purpose has been demonstrated to the world twice in recent years: first, in the 1956 Suez crisis and the 10-day Sinai War (October 29–November 7), and again in the 6-day Arab-Israeli War (June 5–11) in 1967.[3]

**Masters of Lightning War:** It is well known that Israeli plans have been built on the concept of a short war as its principal or only hope of victory over its adjoining Arab antagonists. One military analyst, as a student of the Israeli Defense Forces, wrote in early 1967 that if war should erupt "the Israeli defense forces are organized and equipped

---

[1] War ended January 1949 by mediation supervised first by the well-known internationalist Count Folke Bernadotte (1895–1948) of Sweden who was assassinated in Jerusalem in 1948. Ralph Bunche of the United States concluded the mediation.

[2] Israel, with about 2.5 million people, was faced by a coalition of twelve Arab countries with a combined population of about 110 million. (Egypt, Jordan, and Syria, which border on Israel, have a combined population of about 40 million. Israel's armed forces are drawn only from its 2.24 million Jews. Most of the remainder of the population are Arabs.

[3] Although formal hostilities covered 10 and 6 days, respectively, the duration of actual fighting was shorter. In 1956, hostilities broke out on Monday, October 29, at 1700 hours; the capture of Sharm El Sheikh on the Gulf of Aqaba, at the Straits of Tiran, on Monday morning, November 5—a few hours short of a week—ended the war. The Suez campaign lasted 100 hours, ending November 2; the Gaza campaign, Friday-Saturday, November 2, 3 (Maj. Gen. Moshe Dayan, *Diary of the Sinai Campaign*, 1956, pp. 222–226). In 1967 three days of lightning war knocked out the Egyptian army in the Sinai Desert. The armies of two Arab countries were defeated by the fourth day of the war—the Egyptians in Sinai and the Gaza Strip, and the Jordanians in their eastern region. (*U.S. News & World Report*, June 19, 1967, p. 33.) Jordan quit the war in its third day after 60 hours of active combat operations. (*Newsweek*, June 19, 1967, pp. 28–29.) The Syrian campaign, flaring later, was over in 2 days of fighting.

to see a quick and decisive victory."[1] He also pointed out that the Israeli reserve system provides for combat-ready forces available on hours' notice.

What astonished the world most in June 1967 was Israel's demonstrated mastery of lightning war, its phenomenal simultaneous successes against each of its three bordering enemy countries, and the rapid mobilization of its reserve. It is reported that, starting with its 50,000-man standing army, Israel was able to field a fighting force of 235,000 men within 48 hours. By pre-planning, civilian trucks, buses, taxis, and private cars were organized to deliver reservists to assigned rendezvous points, and each was assigned to specific battle sectors and objectives. As an example, several reports cite that, after the alert, Israeli tanks, each manned by a single regular, were waiting in tank parks for the two or three reservists required to complete its crew.[2]

**A New Classic in Military Operations:** What the Germans started as a blitzkrieg in Europe, and the Japanese used at Pearl Harbor and elsewhere in the Pacific area early in World War II, the Israelis demonstrated in the Middle East, first in 1956 and again even more effectively in 1967. The latter war also earned a place among the great, classic military campaigns of history. A whole lexicon of superlatives has been used to describe the actions. Among these were "superb tactics," "absolute mastery of the air," "six amazing days," "overwhelming triumph," "stunning successes," "audacious use of air power—and surprise," "superbly organized and functioning system," "professionalism in the martial arts," "the Israeli military virtues of superb tactics and timing," "destroyed almost totally the air forces of four countries" (Egypt, Jordan, Syria, Iraq), "turned an Arab defeat into a classic rout likely to be studied with admiration at war colleges the world over," "history's most outstanding air battle," and "the amazing Israeli military machine."[3]

Much can be and has been learned by studying the successes and failures in past wars. Few armies have had the chance to fight the same general situation a second time, and yet the Sinai battle in 1967 was basically a rerun of the 1956 operation. The Israelis learned

---

[1] Colonel Irving Heymont, USA (Ret.). Article in *Military Review*, Feb. 1967, p. 45. Additional information obtained in personal consultation with the author.
[2] *Time*, June 16, 1967, p. 23.
[3] Quotations from news magazines: *Time*, June 16, 1967; *Newsweek* and *U.S. News & World Report*, June 19, 1967 issues.

much from their earlier success; the Egyptians seemed to learn little from their failure.

Although the old classic principles of war remained unchanged, in the Arab-Israeli War these principles were more effectively employed. (2-14) Superior military intelligence and communications, brilliant leadership and command, meticulous mobilization, logistics and operational planning, thorough training and superb performance, up-to-date and improved equipment, all were utilized to full potential with superior coordination by troops with the highest possible morale and dedication; all these were important factors.

**Israeli War Gaming:** The Israeli forces have given serious attention to war gaming and made extensive use of war games. Colonel Yitzchak Ya'Acov, an Israeli officer whose duties involve operations research, including application of war gaming, in commenting on Israeli war-gaming activity, mentioned the difficulty of finding officers in the Israeli forces who could take the part of the opponents and think, act, and react as those other nationals. This was true, he added, even among Israeli officers who had been citizens of these particular other countries before emigrating to Israel.[1]

War games entered into the formulation and testing of the Israeli war plans and presaged their successes, at least in kind, if not in their astounding degree. Israeli concepts, tactics, and plans were worked out and thoroughly tested in a laboratory crucible of war gaming, and in the experience and desert dust of field exercises prior to the 1967 Arab-Israeli conflict.

## GAMING FOR RESEARCH

The term *research* is used here in a broad sense to cover the spectrum of any efforts to find out something not yet known or to validate a finding of prior research that is challenged or not fully accepted. The focus of this section is on applied research and the effort to provide information needed in the study of warfare and in other matters of military concern.[2]

The intent is not to denigrate basic research, the discovery of new

---

[1] It will be recalled that the Japanese, in World War II, used in their war games Japanese officers who had lived long in countries of their enemies.
[2] Alexander M. Mood, *War Gaming as a Technique of Analysis*, p. 1, Santa Monica, Calif., The RAND Corporation, RAND paper P-899, Sept. 1954.

truth for its own sake rather than for some preconceived utility, but to put the primary emphasis on research for definite military purposes. Basic truths emerge as by-products of applied research and are welcomed when identified. For example, an earlier operations research study explored the problem of how much stress and fatigue soldiers can stand, how much time is required to recoup, and what can be done to enable them to stand up under more severe or prolonged stress and fatigue. In this study certain physiological measures were employed as indirect measures of troop stress, strain, fatigue, and recuperation.[1] An entirely unpremeditated finding was that the eosinophil count of the blood and the 17-ketosteroids excreted in the urine revealed an effect of combat fatigue and stress much like a diabetic shock, a reaction not previously known and not an objective of the study.

In the advancement of the art of warfare and in the development of military forces throughout the world, laboratories are needed for the study of new and improved means of conducting warfare, as well as proving grounds for testing concepts and equipment and the training of personnel. Basic is the need of *simulating warfare*, that is, setting up imitation or artificial warfare, resembling and including, as nearly as possible, all the elements of real war.[2]

War gaming at the higher echelons of command and at strategic levels involving political-economic considerations certainly fits into the broad concept represented by operations research and systems analysis. The purist might argue that completely mathematical or computerized simulations fit the earlier definition of operations research. However, in this book both terms, operations research and systems analysis, are used in the broad and more inclusive context associated with the term *systems analysis*. War gaming, as used here, is a technique employed in operations research or systems analysis for the purpose of studying selected military problems in the context of their normal interplay among the multiplicity of actions involved in actual military operations.

Very simple games may be suitable for selected training purposes and for the study of problems that do not require the interactions of a large number of different units. But for most research and command decisions a complete, interacting system is essential. A case history taken from the files of the first U.S. Operations Research Group illustrates this thesis.

---

[1] S. W. Davis and J. G. Taylor, *Stress in Infantry Combat*, Chevy Chase, Md., Operations Research Office, ORO-T-296, 1954.
[2] Francis J. McHugh, *Fundamentals of War Gaming*, 2d ed., p. 1-1, Newport, R.I. The United States Naval War College, Nov. 1961.

## First U.S. Operations Research Group

The U.S. Navy established the Naval Ordnance Laboratory[1] to develop more effective naval weapons. At the onset of war with Japan, activity at this installation was intensified. The NOL scientific program included such activities as countermeasures against enemy weapons and development of new advances of magnitude in U.S. weaponry. Among its problems were mine development and a wide range of means of detecting underwater threats, as by mine locators, magnetic airborne detectors, submarine detectors, and the degaussing of U.S. ships—merchant, troop, and men-of-war—to protect them against Japanese and German "influence mines" activated by the magnetic fields of ships.

A staff of scientists, recruited from 1940 to 1942, was organized into research groups. Dr. Ellis A. Johnson headed the countermeasures section, concerned with mine development and mine countermeasures. To facilitate study, Johnson established an internal seminar in 1941.[2] Some fifty senior scientists of the NOL staff undertook a systematic study of all aspects of mine warfare: technological, tactical, and strategic. The researchers discovered that what they were doing was now called *operations research*. Dr. James B. Conant, as Chairman of the National Defense Research Committee, had visited England in late 1940, learned of operations research (OR) there, and advocated its use in the United States. (2-15) Accordingly, the NOL Operations Research Group, patterned upon the British experience, was formalized on March 1, 1942. Soon OR groups in other U.S. military services were formed.

The NOL group now had a name, operations research; a technique; and a set of problems. Many of the problems had tactical and even strategic aspects as well as technical elements. Because of the nature of the problems and the methodology for study to be developed, this group is believed to have pioneered operations research in the United States. In particular, it was foremost in including strategic aspects in

---

[1] First established as the Mine Laboratory (1918) and later (1929) designated the Naval Ordnance Laboratory (NOL). At inception this facility was located at the Washington Navy Yard, site of the Naval Gun Factory. With the advent of World War II, expansion of the facility was necessary, and a new plant, the home of present operations, was constructed at White Oak, Md., a Washington suburb. The Naval Research Laboratory, created in World War I at the urging of Thomas A. Edison while he was a scientific advisor to the Navy, continued to function in broader research fields. It too is located in the Washington area and is situated only a short distance down the Potomac River from the Navy Yard.

[2] For want of a name, this seminar became known by its research objective, "How to Sink a Ship."

its studies. An important early operations research study conducted by this group is described next.

### Operations Research Case Study: Mine Warfare Against Japan

After Pearl Harbor it was imperative that the United States destroy as much as possible of Japan's Navy and merchant shipping in order to cut its lifeline and thereby starve or cripple the home islands. The ability of Japan to deploy forces had to be negated. The Japanese attack on Pearl Harbor had not only brought World War II to the United States and crippled the U.S. Pacific Fleet; it established the superiority of the Japanese fleet in the Pacific, giving the Japanese an open door to the conquest of the Pacific Islands and East Asia. The effect on the United States was to challenge its every resource in an attempt to circumvent total defeat and to overthrow the Japanese. High priority was given to a study of mine warfare against Japan.

The OR group, building upon experience in mine development and countermeasures, undertook the study. The group soon realized it knew too little about many aspects of the undertaking and that it lacked data to deal with certain specific problems. Much tactical and technological data existed, but there were serious gaps, particularly in factors not suited to laboratory or field experimentation. The NOL seminar was used as a "think tank" to generate ideas and suggest means of providing the missing information and insights. On this, Dr. Johnson reported, "The seminar made very limited progress until [he] reinvented war gaming.[1] As a geophysicist it was not at all surprising" he said, that he "was in complete ignorance of the long and illustrious history of war gaming." (2-16) War gaming of the mining problem held promise; for example, a doctrine to guide the development of mines for offensive operations did not exist. By use of war gaming, models of possible mining operations were studied, various tactics and weapons characteristics were tested, and the strategic implications assessed. Participants are shown in Fig. 2-2.

---

[1] It was not until later that Dr. Johnson learned that the technique he visualized was well known in military circles but was used primarily for training and testing of plans rather than as a research tool.

FIG. 2-2. NOL gamers and operations research analysts, World War II. *Left to right: top to bottom:* Doctors Ellis A. Johnson (1), John Bardeen, Ralph Bennett, Lee H. Hoisington, Charles Kittell (2), Walter C. Michels (3), Thornton L. Page, Lyman G. Parratt, Lynn H. Rumbaugh (4), George Shortley (5), and Foster Weldon (6). *(Photo credits: U.S. Navy, Naval Ordnance Laboratory:* 1, 2, 4. *Bryn Mawr College:* 3. *Brooks,* 5. *ORO,* 6.)

The Role of Gaming    41

First the games were conducted on the tactical level. Selected ports were hypothetically mined in various patterns, using maps and charts of actual areas—San Francisco, Port Darwin, and the Shimonoseki Strait in Japan, among others.

Once the games were organized, ten to twelve gaming teams were set up and about twenty plays were run. Each team consisted of five or six research people, augmented by naval officers as required. One team was the Mining Team. This team placed the mines in any pattern it wished. The opposing team was the Sweeping Team. The task of the opposition was to sweep the harbor clear of mines, or at least to open safe corridors. No holds were barred as to how the mines might be placed or how the sweeping operations were to be conducted. The games encouraged innovation.

The gaming effort was barely under way when it became clear that games of greater scope were required to deal with economic and strategic factors. The games and related analytical studies then were expanded. To pursue this program Dr. Johnson set up a part of his NOL Research Group—now about 200 strong—as the Navy's Mine Warfare Operations Research Group (NMWORG). The NMWORG made the broad analytical studies. The games generated data: on tactics, and on the technological, strategic, and economic requirements and capabilities. The games also assessed the strategic impact of actions proposed. *In this case an operations research study created the need for war gaming, and war games produced needed data not otherwise available for operations research. But this case goes further.*

Dr. Johnson next went into active military service. As a naval officer he was assigned the job of planning and conducting actual mining operations against Japan. (2-17) His assignment in effect was to carry into practice actions demonstrated as being feasible by gaming and operations research. In this connection Dr. Johnson later reported: "As an officer, when I actually conducted the mining campaign against Japan that I had helped to war-game earlier, I found very often that almost every countermove the Japanese made to a powerful mining attack had also occurred in one or more of the war game situations. In fact, the over-all results of the campaign followed closely the results predicted by war gaming."[1]

The tactical and strategic results of this campaign were successful, as attested by the fact that 670 Japanese ships, including some 65 men-of-war, were mine casualties. Sixty percent of Japanese ship

---

[1] Ellis A. Johnson, "The History and Future of War Gaming in Operations Research," a paper presented at the Third War Games Symposium at the University of Michigan, Oct. 6, 1960.

casualties during the last year of the war were the result of mines. Japanese war records indicate the blockade of Japan was almost absolute. Prince Konoye stated that the aerial mining of Japanese waters by the USAF's XXI Bomber Command had an effect on Japan comparable to that achieved by the Command's high-explosive and incendiary bombings. Noteworthy is the fact that only 5.7 percent of this Command's efforts were devoted to mining Japanese waters.[1]

**A Classic Example:** The case is probably unique in World War II; the mining of Japanese waters represents a complete cycle *carried through by OR personnel,* from problem to solution to application and assessed results in actual warfare. Certainly, it is a classic example of the significance of the interrelations between war gaming and operations research. There were four distinct phases of this mining operation.

1. Technical war games, aimed at identifying the technical characteristics of mines and mine placement that yield maximum effectiveness.
2. Tactical war games employing various concepts of mining and sweeping tactics directed to specific harbors.
3. Strategic war games focused on the mining of the home islands of Japan.
4. Finally, actual military operations carrying into practice what had been learned from the NOL seminar, the subsequent series of war games, and the operations research analyses which were carried through. These operations constituted one highly successful aspect of the naval war against Japan. (2-18)

With a passion for modernity, the Japanese have now become active in the field of operations research. With a national organization, a journal, and a growing reservoir of trained operations analysts, the practitioners of this new discipline are finding places in Japan's industry and in new national defense forces. In this connection, small war-gaming groups have been set up in the Ground, Maritime, and Air Defense Forces with three direct objectives:[2]

---

[1] Information to the author from Dr. Johnson.
[2] Kazuo Tada and Macon Fry, reporting on operations research in Japan, "Military Operations Research in Japan," *First International Conference on Operations Research Proceedings,* pp. 432–439, published by the Operations Research Society of America, 1957. Notes from p. 434.
   Recently the Operations Research Group of the Ground Staff, Japanese Defense Agency, has been developing computer-assisted games and computer simulations at division level, based upon certain models developed for the U.S. Army. Communication from Col. S. Kashiwai, Secretary of the General Staff, received Oct. 25, 1966.

1. Evaluation of the nation's defense plans
2. Evaluation of tactics
3. Evaluation of weapons systems

A stated group purpose is to provide "valuable staff training as a by-product."

Military Participants and Computer Support

Research gaming to provide information on which important military decisions are to be made requires a high order of command experience and military expertise. This calls for direct participation in game operations by military commanders possessing the necessary level and quality of experience. If the supporting data are to be developed in sufficient detail to permit systematic and meaningful analysis, and if results are to be made available within a reasonable time, the game must employ rapid data processing achieved only by the use of electronic computers. These two elements must be augmented by a group of experienced and sophisticated operations analysts or game scientists, supported by clerical and administrative personnel able to translate a complex of data, rules, procedures, and the doctrine and practices of the military art into a set of models programmed for the computer. These models must be designed to yield the type of data needed for subsequent analysis if they are to provide the necessary basis for study of the specific problem(s). Such gaming operations normally involve from ten to sixty persons on a full-time basis, most of whom are drawn from higher levels of specialized competence, experience, maturity, and expertise.

Research and analytical games are so complicated, and require so much time, that the games must be conducted by a full-time organized group, with experts functioning on a continuing, rather than an ad hoc, basis. For adequate study the questions associated with one central problem may require continuous gaming and analysis by a gaming group working full time for as much as a year.

## GAMING OF NEW CONCEPTS

What are the research uses to which gaming may be directed? Most of the research problems which employ war gaming as a principal means of investigation are necessarily shrouded under security restrictions. The results of some have taken tangible form and now are common knowledge, as in the case of the World War II mining of Japanese waters. Examples of other purposes of gaming were:

1. The development of doctrine for the use of nuclear weapons
2. The study of problems encountered in the transition from the use of conventional weapons only to the use of a mix of conventional and nuclear weapons in tactical warfare
3. The study of the airmobile and the air-assault concepts

Nuclear Weapons

The introduction of nuclear weapons made new and exacting demands of gamers. Tactical and strategic doctrines both were affected by this technological advance in weaponry. See Fig. 2-3.

**Doctrine for Tactical Nuclear Weapons:** When tactical nuclear weapons became a reality, and potential antagonists had the capability to use them on the battlefield, the possible employment of these weapons caused a reexamination of tactical doctrine.[1] Through field tests of atomic weapons held principally at the Atomic Weapons Testing Grounds, Frenchman's Flats, Nevada; Camp Mercury of the Atomic Energy Commission; and the U.S. Army's Camp Desert Rock (all north of Las Vegas, Nevada), many of the effects of these weapons were learned. Significantly, however, tactical nuclear weapons had never been employed in actual combat, and experience on which to build tactical nuclear doctrine did not exist.[2] War gaming in a simulation of nuclear tactical warfare was the only practical method that could investigate this problem. It is not surprising that the U.S. Army's war-gaming facilities were employed to provide the needed information.

**Doctrine for Strategic Use of Nuclear Weapons:** For some years many simulations were developed and used in war games to study problems involved in massive nuclear strikes and counterstrikes.

---

[1] The term *doctrine* as used in the military refers to "Principles, policies and concepts applicable to a subject, which are derived from experience or theory, compiled and taught for guidance. It represents the best available thought that can be defended by reason." *Dictionary of U.S. Army Terms*, p. 201, D/A, AR 320-5, Jan. 1961.
[2] Two nuclear weapons were used in actual warfare: the first on Hiroshima on Aug. 6, 1945; the second on Nagasaki on Aug. 9, 1945. These were strategic weapons and not used in the tactical sense of supporting surface forces engaged in combat at that place and time.

The capability to withstand attacks by manned bombers and by missiles, to defend against them, to assess large-scale damage, and to protect military and civilian targets, population centers, and industrial centers had to be learned. It was this general problem of large-scale nuclear attack and chaos that stimulated a correspondingly large-scale construction and use of computer simulations to game such problems.

### The Airmobile and Air-assault Concepts

A final example is drawn from the studies conducted to test and evaluate the concepts of using aircraft to substantially improve and increase the Army's mobility under operational conditions, and to explore the possibility of creating an airmobile force with sufficient firepower and staying capability to use as a deep-penetration assault force. Airborne divisions with their paratroops had demonstrated capability to raid, disrupt, and destroy (communications centers, installations, bridges, etc.); to block and delay advancing columns; and to seize and temporarily hold central points (high ground, road junctions, etc.) until heavier forces could arrive. The air-assault concept was to add sufficient heavy, airmobile firepower such as armor—artillery, rockets, and missiles—to substitute for equipment that heretofore had been tied to surface movement.

War games of a high degree of sophistication were applied to airmobility and air-assault concepts and were validated later in Vietnam and in the 1967 Arab-Israeli War.

### Vietnam and the Airmobile Division

The increasing involvement of the U.S. Army in Vietnam gave impetus to interest in the possibilities of putting aircraft to greater use to enhance mobility and firepower. Under Maj. Gen. Hamilton Howze the Army's Tactical Mobility Requirements Board was established to study the problem. A program of war gaming was initiated to investigate certain aspects of the mobile concept; evaluate the operational requirements, characteristics, and capabilities of a proposed Air As-

---

FIG. 2-3. Nuclear-weapon and missiles delivery systems. War gaming has been used extensively to explore the many problems associated with the advent of nuclear weapons and their employment in defense and offense. (Photos: upper and lower right, U.S. Navy; lower left, U.S. Army.)

sault Division; and to identify specific strengths and weaknesses. This war-gaming program began in early 1962 and continued through 1963. (2-19) Results of the game and associated analytical studies indicated the validity of the airmobile concept and suggested certain modifications in the proposed Air Assault Division. Concurrently an experimental Air Assault Division, under the command of Maj. Gen. Harry W. O. Kinnard, was established at Fort Benning, Georgia—the home center for U.S. Army Infantry—and a series of "more than 80 tests, war games and study efforts" were conducted there and elsewhere.[1]

The final result of these studies, to which gaming made an important contribution, was the Pentagon decision:

1. Approving the airmobile concept
2. Authorizing formation of a *modified* Air Assault Division
3. Reorganizing the 1st Cavalry Division to become the world's first airmobile fighting force

The division, redesignated 1st Cavalry Division (Airmobile), was reequipped and went into intensive training prior to deployment to South Vietnam in August 1965. (2-20) It was soon involved in combat against the elusive Viet Cong guerrillas and units of the North Vietnamese forces, and the value of the increased responsiveness (mobility and firepower) of the helicopter-borne forces was quickly demonstrated.[2]

Effectiveness in combat is the ultimate measure of performance of a military unit. Only the most outstanding achievement merits the highest possible honors. Nobel Prizes represent such outstanding performance in the fields of science and medicine. In war, the Congressional Medal of Honor, awarded to individuals, and the Presidential Citation, awarded to military units, are the corresponding highest awards for outstanding performance in the Armed Forces of the United States. The 1st Cavalry Division (Airmobile) was the first division-size unit to receive the Presidential Unit Citation for out-

*(Text continues on p. 54.)*

---

[1] Major General Edward L. Rowny, "After the Air Mobile Tests," *Army*, May 1965, p. 36. General Rowny then was Deputy Assistant Chief of Staff for Force Development. In 1962 he was Chief of Field Test Operations of the Howze Board.

[2] Major General Kinnard, "Activation to Combat—in 90 Days," *Army Information Digest*, April 1966, pp. 24–29. [General Kinnard reported the 1st Cavalry Division was fortunate in that it "had been war gaming with studies based on the very area of Vietnam to which it was ultimately deployed." (p. 26.)]

## The Role of Gaming 49

FIG. 2-4. Large-scale replica of Pearl Harbor, constructed by the Japanese Imperial Navy on an "off-limits" evacuated island in the Kurile Chain. Used for testing of war-gamed plans and for training of naval aviators prior to the Pearl Harbor attack. (Japanese official photo, captured by U.S. Navy. Courtesy of Vice Admiral John F. Shafroth, USN.)

FIG. 2-5. The Japanese attack on Pearl Harbor. Ford Island and Battleship Row. Attack under way. Note: *upper center*, Hickam Field, Army Air Field, already hit; *lower center*, torpedo hit on battleship. (*Captured official Japanese photograph. U.S. Navy.*)

FIG. 2-6. Ford Island early in attack. Note two Japanese planes in air: one over battleships, the other at upper right near oil storage tanks. (*Official Japanese photograph, captured by U.S. Navy.*)

FIG. 2-7. Torpedo hits on two battleships. (*Captured official Japanese photograph. U.S. Navy.*)

The Role of Gaming   51

FIG. 2-8.   Battleships under attack.   *Left to right: Oklahoma* type already sunk: direct hits on another *Oklahoma* type, a *California* type, and a *Maryland* type. (Captured Japanese photograph.   U.S. Navy.)

FIG. 2-9.   Battleships burning while attack continued.   (Captured Japanese photograph.   U.S. Navy.)

FIG. 2-10. U.S. antiaircraft defense against Japanese attack. Several Japanese attack aircraft visible. (*Official U.S. Navy photograph.*)

FIG. 2-11. Capsized U.S. battleship at left, others burning. (*Official U.S. Navy photograph.*)

The Role of Gaming    53

FIG. 2-12.    U.S.S. *West Virginia* and U.S.S. *Tennessee* burning.    Rescue boat picking up survivors from the water.    (*Official U.S. Navy photograph.*)


FIG. 2-13.    U.S.S. *Arizona* after the attack.    (*Official U.S. Navy photograph.*)

54  Venture Simulation in War, Business, and Politics

standing performance in Vietnam. The award was made in September 1967, with special reference to "numerous victories" against an invading North Vietnam division in Pleiku Province between October 23 and November 26, 1965. (2-21)

## Chapter Notes

2-1. The early Indian maneuvers were reported to the author by S. S. Srivastava of the Indian Ministry of Defence in a letter from New Delhi, dated June 13, 1966. Writing in response to the author's query about ancient war games in Asian countries, Dr. Srivastava replied under the caption "Ancient Games, 1500 B.C. to A.D. 400" and related: "The first indirect evidence of the organization of some form of war games is available from the great Indian Epic Mahabharata about 500 B.C." He continues, describing the great sham battle, as summarized in the text. Dr. Srivastava searched the literature available to him in India and reported several additional sources to the author.

2-2. The curiosity of Brownlee Haydon of the RAND Corporation was aroused after the statistics of war "quotation" had appeared in an internal publication of RAND in the spring of 1961. He found others had seen the quotation in essentially similar form in a number of other sources. In following these and other leads, Haydon found the statement was written by Norman Cousins, under the title, "A Report of an Imaginary Experiment" in the 75th anniversary edition of the *St. Louis Post Dispatch*, published Dec. 13, 1953. Cousins, himself, used it subsequently in other writings but never intended anything more than an imaginary or fanciful set of statistics. The widespread requoting extended to Brazil, Canada, Ireland, and the worldwide news dispatches of a number of religious organizations. See Haydon's "The Great Statistics of War Hoax" in the November 1964 issue (pp. 94–96) of the magazine *Air Force*.

Not fictious, the information about graduates of the U.S. Military Academy, mentioned in this chapter, came from the Chief, Archives and History Section, USMA Library, Sept. 28, 1965, and from the 1964 *Register of Graduates*, p. 760, the West Point Alumni Foundation, Inc., West Point, N.Y.

2-3. This is not to suggest war games are a substitute for laboratory or proving-ground tests of equipment such as are done at Aberdeen Proving Ground at Aberdeen, Md., or for field experiments—testing troop and equipment capabilities—as are done at the Combat Developments Experimentation Center of the U.S. Army Combat developments Command at Fort Ord, Calif., or the field tests this

Command conducts at the Hunter Liggett Military Reservation in California. These latter types of testing, trials, and experimentation are needed to provide some of the basic data—the life blood— of meaningful war games.

2-4. *War Games*, p. 83, U.S. Army Historical Document MS P-094, by Rudolf Hofmann, General der Infanterie a. D., with a foreword by General Oberst a. D. Franz Halder, translated by P. Nuetzkendorf, and prepared in the Historical Division, Headquarters, United States Army, Europe, 1952; published by Department of the Army, Office of the Chief of Military History.

2-5. Alan Clark wrote a book on the actual operation rather than the gaming preceding the operation. The book, entitled *BARBAROSSA, The Russian-German Conflict*, 1941–1945, was published in 1965 by William Morrow & Company, Inc.

2-6. Hugh M. Cole, in his book, *The Ardennes: Battle of The Bulge*, published in 1965 in the official-history series of the U.S. Army in World War II, gives greater detail and hour-by-hour or day-to-day developments in the operations.

General Omar N. Bradley also described these operations in his book, *A Soldier's Story*, New York, Henry Holt and Company, Inc., 1951. His chap. 21, "Countermeasures" (pp. 451–489), gives background and insights as well as an account of the operations.

See Dwight D. Eisenhower, *Crusade in Europe*, chap. 18, "Hitler's Last Bid," pp. 342–365, Garden City, Doubleday & Company, Inc., 1948.

Field Marshal Erich von Manstein, in his book *Lost Victories* (English translation published in Chicago by Henry Regnery Company, 1958), reveals what was going on in the German Command. He also throws light on the attitudes and personalities, as well as the actions, of the principal German commanders and political leaders. Von Manstein served as Chief of Staff to Field Marshal von Rundstedt in the early phase of World War II and operated war games for the Command. He mentions the Western Front plans and adds: "Our view was corroborated by a sand-table exercise in Coblenz on 7th February at which we ran through the advance of 19 Panzer Corps and the two armies of our Army Group . . ." (p. 119). He then added (p. 120), "At the close of the above-mentioned sand-table exercise, which I had helped to run, von Rundstedt thanked me . . . for all I had done as his Chief of Staff." Two days later (Feb. 9, 1940) Von Manstein left Coblenz to assume his new duties as Commanding General of 38 Corps.

2-7. Grand Admiral Karl Doenitz, among all the principals in the Third Reich who were tried at Nuremberg before the International Military Tribunal (Nazi War Crimes Trials), drew the lightest sentence— 10 years in prison, which he served. As for the others, Hitler (born

April 20, 1889 in Austria), Goebbels, Himmler, and Goering committed suicide April 30, May 1, May 23, and Oct. 15, 1946, respectively; Ribbentrop, Keitel, Kaltenbrunner, Rosenberg, Frank, Frick, Streicher, Seyss-Inquart, Sauckel, and Jodl were all hanged in the sequence named on Oct. 16, 1946; Hess, Raeder, and Funk drew life sentences, Speer and Schirach got 20 years, Neurath 15. For an account of the whole period of the Third ("Thousand-year") Reich, from its birth on Jan. 30, 1933 (p. 5) to its end in unconditional surrender at 2:41 a.m. May 7, 1945 (hostilities ceased on midnight May 8–9), see William L. Shirer's *The Rise and Fall of the Third Reich*.

2-8.  *a*. Samuel Eliot Morison, *History of United States Naval Operations in World War II*, vol. III, pp. 82–85, Boston, Little, Brown and Company, 1948.

*b*. Ellis M. Zacharias (Capt., USN, later Adm.), *Secret Missions—The Story of an Intelligence Officer*, pp. 113–114, 243–245, Putnam, 1946. Zacharias attributes to Commander Minobi, in a German translation of his 1942 secret Japanese-language book intercepted during World War II, that the Japanese island of Skioku was evacuated about 1927–1928 and a full-scale target area for carrier-based bombers was constructed. Only after the decision to attack was made on Sept. 13, 1941 in Tokyo was this area identified to the pilots as the Hawaiian island of Oahu, and the area as Pearl Harbor. There is some ambiguity as to whether the smaller-scale mock-up of Pearl Harbor was a part of the Skioku setup or at a different location.

*c*. Roberta Wohlstetter, *Pearl Harbor, Warning and Decision*, pp. 371–373, 381, Stanford University Press, 1962.

*d*. R. D. Specht, "War Games," in M. Davis and M. Verhulst (eds.), *Operations Research in Practice*, pp. 144–145, Report of a NATO Conference, New York, Pergamon Press, 1958.

*e*. John P. Young, ORO-SP-98, 1960, p. 86.

*f*. Original data on these war games and plans were reported in the very extensive Tokyo War Crimes Trial (1946–47) Documents, comprising more than 50,000 pages and on deposit in the U.S. National Archives and the Library of Congress. An index to these papers was prepared by Paul S. Dull and Michael Takaaki Umemura, the latter a nisei, under the title *The Tokyo Trials—A Functional Index to the Proceedings of the International Military Tribunal for the Far East* and published by the Center for Japanese Studies, University of Michigan Press, in 1957 as Occasional Papers No. 6. The first few references cited above are briefer and more readily available.

2-9. Admiral Yamamoto, Commander of the Japanese Combined Fleet, who had served as Naval Attaché of the Japanese Embassy in Washington many years earlier, steadfastly held for a surprise attack on Pearl

Harbor as a means of removing the U.S. fleet's threat to the Japanese Navy's operations in the Pacific waters in conducting Japan's Southeast Asian plan of conquest. Finally his will prevailed over the misgivings of other admirals of the Naval General Staff.

a. Office of the Chief of Military History, *Command Decision*, pp. 63–87, 419–422; particularly pp. 70, 73, 78, 82, 83, New York, Harcourt, Brace and Company, Inc., 1951.

b. Mitsuo Fuchida and Okumiya Masatake, *Midway, The Battle That Doomed Japan* (detailed account), pp. 20–25, 52, 94–97, Annapolis, Md., U.S. Naval Institute Publication, 1955.

c. John Deane Potter, *Yamamoto, The Man Who Menaced America* (detailed account), pp. 46–51, 59, 63, 141, 183, 252, New York, The Viking Press, Inc., 1965.

d. Roberta Wohlstetter, *Pearl Harbor, Warning and Decision*, Foreword by Thomas C. Schelling, and particularly pp. 354–355, 368, 371–373.

e. Sidney F. Giffin, *The Crisis Game. Simulating International Conflict*, pp. 59–61, Garden City, N.Y., Doubleday & Company, Inc., 1965.

f. *War Games*, p. 121, U.S. Army Historical Document MS P-094, by Rudolf Hofmann, General der Infanterie a. D., with a foreword by General Oberst a. D. Franz Halder, translated by P. Nuetzkendorf, and prepared in the Historical Division, Headquarters, United States Army, Europe, 1952. Published by Department of the Army, Office of the Chief of Military History.

2-10. Roberta Wohlstetter, *Pearl Harbor, Warning and Decision*, pp. 371, 380–381; Fuchida and Masatake, *The Battle That Doomed Japan*, pp. 20–21, 24.

For a detailed account of the Pearl Harbor attack see Roberta Wohlstetter's book that drew upon official documents, many of which were Tokyo war-crime trial documents.

For a Japanese viewpoint, see Fuchida and Masatake, *Midway, The Battle that Doomed Japan*, composed of reports by two Japanese officers, Mitsuo Fuchida, former Captain, and Masatake Okumiya, former Commander, Imperial Japanese Navy. Roger Pineau, a former U.S. Naval Intelligence Officer, translated the Japanese accounts into English and edited the book. See particularly pp. 25–33 for details of the Japanese strike at Pearl Harbor.

Samuel Eliot Morison, in *History of United States Naval Operations in World War II*, gives a very informative and lucid account of the Pearl Harbor attack. See vol. III, chap. V, pp. 80–171.

John Deane Potter, *Yamamoto, The Man Who Menaced America*, pp. 46–83, 140–141, 182–185, 252–253, and 301–313.

2-11. The Imperial Japanese Naval Officers were Capt. Schiro Yokoyama, Naval Attaché in Washington, Lt. Yoshimori Terai, his assistant, and Capt. Tsunego Wachi, Naval Attaché in Mexico City.

2-12. This information was obtained in separate interviews with the Japanese officers involved, in Tokyo, after the defeat of Japan, by Lt. Roger Pineau, USNR.

2-13. Samuel Eliot Morison (Rear Adm., USNR), the eminent historian and author of the fifteen-volume *History of United States Naval Operations in World War II*, gives an authoritative account of the naval war in the Pacific. Another pertinent account, written by Professor Potter, of the U.S. Naval Academy faculty, appears in *Yamamoto, The Man Who Menaced America*.

2-14. The principles of war as taught in U.S. service schools are listed and described in the U.S. Army Field Manual 100-5, *Field Service Regulations, Operations*. They are listed as follows: Objective, Offensive, Mass, Economy of Force, Maneuver, Unity of Command, Security, Surprise, Simplicity. Variously stated, they have been enunciated by J. F. C. Fuller and many others and are rather universally taught in the military services of all nations with modern military forces. (Fuller, *The Foundations of the Science of War*. Debatable as "principles," nevertheless they are widely accepted as guides or rules for the conduct of war. Nor are they immutable nor complete for all situations. For example, vulnerability, obscurity (as in jungle fighting in Vietnam); intelligence, timing, training, and determination or morale (as in the Arab-Israeli War of 1967); and endurance (which Fuller lists, as well as "determinations") are other basic factors. For a discussion of some of these factors see Bretnor, "Vulnerability and the Military Equation" in *Military Review*, Sept. 1966, pp. 18–26. Bretnor expresses the opinion that these "principles" were in a great measure responsible for the vast superiority of tactics and techniques in World War II over World War I (e.g., trench warfare, etc.) counterparts. (p. 19.)

Mao Tse-tung, as the architect of Communist Revolutionary War, also has laid down a set of "10 Basic Principles of [revolutionary or guerrilla] War." These are briefly listed as: aims, mobile concentration, annihilation, fighting on the move, the offensive, surprise attack, continuous attack, autonomy, unity, the military spirit. See Kenmin Ho's summary article, "Mao's 10 Principles of War," *Military Review*, July 1967. For a more extended discussion see Mao's own writings, particularly *On Guerrilla Warfare*.

Another viewpoint is that there are no principles of war if by principle is meant a law or truth which always applies with regularity and exactitude to any situation. In this view "the objective" (or Mao's "aims") is of primary importance; everything else should be considered in terms of how it contributes to the achievement of that objective.

2-15. Credit for introducing the concept of operations research in the United States is given to Dr. Conant, as in Florence Trefethen, "A History of

Operations Research," chap. I of McCloskey and Trefethen, *Operations Research for Management*, p. 12, citing Parker and Parker, "Operations Research for the Army," *Combat Forces Journal*, May 1951, p. 5. Dr. Shirley Quimby of Columbia University worked with the British OR Group in England and is also credited with introducing the practice in the United States in his later association with mining operations against Japan in the Pacific and the China-Burma-India theater. This view has been expressed by Ellis A. Johnson and, more recently, by Raymond H. Milkman, "Operations Research in World War II," in the *United States Naval Institute Proceedings*, May 1968, p. 78, quoting Charles Kittel of MIT. Kittel was associated with Johnson in the first U.S. OR Group at the U.S. Naval Ordnance Laboratory.

2-16. Ellis A. Johnson, "The History and Future of War Gaming in Operations Research" a paper presented at the Third War Games Symposium at the University of Michigan, Oct. 6, 1960. Also in "The Application of Operations Research to Industry," pp. 31–33, a paper presented at the Fifth Annual Industrial Engineering Institute, University of California, Berkeley, Jan. 31 and Feb. 3, 1953.

2-17. Lieutenant Commander (later Commander) Ellis A. Johnson, USNR, served successively on the staff, Chief of Naval Operations, as mining officer in various naval commands in the Pacific, as the Naval Mining Officer, Twentieth Air Force, and as Director of Mining, XXI Bomber Command, USAF.

2-18. An account of these activities was given by Dr. George Shortley in his presidential address to the Operations Research Society of America at the meeting in Santa Monica, Calif., May 18, 1966. It was subsequently published in *Operations Research*, 15(1):1–10, Jan.–Feb. 1967. Some additional details can be found in another unclassified source in Florence Trefethen's chapter on the history of operations research, McCloskey and Trefethen, *Operations Research for Management*, pp. 14–17.

2-19. Lieutenant General Dwight E. Beach, USA, Commanding General, Combat Developments Command, Fort Belvoir, Va., writing in *The Army-Navy-Air Force Journal and Register*, June 13, 1964, pp. 20–22, refers to the use of "studies, war games, and field experiments" in the Air Assault Division. More details of "The Army's Tactical Mobility Concept" were reported under that title by Brig. Gen. George B. Pickett, Jr., USA, Chief of Staff, USCDC, in the Nov. 1964 issue of *U.S. Army Aviation Digest*.

2-20. Major General Harry W. O. Kinnard was chosen to command the first experimental airmobile division, the 11th Air Assault Division (Test) at Fort Benning, Ga., in 1962 and to conduct extensive studies and tests to evaluate the concept. He summarized the transition from test status to activation, operational readiness, deployment to Vietnam, and operational employment in the Vietnam War in an article, "Activation to Combat—in 90 Days. The Story of the 1st Cavalry Division (Airmobile)," *Army Information Digest*, April 1966, pp. 24–31.

2-21. For a detailed account of the Pleiku operations see the article by Lieutenant General Kinnard, "A Victory in the Ia Drang: The Triumph of a Concept. The 1st Air Cavalry Division in Battle," *Army*, 17(9):71–91, Sept. 1967.

# Operations Research and Analytical Gaming

## DEVELOPMENT OF GAMING IN OPERATIONS RESEARCH

In World Wars I and II, science was called into service on an unprecedented scale, particularly in developing weapons of war. Mobilized in support of the war effort, science also developed techniques needed to make effective use of weapons and resources. Operations research (OR) was one of those important techniques.[1]

The post-World War II technological explosion created further need for extensive analytical studies. Nuclear weapons, missiles, satellites, and the progressive conquest of space required enhanced analytic capabilities. Moreover, improved means, such as electronic computers, now existed to aid in the study of optimum applications of new weapons systems, tactical and strategic concepts, and projections of new generations of both.

Because operations research and analytical gaming arose from military needs, their development will be traced in a military context. Gaming as a contributing technique in operations research and its relation to higher-level sponsoring organizations are illustrated in Fig. 3-1. Also shown is the tie-in between war games and field studies.

---

[1] One of the earliest books on operations research gives the following definition: "Operations research is a scientific method of providing executive departments with a quantitative basis for decisions regarding the operations under their control." Philip M. Morse and George E. Kimball, *Methods of Operations Research*, p. 1, New York, John Wiley & Sons, Inc., 1951.

62  Venture Simulation in War, Business, and Politics

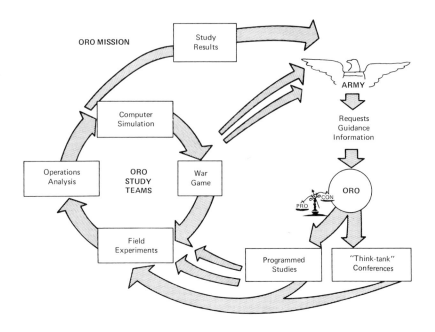

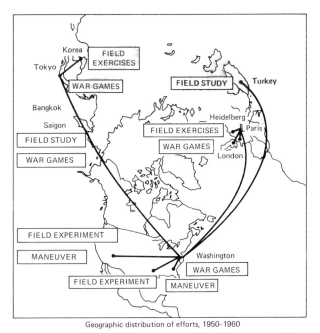

Geographic distribution of efforts, 1950-1960

FIG. 3-1. The place of war games in an operations research organization. ORO as an example.

The latter involve the generation of data needed both as inputs for the simulations and games and for the testing and validation of the results of games. Often field experiments were designed expressly to yield the kind of data on which games and simulations must depend, e.g., on hit probabilities in tank battles. Field exercises and maneuvers often are necessary to assure the practicality of results indicated in games, e.g., the operational limits and problems encountered in the use of helicopters in tactical situations.

In early operations research, analyses for military decisions were focused on specific problems for which precise mathematical solutions were well suited. Later, operations research came to be applied to broader problems of national security. An exclusively quantitative approach no longer would extend over this broad field. Instead of study of one phase of a military problem, or one specific aspect of an operation, such as optimum fuse settings for depth bombs in antisubmarine warfare, entire weapons systems became the subject of OR studies. To match this widening domain, operations research as a field of study came to embrace an approach to complex problems that often require decisions or choices, at times involving multiple objectives, some of which may be conflicting, and analysis which must allow for a larger element of judgment. Along with this development, other terms sometimes were substituted for broader or narrower aspects of OR. Among such terms in contemporary use are operations analysis, systems engineering, systems analysis, cost-effectiveness analysis, management science.[1]

Analytical War Gaming

Analytical war gaming emerged as a recognized activity in World War II. The British and the Germans had developed the precursor of analytical war gaming in experimentation with various tactical concepts, notably armored warfare. The British also had initiated a type of analytical study of military operations which they called "operational research." In 1942 the first operations research and analytical type war gaming in the United States began at the Naval Ordnance Laboratory under the direction of Dr. Ellis A. Johnson, as mentioned in the preceding chapter.

After World War II, on the initiative of Dr. Johnson, an operations

---

[1] E. S. Quade (ed.), *Analysis for Military Decisions,* p. v; "Orientation," p. 1, and "Introduction," pp. 2–12, Chicago, Rand McNally & Company, 1964.

FIG. 3-2. Originators of war gaming at Operations Research Office. *Left to right:* Dr. Ellis A. Johnson, Director; Dr. George Gamow, Consultant; Richard E. Zimmerman, Analyst. *(Gamow photo by Bishop.)*

research organization was set up in Washington principally to serve the needs of the U.S. Army. First called the General Research Office, the name soon was changed to Operations Research Office (ORO). The organization was an adjunct of The Johns Hopkins University, and Dr. Johnson was appointed Director. Dr. Johnson's earlier distinguished work in operations research on mining operations against Japan had left him an enthusiast for the potential that war gaming held in the study of operational problems.

Among the scientists who came under Dr. Johnson's influence was the late Dr. George A. Gamow, erstwhile professor of physics at George Washington University. As a consultant to ORO in 1950, Gamow was spurred by the intellectual challenge of Johnson's ideas of using Monte Carlo methods as an analytical technique on war gaming, and by the problems of building mathematical models to represent the dynamics of modern warfare. Gamow found time among his several connections and various assignments to develop, at ORO late in 1950, a simple hand-played game known as TIN SOLDIER. See Fig. 3-2.

**TIN SOLDIER:**[1] Dr. Gamow had worked on Monte Carlo[2] methods at the Los Alamos Scientific Laboratory, New Mexico, applying these methods to problems associated with the atomic bomb. In applying these same methods to war gaming, he developed a simple model that

---

[1] TIN SOLDIER became operable in 1952 and is believed to have been the first war game designed as a mathematical model for use in analytical research studies.
[2] Monte Carlo refers to play in which there is a considerable element of chance, as in the spin of a roulette wheel.

corresponded to an engagement between two tank forces on partially wooded, flat terrain. He used a lattice of hexagonal areas, crosshatched for wooded areas and unshaded for open fields. See Fig. 3-3. Gamow's description of the scheme of the game follows.[1]

> The two opposing tank forces, ten units each, are originally located at the rear lines of the battlefield, and *a move* on each side consists in displacement of each of the tanks to one of the adjoining hexagonal fields (although not all tanks must necessarily be moved).
>
> If two opposing tanks come to adjoining white fields, *a battle* is announced, and its outcome is decided by tossing a coin or a die. If, as may happen, a moving tank comes in contact with two enemy tanks simultaneously, it must "shoot it out" first with one of the tanks and, if victorious, with the other. (More realistic rules can be introduced in that case.)
>
> If a tank on a white field is in contact with an enemy tank in a crosshatched field (and considered concealed), the first tank is always killed (or given a much higher probability of being killed in the dice-tossing process). If both tanks are in the woods, a battle is announced only if one of them moves into the field occupied

---

[1] Adapted from unpublished notes by George A. Gamow at ORO, April 1, 1952.

---

FIG. 3-3. Schematic battlefield of TIN SOLDIER. Two opposing tank forces (Blue and Red), of ten tanks each, face each other at start of battle. Tanks are shown as black and white circles. Crosshatching represents areas of the battlefield in which tanks would be concealed from enemy sightings by overhanging trees.

by the other (half the normal visibility distance), and the outcome is again decided by the toss of a die.

The objective of the game may be the destruction of a maximum number of enemy tanks with least losses to one's own forces, the destruction of some objective located at the rear line of the enemy forces, or still some other purpose.

For exploratory purposes the game was hand-played, and moves were controlled by human decision, but the ultimate intent of the game was to provide for *random actions*. In "intelligent" playing, different strategies will be used by players. The strategy of a game incorporating random action must be built into the rules. Thus, for example, the strategy of *massing all tanks* or *dispersing all tanks* over the field may be introduced. The motion of each tank (random action) of both forces then is decided by tossing a die for each such random tank movement. Playing this kind of random game by hand is a lengthy process, but if played by an electronic computer, as intended, a few hundred games can be played per hour.

Among a large sample of such random games played under the same rules and with the same initial conditions, there will be a large percentage of even-exchange outcomes. There also will be games in which one or the other side achieves a decisive victory. By plotting the results, distribution curves are obtained that can be used in making analytical studies of various tank tactics and strategies. A degree of clustering tendency, for example, may be given to the black tanks and a degree of dispersing tendency to the white. Since clustered, or bunched, tanks present a more lucrative target for antitank weapons, such results may furnish noteworthy indications. Also by random games one can test, for example, the relative advantages of the higher speed of lighter tanks, with their greater vulnerability because of thinner armor, against the advantages of the heavier and less vulnerable tanks. Appropriate strategies for tank utilization differ in relation to tank characteristics. Random games can furnish insights in an evaluation of strategic concepts.

The TIN SOLDIER operation can be described as follows: Imagine a contour map marked to indicate the pre-battle positions of two opposing tank units. Call the forces the Blue and Red sides, respectively. Suppose one man directs the Blue side, and another directs the Red through an imaginary battle. Organize the progress of this battle by assuming the two men agree to the set of rules that follow:[1]

---

[1] From unpublished notes by Richard E. Zimmerman at ORO, dated Nov. 24, 1952.

1. Each man, in his turn, is permitted to move his tanks a distance that would correspond to 10 seconds elapsed time on real terrain, consistent with the mobility capabilities of the tank.
2. After each man moves his tanks according to his military good sense, tanks within range are assumed to bring fire on the opposition. The winner of tank duels is decided by flipping a loaded coin that is supposed to express the odds on the battle outcome in actual fighting between these tanks. For example, if the Blue tanks are M-46's armed with 90-millimeter guns and the Red tanks are the German Mark III's armed with 50-millimeter guns, then it might make military sense to give the M-46 4 to 1 odds over the Mark III. The loaded coin should therefore name the M-46 the winner 80 percent of the time and the Mark III the winner 20 percent of the time.
3. The battle is over as soon as all tanks on either side have been knocked out or have made successful withdrawal from the battlefield.

The above rules follow those of map exercises long used for the training of officers as, for example, in KRIEGSSPIEL. Rule 1 is so phrased that the men playing the game must be familiar with the use of tanks. Military judgment is exercised at every move. Contrariwise, compliance with Rules 2 and 3 does not require military experience. Military men, however, are required to formulate the rules.

The intent of the basic scheme for TIN SOLDIER was to have militarily experienced men, or trained gamers, replace the vagueness of Rule 1 with a set of specific rules. The assumption was that, if this could be done and if the resultant rules made military sense, direct placement of rules in mathematical form could be accomplished and the entire "battle" run through on desk calculators by semiskilled technicians, or the battle could be fought many hundreds of times more quickly on an electronic computer. From TIN SOLDIER it was concluded that the procedure did make military sense and was feasible.

**The MAXIMUM COMPLEXITY COMPUTER BATTLE:** The next step beyond the hand-played TIN SOLDIER tank-battle model was a computerized Monte Carlo version (see footnote 2 on p. 64) capable of dealing with more of the variables of actual battle. To work on this problem, with assistance from Gamow, two of Gamow's Los Alamos colleagues, Richard E. Zimmerman and Warren W. Nicholas, joined the ORO staff in 1952. In the meantime Dr. Nicholas M. Smith and Dr. Joseph O. Harrison, Jr., of the ORO staff, arranged for an electronic computer installation at ORO. Dr. Harrison contributed to the design of the computer model and programs. The game became operable in 1954. The MAXIMUM COMPLEXITY COMPUTER BATTLE was the first *computerized* analytical war game intended for research use in operational research studies.

Richard Zimmerman considered the Monte Carlo application technically feasible for study of small-unit actions. Zimmerman also concluded that all significant factors affecting combat actions that have yielded to measurement and quantification can be included in this type of computer battle and that the technique can be extended to larger combat units. Additional factors may be included when such factors become measurable.[1]

The Rise and Spread of Analytical War Gaming

At a conference held at Headquarters, Army Field Forces, Fort Monroe, Virginia, in August 1952, attention was focused on what operations research, in general, and ORO in particular, could do to assist the Army in improving its basic capability and its research and development (R&D) program. At this conference, war gaming, holding promise of providing the kinds of situations needed to assess the feasibility and the advantages and disadvantages of certain proposed innovations, came under careful consideration.

**Concept Crystallized:** Two significant ideas emerged from the Fort Monroe conference. The first was that war gaming could be used as a research instrument. The concept of analytical war gaming crystallized; possibly the idea was born at this conference. The second idea was that a series of games pegged to different levels could study the whole spectrum of warfare and should be developed to probe into each of these levels with some intensity.

An example. Although it is common knowledge that both the United States and the U.S.S.R. back up infantry divisions with a certain amount of corps artillery, no one was prepared to say what ratios of tanks or numbers of field artillery pieces could or would be

---

[1] Unpublished notes of R. E. Zimmerman at ORO, Jan. 25, 1955.

employed in an actual situation, i.e., in meeting an attack by a superior force, with specified numbers of divisions available on each side. This kind of question could be studied by war gaming. Further, it was not possible, solely by the analytical methods then used, to integrate all required details into a final, acceptable solution. A method was needed to bring all factors and interactions into play concurrently. Only war gaming held promise for studying the entire problem.

At the Fort Monroe conference certain high-priority problems were formulated with the implication that these were the types of problems operations research organizations (specifically ORO) should study to make a major contribution to the Army. In essence, and by explicit group agreement, the job was to be done by war gaming.

Next, ways of organizing a war-gaming attack to provide comparative data and the means by which gaming could be supported were considered. The possible assignment of active-duty officers to provide military expertise in dealing with new concepts applied to familiar military problems was considered. Problems of developing and operating war games were discussed: the number of personnel required, the acquisition of input data, the requirement for rigid rules, and the time-consuming and onerous tasks inherent in preparing games for research purposes were pinpointed. There was deep concern over the urgent need to determine effective ways of employing innovations for tactical advantage prior to a possible enemy's success in development and use of such methods. Cases were cited to illustrate how in World War II the Allies missed many guesses on enemy capabilities and suffered costly consequences. Among items illustrated were development and use of magnetic sea mines, homing torpedoes, V-1 missiles, V-2 rockets, jets, and certain toxic gases and biologicals, in addition to a German effort to develop atomic weapons. Only in the fields of radar, the proximity fuse, and the atomic bomb did United States development precede that of an enemy.

**Early Gaming at CORG, CONARC:** The Combat Operations Research Group (CORG) at the U.S. Continental Army Command (CONARC) was an ORO field office from 1953 until October 1955 when ORO suspended its contract and Technical Operations Incorporated (TOI) took over. An ironic event attending the early operations research studies at CORG was that a fire destroyed all records and about twenty-five of the first staff memos produced. The only source now available for some of this information is the memory of the operations analysts and military officers who participated in the studies. Some of the early CORG gamers are shown in Fig. 3-4.

FIG. 3-4. Pioneer gamers at Combat Operations Research Group (CORG), U.S. Army Continental Command (CONARC) Headquarters, Fort Monroe, Virginia. ORO and/or TOI (Technical Operations, Inc.) personnel (left to right, top to bottom): Dr. William Whitson, Dr. Franklin Brooks, Archie Colby, Norman Parsons, Dr. Donald Meals, Allen Hulse, William Archer, John C. Flannagan, E. R. Williams.

The early CORG game was fabricated entirely on concepts and models formulated by ORO field-unit personnel. The game was called SYNTAC. The game continued after TOI took over CORG operation. The sequence of subsequent game development follows: (1) A tank-antitank combat model for gaming at the battalion level was needed. Accordingly, a submodel of SYNTAC known as the

TABWAG game was developed in late 1954 and early 1955. Both SYNTAC and TABWAG required manual play. (2) After TOI took over, a computerized version of TABWAG was developed and became known as AUTOTAG.

An interesting aspect of the original SYNTAC game grew out of a need for game data that did not exist. Donald W. Meals and Joseph Bruner reported that nothing was known about the problem of pinpointing fire from an antitank weapon. This capability was necessary to disclose the position of an attacking weapon to a tank crew under attack. A field experiment was set up to provide such data. The ORO report[1] of this study became a classic and has yielded data for all war gamers in the United States and the United Kingdom. Here was an example of a game that established a requirement for field tests. The field tests in turn yielded data that enabled the game to solve problems not previously within the capability of operations research analytical study. See Fig. 3-5.

British Analytical Games

The earliest analytical war game in the United Kingdom started in 1953 at the then Army Operational Research Group (AORG), later the Army Operational Research Establishment (AORE). G. Neville Gadsby conceived a physical simulation to study interactions of several mathematical models emerging from conventional analytical studies.[2] Graham F. Komlosy, under the encouragement of A. W. Ross,[3] Chief Superintendent, and assisted by E. Treadwell, worked on the development of simulation techniques to study logistic supply problems for British Army bases on the European Continent and the Forward supply depots of the British Army of the Rhine (BAOR). Ross foresaw the possibility of developing a more complete and sophisticated war game for analytical studies of tactical problems and discussed this idea with Edward Benn, a senior operational analyst at AORG (latter Deputy Director).

Edward Benn and R. W. Shephard, contemporaries working in two different Establishments in the United Kingdom, became the prime

---

[1] ORO-T-362, Project Pinpoint (disclosure of antitank weapons to overwatching tanks). Also direct information from Dr. Meals.
[2] Information to the author in a letter, dated Dec. 12, 1966, from G. N. Gadsby.
[3] Ross had visited RAND in 1954 where Paxson's war-gaming efforts stimulated and encouraged his interest in the technique. Letter dated Feb. 7, 1967 to author from A. W. Ross.

72     Venture Simulation in War, Business, and Politics

movers in the development of an analytical tactical war game at each of the two Establishments. Earlier each of these men had been skeptical and negative concerning the potential for war games in analytical studies. Pressures by others, however, forced Benn and Shephard to serious consideration of gaming. In the process they became the principal developers of analytical war gaming in the United Kingdom. Shephard believes one significant idea converted him to the use of analytical gaming as an operations research tool in tactical studies. That idea was the use of colored lines on the terrain board to mark ridge and valley lines,[1] thereby solving the problem of weapon-to-weapon (e.g., individual tanks) intervisibility, and enabling formulation of pertinent rules. From that point he could see the possibilities of analytical games.

In 1954 Benn and Lt. Col. G. W. H. Field of the British Army developed concepts of gaming on which the bulk of subsequent analytical studies in AORG/AORE were based. These concepts, originated in the earlier Aldershot-type training games, were updated with rules and techniques amenable to analytical studies. Aldershot of 1872 vintage, like E. Baring's translation in 1872 of the Tschischiwitz KRIEGSSPIEL games, provided the foundation for, but was not suitable per se for, analytical studies of the nuclear age. Benn reports that Colonel Field fostered many of the ideas and techniques for the physical representations and physical displays involved in the game and, most important, had much to do with the development of the communications and control systems and procedures. In the critical early years of AORG/AORE gaming Colonel Field acted as Chief Controller and, more generally, as Project Manager. He also shared

---

[1] Shephard regarded this concept as the "breakthrough" that opened up analytical techniques in tactical-level war games. By this device the analyst could relate the positions of small units, individual weapon teams, vehicles, etc., to command and analytical parameters, all requirements of the Royal Armament Research and Development Establishment (RARDE) studies.

---

FIG. 3-5. Field experimentation sponsored by the U.S. Army Continental Army Command (CONARC) to yield tank-battle data required for war gaming at Combat Operations Research Group (CORG) in the SYNTAC and AUTOTAG games. *Upper:* analysts interviewing tank crews after firing. *Lower:* Field Data Analysis Center, automatic recording equipment and other data records. (*U.S. Army photos.*)

with Benn the rather onerous task of producing and revising the first sets of gaming rules.

The basic problem, as in all meaningful war games, was to get valid data as the foundation for reference tables (input data) and rules. Benn reports they "were able to get reasonable estimates of many tactical parameters from the large-scale maneuvers held in BAOR and this was a big help. (Unfortunately for later workers these large and expensive maneuvers were discontinued late in the 1950's.)"[1] In looking back on the pioneering efforts to develop the AORG game as an analytical tool, Benn assessed the principal achievement as "the blending together of rules and concepts in a way that seemed to give the right feel of battle, which did represent a new advance. This at any rate was the impression of our military colleagues."[1]

Further problems were met in matching detailed simulations of many components *to produce a physical presentation and a demand in the game for command activity and decisions in forms similar to those which commanders would meet and face in field headquarters.* Another significant feature was "the widespread introduction of probabilistic parameters in fire, movement, planning, casualties, etc., *to produce appropriate uncertainty* in the battle simulation."[2]

**The AORE Games:** Aore developed three war games. The first was a cross-channel game with a schematic representation of day-to-day troop and supply movement from the United Kingdom to Continental ports. Some allowance was made for the treatment of movement of supplies to and from the ports involved.

The second game was a *map maneuver* that permitted calculation of the *movement of supplies* from the Continental ports forward to the division supply points. The third, and perhaps principal, game was a *map maneuver* played on a three-dimensional terrain board with the *combat actions of the British Army of the Rhine* (RAOR) *versus an aggressor.* The cycle or interval of play was 1 hour of combat. Resolution was primarily at the company level. The games were designed so that all three could be interrelated, thereby generating a feel for the movement of supplies to port, across the channel to division points, and thence to combat actions on the battlefield. Figure 3-6 shows one of the AORE games in operation.

---

[1] Letters to author from Edward Benn.
[2] Letters dated July 1, Aug. 8, Dec. 9, 1966, and Jan. 9, 1967 from E. Benn to author. Italics supplied.

FIG. 3-6. British Army Operational Research Establishment (AORE) war game on a map-maneuver board, Control Room. (*British Crown Copyright.*)

**The RARDE Games:** In addition to the AORE games concerned with movement, supply, and fighting of a field army, a need existed for a lower-tactical-level game. This need was recognized in a rather interesting way at the Royal Armament Research and Development Establishment (RARDE) at Fort Halstead, England. It was customary for RARDE to hold an annual open house. A display had been prepared for the 1954 event by an Operational Research Group headed by R. W. Shephard.

This display attempted to show military units on a section of simulated terrain and attracted the attention of the many military officers attending the exhibit. Among the officers were four brigadiers who almost immediately fell to discussing and arguing the tactical problems they could represent on this terrain. Shephard's idea of using gaming for analytical studies at RARDE then and there gained considerable interest, following which Shephard was able to get prompt sanction, including authorization of funds to proceed. In 1954 Shephard initiated work on the RARDE game. Some months later he invited the Director of the Royal Armoured Corps and the Director of Infantry to see the game. The visit resulted in increased support for the game on a permanent basis. The RARDE game continues in operation today. See Figs. 3-7 and 3-8.

Shephard relates:[1]

The final stimulus for the RARDE game did not arise from the need for an attack on problems raised by tactical nuclear weapons as did the AORE game. Rather it arose primarily from a dissatisfaction that I felt with some of the methods of weapon assessment that we were using at RARDE at that time. There had been a tendency to develop rather simple and straightforward analytical models which ignored to a large extent the effects of tactics and terrain on weapon effectiveness. War gaming seemed to be a technique that gave an opportunity of including these factors in a realistic way and more particularly, of taking account of the many interactions that there were between weapons of different types in a small unit action. In other words, although like Benn, I turned to war gaming because there was a need to solve real problems in a manner that was better than any used previously, the detailed requirements were really rather different.

The RARDE game, pegged to operational units down to sections

---

[1] Letter dated Nov. 28, 1966 to author from R. W. Shephard.

FIG. 3-7. British Royal Armoured Research and Development Establishment (RARDE) war game on a terrain board, Control Room. (*British Crown Copyright.*)

Operations Research and Analytical Gaming 77

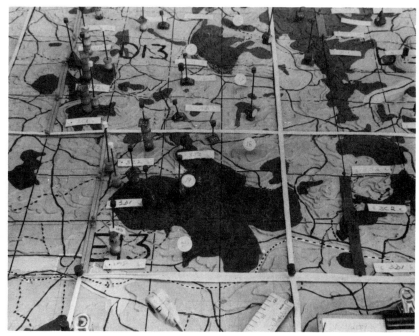

FIG. 3-8. Close-up of a map-maneuver board, British (RARDE) war game. (*British Crown Copyright.*)

(squads) of men, gives individual representation of tanks, antitank guns, and artillery weapons. The game could play up to battalion-size units. Shephard reports: "Our rules were based in the main on the results of analytical studies that we had carried out previously in the Weapons Assessment Branch at RARDE, coupled with the advice we had obtained from various Army Schools."[1]

Although models, rules, and data were developed independently, there was some interchange of information among war gamers in the United Kingdom and in the United States. Although these exchanges did much to share methodology and raw data, each game was built on its own set of concepts and models. The result was that, although the games were similar in features, models were not interchangeable.[2]

---

[1] *Ibid.*
[2] The games at ORO—CARMONETTE, INDIGO, ZIGSPIEL, etc. (Washington, D.C. area)—SYNTAC at CORG (Fort Monroe, Va.), and SIERRA at RAND (Santa Monica, Calif.) were developed concurrently, followed soon by many games and simulations in other research and military establishments. Even if interchangeability of component simulations and models had been an objective, it is doubtful that researchers working at such distance from each other could have made much progress toward this goal at that time.

Concepts of interchangeability were evolving but means of representing and applying these concepts had to be invented or engineered. Some leading British gamers are shown in Fig. 3-9.

**Comparative Gaming:** A start on comparative gaming was made within the United Kingdom in 1964 when a series of six battles at company level were played in the AORE game, and the same six battles were played in the RARDE game. Each game used the same locations and forces and produced similar battle outcomes.

This AORE/RARDE comparative gaming was made possible by the similarity of the games; beyond that, both games were designed to play the same type of situations for the British Army of the Rhine as a part of the NATO defense forces stationed in West Germany. The substance of the games is immaterial for this discussion; the significance here is that there is promise in comparative gaming at different facilities. Technical difficulties are not involved in comparative gaming at the same facility, either with the same or paired teams.[1] However, the practice is uncommon.

International cooperation in comparative gaming had been favorably considered at the triennial Tripartite Conferences on Army Operations Research, in which the OR and war-gaming personnel of the United Kingdom, Canada, and the United States participated. The idea was carried into action, with the CARMONETTE game played at the Research Analysis Corporation in the United States in 1965 and the CAORE game played in Canada in 1966.

Extension of Analytical Gaming

The decade of the 1950s was a period of great expansion of military operations research in the United States and Canada.[2] Operations research groups, by whatever name, were set up in all military services at a large number of headquarters and installations. Problems calling for solution flooded into these OR groups. Many of the problems could not be studied by traditional analytical techniques

---

[1] The results of comparative gaming can be useful in gross outcomes but the details are not comparable in human-decision games. Automatic simulations avoid the variability of human decisions but also bypass the role of the experienced (and astute) commander.

[2] The United Kingdom was already firmly committed to OR, particularly in the military establishments where it originated. Although OR was introduced in a number of European countries and in Japan in this period, the expansion of military OR activities in most countries was to follow post-World War rehabilitation, a process far from complete in this decade.

FIG. 3-9. Some war gamers of the United Kingdom Operational Research Establishments. *Left to right, top to bottom:* E. Benn, J. E. Beswick, G. W. H. Field, G. N. Gadsby, G. F. Komlosy, A. W. Ross, R. W. Shephard, E. Treadwell, D. Williams.

alone. Analytical war gaming had demonstrated usefulness in the few situations to which it had been applied. Now it became widely recognized as a promising means of contributing to operational studies beyond the scope of traditional methods.

Proliferation of Games and Simulations

By 1962 one military directory listed some sixty organizations engaged in war games and simulations of direct interest to the military

80    Venture Simulation in War, Business, and Politics

Table 3-1. Summary of Existing War Games, 1962

| Purpose of game | No. of organizations involved* |
|---|---|
| Training | 5 |
| Operational† | 16 |
| Research and development | 47 |

* All organizations conducted games for more than one purpose.
† Games designed and played to test military operations plans.

services.[1] Fifty-seven of these organizations were in the United States, two in England, and one in Canada. This directory also reported the purposes for which the games were conducted. A summary tabulation is given in Table 3-1.

Some organizations had several different games and simulations in operation. Almost all the research and development games listed were analytical in nature, as were many of the operational games; thus the total of analytical-type games exceeded sixty. A survey conducted 3 years later again took stock of the games and simulations in use. In this inventory the models were almost exclusively analytical and numbered more than 100.[2]

Starting about 1956, business and management games increased rapidly. A tabulation of such games appeared in books published in 1961 and 1962. One book listed ninety-five, and another listed eighty-nine organizations that had developed and used a management game or simulation.[3] The games (usually company-oriented) were intended largely for in-service training and education to stimulate and improve insights and understanding of business and management problems. A secondary purpose was to contribute to informal analytical considerations, that is, to induce questioning and provide incentive and substantive leads for further study. See Chap. 8.

American business and industry had discovered a new analytical tool—one with appeal sufficient to warrant substantial financial sup-

---

[1] STAG—U.S. Army Strategy and Tactics Analysis Group.
[2] Survey conducted by the Joint War Games Agency. Information furnished by Rear Adm. C. J. Van Arsdall, Jr., Chief, Joint War Games Agency, 1966, and Joseph O. Harrison, Jr., RAC-TP-133, 1964, p. 17.
[3] Kibbee et al., *Management Games*, 1961; Greenlaw et al., *Business Simulation in Industrial and University Education*, 1962. One hundred and eighty-two business games are listed in the book by Robert G. Graham and Clifford F. Gray, *Business Games Handbook*, New York, American Management Association, 1969.

port by the profit-making world. Extensive simulation studies now are under way in graduate schools of business, and long-range projects that include simulation models and extensive analyses of business applications have been initiated in the financial world.[1] The use of war-gaming and simulation techniques within the military establishment had reached a stage of great usefulness and was spreading into many other fields.

---

[1] For example, large brokerage houses have provided funds for simulation studies of the stock market at leading universities and research organizations.

# Essentials for Gaming

### RUDIMENTS OF WAR GAMES

War games differ in many respects, yet basic characteristics are present in all. Among salient similarities are:

1. Every war game simulates a military operation (irrespective of phase or manner of gaming).
2. Each game involves two or more opposing forces.
3. Each war game is conducted in accordance with data, rules, and procedures acceptable to the military profession.
4. Every war game represents an actual or assumed real-life situation.

By common acceptance, one of the opposing forces in a war game is designated the Blue force, or Blue; the other, the Red force, or Red. Each force, Blue or Red, may represent military units of a single country or political entity, or a coalition of forces of more than one nation. Thus, if the game should be set in Europe in the present or future time frame, the Blue force might represent units of NATO forces; the Red force might represent units of the Warsaw Pact forces.

An activity cannot be termed a war game unless military forces are involved in movements or operations accompanied by the clash of arms or the threat of such a clash. War games are played only under conditions of simulated warfare, i.e., shooting, attack, or invasion by one force upon the territory of another. War games rarely consider non-shooting situations unless the situation is an interlude during which forces get into the position or range to fight.[1]

1. War games are based on reasonable military capabilities. Effects for most part are drawn from actual performance records and measurements obtained under field-test situations or from records of actual warfare.
2. Capabilities and effects, adjudged to be militarily sound, are built into war-game data, rules, and procedures.
3. The adherence to prescribed data, rules, and procedures continues throughout the game. The application of specific rules or the unbiased decisions of a knowledgeable Umpire give consistency to the game.

A war game is built upon a situation such as exists or realistically could exist in some locale under certain assumed conditions. The situation is described in detail sufficient for the Commander to visualize the conditions under which he must conduct military operations.

Basic Requirements for War Games

Reduced to minimum essentials, there are four ingredients required for all war games. These prime ingredients are:

1. Knowledgeable personnel
2. Special facilities and equipment
3. Standard data, rules, and procedures
4. A description of the situation and missions to be represented in the game

**Personnel:** War games conducted for serious military purposes need to be staffed with highly qualified military officers. Senior commanders on active duty, or recently retired, are the most appropriate type of resourceful and knowledgeable personnel. Senior military specialists are required when the game includes specialized areas, such as nuclear, electronic, or chemical capabilities.

---

[1] Recent efforts have been made to game cold war, political negotiations, and economic competition. Such moves as blockades, shows of force, protective occupations, and political confrontations, e.g., the Cuban missile crisis, could be gamed but such applications of gaming would not be within the current normal meaning of war game.

Scientists employed in gaming should be militarily oriented. Game scientists possess the mathematical, statistical, and related technological knowledge and skills to design game models. These scientists must translate military characteristics and actions into models and data that may be recorded and tabulated. They must then compute the interactions. They must assure that chance events and results are treated in a sound mathematical and statistical manner to represent realistic probabilities. In supervising the recording and processing of game data, and in the interpretation of results, proper allowance must be made for the effect of assumptions, approximations, and chance results. All elements must be managed within the limits of reliability imposed by the original input data and data-processing methods.

If the game is supported by an electronic computer, some of the scientists must be skilled in computer technology. This skill includes designing and programming, modifying, and troubleshooting the models processed in the computer; the supervision of the computer input preparation; and the interpretation of computer outputs in each cycle of game play. In addition, persons are needed who possess clerical and administrative support skills to perform record-processing duties.

**Facilities and Equipment:** Two or three adjoining rooms, each having access to a connecting corridor and each equipped with a display board on which the forces involved in the game can be represented, are the usual requirement. The display board, most commonly, is a large table on which maps can be fastened or upon which terrain models can be laid. For some games vertical display boards are suitable. These may be maps fastened to walls or projection screens on which maps and overlays may be shown. The display board, horizontal or vertical, must provide an area to place and show markers or "pieces" to identify and represent individual units in their relative positions of deployment. The markers can be a variety of types. On a horizontal display board the pieces may be scale models of tanks, vehicles, and individual soldiers, or symbolic markers representing units up to and including divisions. Markers sufficient to represent every unit involved in the game are a necessary part of the supplies and equipment.

In business and management games the display boards may be used for charts of business operations such as Gantt charts, financial statements, production and sales charts, etc. In short, the display boards show whatever data, conditions, and trends are the significant indicators of status, plans, and results of efforts expended in the situation being gamed.

A word should be added here about computers. Desk computers may serve the needs of game participants in less complex games, but for large complicated simulations and games involving many interacting factors and units and for games for analytical purposes sophisticated modern electronic computers are a necessity.

Other facilities and equipment are involved in games, but only basic requirements are mentioned here.

**Standard Rules, Data, and Procedures:** Standard rules, data, and procedures are provided in the form of a game manual, handbook, or set of game rules. In addition to the rules, a set of reference data (detailed tables) is required. The tables provide data such as the distances different types of military equipment or personnel can move over different types of terrain under different weather conditions, by day or by night, and with or without opposition. As details of the game increase, more and more elaborate sets of rules and tables are required for game operation.

**Description of the Situation:** The statement of the situation required for initiating and conducting any game may be presented in a briefing, as in the case of a simple game. For complicated games, a detailed game directive is used. The statement of situation must include all the "givens" for the game. These "givens" include a statement of the geographic locale, the forces and weapons available to each side, the time frame in which the action is set, and the political and other constraints applicable to the game. Also, "givens" customarily include a résumé of conditions leading to and resulting in the initiation of conflict, the starting deployment of forces involved, and, above all, orders or instructions to each commander stating the purpose and objectives of the game, his special mission, and related details. These statements of the situation may be called the *game directive*, the *scenario*, the *general* and *special situations*, or *starting conditions*. (See Gaming Glossary.) These terms are used in connection with the particular games to which they relate.

The games discussed are those developed to a state of operational readiness. However, before any "operational" game can be employed for a particular purpose, it must be "adapted" or modified to meet the special requirements of intended use. Many distinct games exist in a state of operational readiness. These continuing, operational games usually carry a generic or family name. KRIEGSSPIEL is an example. Over the years KRIEGSSPIEL was modified and adapted to many specialized forms or varieties, of which American KRIEGSSPIEL was one. Although the basic characteristics of these games were similar, usually each variety appeared under its own distinctive "given name."

Family names of operational games have proliferated over the years. A few of the more widely known family-name games are NEWS, SYNTAC, POLEX, TACSPIEL, THEATERSPIEL, Inter-Nation Simulation, VALOR, TEMPER, and CARMONETTE.

Periods in Game Sequence

There are three major periods or phases in the conduct of a war game. These are the pre-play preparatory period; the playing period, during which the game is run through its series of actions; and the post-play period in which the analysis is conducted and the report prepared. The pre-play and post-play periods are each likely to require as much or more real time than is required for the game-playing period. To gain an adequate understanding of the gaming process it is important to look into the work that must be accomplished during the non-playing periods. See Fig. 4-1.

**Pre-play Period:** In the pre-play period[1] preparations are made to initiate the play of the game. Important aspects of a game must be formalized at this time. Reduced to basic essentials, these aspects include:

1. *Problem analysis.* Here a determination and statement of the purposes and objectives appropriate to the game are made. These items are analyzed and categorized to specific topics to be investigated or to questions to be answered by the game.
2. *Definition of conditions.* The definition of the starting conditions, the political antecedents, the locale, and the forces and related military elements to be involved in the game are stated.
3. *Identification of needed data.* The kinds of data that must be generated in the game to clarify the topics or questions to be studied are identified.

---

[1] Frequently called pre-game preparations or period. Because the game includes this period this author prefers to call it the pre-play period.

Essentials for Gaming 87

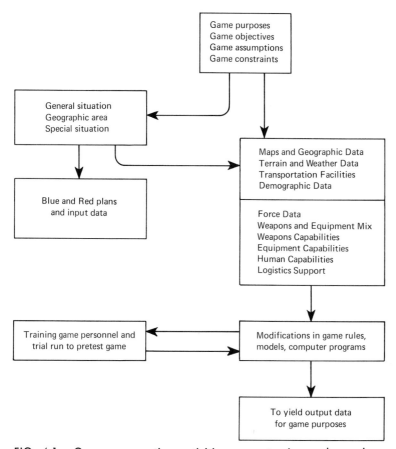

FIG. 4-1. Game preparation activities are extensive and complex.

4. *Input acquisition.* New or specialized input and reference data and tables needed for the game must be generated or acquired.
5. *Rules review.* Augmentation of, or changes in, the standard rules, as may be required by special features and purposes of the game, must be accomplished. A rules review is necessary to ascertain the extent of this effort.
6. *Form design.* The special data-recording forms and work sheets must be designed for the particular game.
7. *Map and chart preparation.* The maps and wall charts to be used must be obtained or prepared.
8. *Team organization.* The players, control and support personnel, and the functions to be performed to conduct the game must be

organized as an integrated whole and arrangements made for the preparation and flow of game information.

9. *Construction of computer routines.* Computer routines required for game operation must be constructed, or reconstructed, and checked out (debugged).
10. *Logistics, including maintenance estimates.* Supplies needed for the game must be estimated, ordered, and stocked. Equipment must be checked out, reconditioned, or replaced, and replaceable parts made available to assure continuous game operation after play begins.
11. *Preparation of game directives.* The game directives, scenario, general and special situations, order-of-battle information, and similar duplicated materials needed as information handouts to the game participants must be prepared.
12. *Preliminary check performance.* The initial work and orientation of players and controllers; the adequacy of models, factors, and basic data; and overall gaming coordination must be checked in an operational test prior to the game. See Fig. 4-1.

**The Playing Period:** The play of a game spreads over a series of cycles. Each cycle represents a fixed period of combat (game) time.[1] In some games the cycle may be 30 minutes of combat actions (game time); in other games a cycle may be a full day (or more) of military operations. A standard procedure is followed during each cycle of play. The Blue and Red Commanders are given a certain number of hours of working time in which to make plans and issue orders for the next cycle (period of game or combat time). See Fig. 4-2.

It is not uncommon for one cycle of play to require from 1 to 3 or more working days. As soon as a cycle is completed, planning begins for the next cycle. This process continues, cycle by cycle, until the objectives of the game have been achieved. The entire series may cover only a few hours or several days of simulated combat (game time) but may require anywhere from approximately a week to several months of working time by the gaming team. (Refer to Fig. 4-2.)

Each Commander's orders are submitted to Control where they are judged and assessed by the Chief Controller or Chief Umpire, who, acting in accordance with the game rules and data, determines what would happen as a result of the orders. The Chief Umpire, in the case

---

[1] Simulated time representing the real time required for a particular stage of a battle or war.

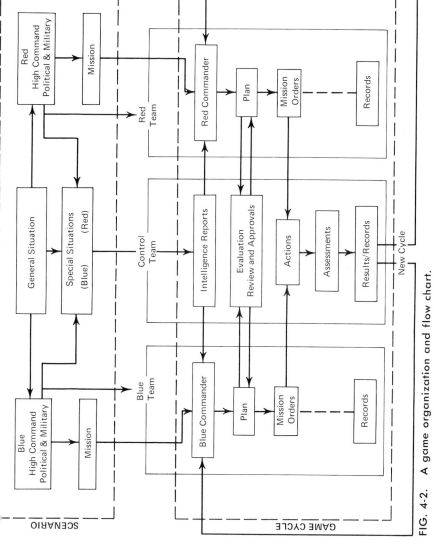

FIG. 4-2. A game organization and flow chart.

of an umpired game, makes a determination in terms of his military judgment. In many games it is necessary that the orders be given in written form. In some highly informal games an oral presentation of orders is acceptable. If the game calls for the Commanders' orders to be written, the Controller's assessments also are written and are delivered to the Commanders. These after-action assessments, cycle by cycle, are called *status reports*.

The Controller, or Control Party, conducts a rather complicated assessment process. The determination of applicable rules, the calculation of reference data pertinent to the particular rule and situation, unit by unit (for all units actively engaged in the operation), and the preparation of the status report on all units of the two forces (Blue and Red) are involved. The Controller's report covers items such as casualties suffered and ground gained or lost, unit by unit, and the new location of units. The Controller then reports these results to the respective Commanders.

**The Post-play Period:** Analysis of game data is involved in the first and also in the final step in the gaming process. As has been indicated, a large amount of information of many different kinds is accumulated and analyzed prior to play. Additional quantities of data are generated or acquired during the play of the game. Each cycle of play involves moves by both Blue and Red. These moves result in certain interactions among components (both conflicting forces—Blue and Red) and among the mutually supporting units of each force. Certain types of units are actively involved at one time and not at another; certain weapons are employed more or less regularly, others less frequently and for special purposes. The dynamics of a war game—as in real warfare—presents a continually shifting set of circumstances to which Commanders must react. Although game objectives or objectives of the Commanders may remain more or less stable, plans and means must be altered as new elements in the situation develop. In accordance with the objectives of the game, the great mass of data generated in the play of the game must be sorted into categories and compiled, and the results interpreted. A number of different analytical techniques may be employed in this process.

The culmination of the analytical process usually is a report. Reports may vary in degree of formality. A critique session, in which the players report on the whys and wherefores of their various decisions and subsequent evaluation of results, may serve the purpose for a training game. Or the Chief Controller or Chief Umpire may report on the lessons learned. Comments are illustrated with excerpts from the game that represent astute, or not so astute, tactical decisions and surprises; the way they were met; and any other features considered significant from an experienced viewpoint.

Formal games attempt to investigate a particular group of questions or problems, obtain an adequate amount of data from the game, and organize all into meaningful comparisons or summaries. It is rare that the gross result of the game—as, for example, Who won?—is a prime consideration. In this respect a war game is unlike a sports contest. In a sense, all war gaming is undertaken for learning purposes, in preparation for leadership and employment of military forces in real war. Therefore, the emphasis is placed on finding ways to improve, or to use, resources for maximum combat effectiveness.

In the foregoing discourse an attempt has been made to outline common characteristics and the basic rudiments that apply to the conduct of war games. War games exist in many forms and are applied to many different purposes, and although there are many similarities each type of game has its own specific set of characteristics.

## COSTS OF WARS, MANEUVERS, AND GAMES

All means of testing plans, providing military training, and furnishing research information on military problems that have been mentioned are expensive, and consideration must be given to the cost factor in relation to the purposes for which the means are employed. A gaming effort using more than forty people may cost as much as a million dollars just to conduct a full game and analyze the game data. This cost does not include the long-term capital investment and costs of developing the gaming capability. The development efforts, plus the design, programming, and check-out of models, and the training of personnel are spread over a period of years. It is not unreasonable to assume that the development of a continuing sophisticated gaming capability may cost in the neighborhood of 2 to 5 million dollars.

Cost of Wars

There are no directly comparable substitutes for gaming, although data from actual war and field maneuvers come to mind. Although rough cost estimates can be attached to the various means of studying warfare, each has certain values and costs beyond those of war gaming. Nevertheless, "ballpark estimates" may point up the magnitude of the money involved and some relative costs.

The cost of war itself cannot be measured adequately by money, even though wars do cost money and official estimates of dollar costs have been made. The figures shown in Table 4-1 report these costs for a number of major United States wars.

Reduced to a daily cost, the United States spent at the rate of 246 million dollars per day for World War II (total expenditures) and an estimated 16 million dollars per day for the Korean War; it was spending about 5.5 million dollars per day in 1965, increased to about 70 million dollars per day in 1968, for the hostilities in Vietnam. The daily cost of operating a game that would be representative of conflict of the type described above amounts to about $500 to $4,000. It is emphasized, however, that neither total dollar costs nor daily costs are comparable measures of the costs of actual warfare or war gaming.

Cost of Maneuvers and Exercises

Field maneuvers also are expensive and usually are budgeted in terms of "add-on" costs. Thus normal operating costs of the military establishment and of the units and equipment involved are not included in maneuver costs. Only such additional expenses as transportation to the maneuver locale, contracts for use of private land and

Table 4-1. Estimated Dollar Cost of Major U.S. Wars* (4-1)†

| | |
|---|---|
| Revolutionary War | $370,000,000 |
| War of 1812 | 92,992,000 |
| Mexican War | 105,756,000 |
| Civil War (Union forces) | 3,055,413,000 |
| Spanish-American War | 444,599,000 |
| World War I | 33,189,864,000 |
| World War II | 331,350,000,000‡ |
| Korean War | 18,082,485,000§ |

* Prior to World War II, the figures include only initial costs, i.e., military expenditures during the periods of active warfare by the U.S. War and Navy Departments.
† These numbers refer to Chapter Notes listed at the end of the chapter.
‡ World War expenditures include money spent by *all federal agencies* on war activities.
§ An example of additional costs not included here is the pensions paid to veterans and their dependents, hospital care, etc. (4-2)

Essentials for Gaming    93

FIG. 4-3. LOGEX-1953. Plotting board. The LOGEX exercises are gamed on a terrain board of this type as part of the Command Post Exercise (CPX) planning and pre-testing process. (*U.S. Army photograph.*)

public facilities in the maneuver area, damage claims arising therefrom, and costs for items such as extra fuel, communications, printing, etc., are included. An example: A 2-week maneuver conducted in 1964 over an area of some 13 million acres of California, Nevada, and Arizona cost some 60 million dollars, or more than 4 million dollars per day. (See Army portion in Table 4-2.) This maneuver, called Joint Exercise Desert Strike, was conducted primarily as a training exercise for the Army–Air Force Strike Command under Gen. Paul D. Adams. The exercise involved some 90,000 Army personnel, 10,000 Air Force personnel, 780 aircraft (operated from twenty-five airfields ranging from Texas to Oregon), 1,000 tanks, and 7,000 wheeled vehicles. Unfortunately, even maneuvers exact a toll in lives. (4-3)

The Command Post Exercise is illustrated by the LOGEX (Logistics Exercise), conducted annually at Fort Lee, Virginia, by the Second U.S. Army Logistics Command. See Fig. 4-3. A LOGEX usually is of 2-week duration and involves 6,000 to 7,000 participants (primarily only Headquarters, staff officers, and supporting per-

Table 4-2. Maneuver and
Exercise Costs (4-4)

| | |
|---|---|
| LOGEX | $541,000 |
| Big Lift | 753,000 |
| Swift Strike III | 9,747,000 |
| Desert Strike | 35,676,000* |

* Army portion only. Air Force and costs bring total to 60 million dollars.

sonnel), with communications networks and subordinate Headquarters. The 1965 LOGEX was budgeted at $541,000.

Budgeted add-on costs for selected maneuvers and exercises are given in Table 4-2.

In recent years the U.S. Army established a West Coast field experimentation laboratory operating out of Fort Ord and employing the terrain of the Hunter Liggett Military Reservation. This laboratory is known as the Combat Developments Experimentation Center, (CDEC), a subordinate command of the U.S. Army Combat Developments Command. The Center is served by a group of military officers and civilian scientists who develop plans for experiments, design specialized instrumentation for measurements, and supervise scientific-data collection. The scientists also analyze the data and prepare the test reports. A battalion of troops is maintained at Hunter Liggett to carry out the experiments. An example of one such experiment was Flash Fire II, conducted in 1965, to test certain concepts for improving night tank gunnery. (4-5) The Center operates on a year-round basis, with an annual budget of about 8 million dollars. (4-6) See Fig. 4-4.

Regular troop training constitutes most of the programmed activity of military units in peacetime and when preparing for combat duties in wartime. This training is a costly process, and these costs, too, are not fully measured in dollar amounts. Injuries occur, lives are lost, and equipment is worn out, damaged, or destroyed. Prior to combat participation, airborne troops must make a miniumm of six practice parachute jumps from aircraft to qualify as paratroopers. Casualties are incurred in these and other training exercises.

The experience of the Japanese in preparing for the attack on Pearl Harbor is a case in point. Captain Ellis M. Zacharias, an intelligence officer of the U.S. Navy who specialized in Japanese affairs, reported construction of a full-scale replica of Oahu Island and Pearl Harbor[1]

---

[1] See official Japanese picture of reduced scale mock-up, Fig. 2-4.

FIG. 4-4. Large-scale terrain model of the Hunter Liggett Military Reservation in use for planning and gaming a field experiment prior to conduct of the experiment with troops in the field. U.S. Army Combat Developments Experimentation Center (CDEC) at Fort Ord, California. *(Photo by Joseph Hazel, Stanford Research Institute.)*

on an evacuated "off-limits" island of Japan as early as 1928. This island was used as a target area for aircraft-carrier bombers. Without knowledge that the island was a replica of any known place, Japanese naval aviators trained for years, perfecting their attack tactics, later employed with devastating effect on December 7, 1941 on the real Pearl Harbor. These facts, Zacharias reported, are contained in a secret book written in Japanese by Commander Minobi (later Admiral) after the Pearl Harbor attack. A copy of this book translated into German was intercepted in 1942, through a German prisoner of war then interned in the United States, by U.S. censorship. In this book Minobi revealed that in the training preparations for Pearl Harbor ". . . 300 planes were lost in two years, partly due to unfavorable weather conditions and partly due to the inexperience of the aviators. . . ." (4-7)[1]

Another example of losses in training from an authoritative Japanese source is reported by Morison, the U.S. Naval historian, who

---

[1] Ellis M. Zacharias, *Secret Missions*, p. 114, New York, G. P. Putnam's Sons, 1946.

wrote: "The Japanese Navy conducted its battle training by preference in northern (Pacific) waters where it would not be observed, and where the men would be hardened by exposure to the elements. It is said that the Combined Fleet would lose between 50 and 100 men, swept overboard or killed by operational accidents, on one of these (annual) exercises; and the press was not allowed to mention it."[1]

Many analysts question the measurement of costs of military operations in terms of dollars. One analyst, who probably reflects the views of many, scoffed at cost effectiveness as a criterion or military measure, with the remark, "Tell me the cost in dollars of a human life. Then I will be able to give you cost-effectiveness figures." But dollar costs are involved in support of military establishments and operations and need to be considered. Even though the means of measuring intangible and non-monetary costs remain elusive, these too warrant serious consideration. This need increases in importance as gaming is applied in more and more intangible arenas as in international relations.

## Chapter Notes

4-1. Chief Program and Analysis Division, U.S. Army Comptroller. Memo to Budget Div., OCA, Comment 2, D/A Feb. 25, 1954.
  (a) *The Military Policy of the United States*, p. 66, by Maj. Gen. Emory Upton, Preface by Elihu Root, Secretary of War, War Department Document 290, 1904. Reprinted, 4th impression, Government Printing Office, Washington, D.C., 1917. Edited by Joseph Sanger.
  (b) Revision by U.S. Treasury Bulletin of Feb. 1952, Table 6B, p. 9, DOD uses 330 billion dollars.
  (c) Information from Office of Comptroller of the Army, November 1965. Prepared for and furnished to Undersecretary of the Army, Aug. 20, 1953.
4-2. Data extracted from a letter, dated Feb. 19, 1952, to Douglas Eddy, Waldorf College, Forest City, Iowa. This letter, prepared by Sol Ruddell, Acting Chief, Program Review and Analysis Division, was included as Comment 2 in a memo to the Budget Division, Office of the Comptroller of the Army, from the Chief of the Program and Analysis Division on or about Feb. 25, 1954.
4-3. These details were reported in *Time* Magazine, June 5, 1964, pp. 24–25.

---

[1] Morison, *History of the United States Naval Operations in World War II*, vol. III, p. 25.

Essentials for Gaming    97

4-4.  Information from U.S. Continental Army Command, Fort Monroe, Va., Oct. 6, 1965.
4-5.  The experiment yielded input data suitable for use in gaming, either to modify or to verify rules formulated in various prior games, on such matters as target acquisition and hit probabilities. Data collected included (1) time of enemy flash; (2) time of acquisition, target identification, loading, and fire; (3) number of target hits at various ranges; and (4) miss distance, with and without reference lights.

In addition, two small-scale feasibility trials were conducted. They were (1) Button-up Flash Fire, consisting of non-live fire, to determine the degradation in aiming accuracy, with and without reference lights, due to the tank being "buttoned up"; and (2) Target Acquisition by Infrared, consisting of non-live fire to determine the increase in flash-localization accuracy through the use of infrared equipment and to acquire an enemy tank as a target by observation of engine exhaust by means of infrared equipment.

4-6.  Information from Budget Office. U.S. Army Combat Developments Command, Fort Belvoir, Va., November 1965.
4-7.  Compare Zacharias, *Secret Missions*, p. 114, and Morison, *History of the United States Naval Operations in World War II*, vol. III, p. 85 and footnote 8.

A joint Japanese-American venture in producing a purportedly accurate documentary motion picture of the attack on Pearl Harbor reported the replica of Pearl Harbor was constructed on an island in the Kurile chain. Reported in the *Washington Post* in November 1967. The film entitled *Tora! Tora! Tora!* was released on the theater circuit in September 1970.

# Models: The Anatomy of Games

## WHY MODELS ARE USED FOR SIMULATIONS

The purpose of a model is to demonstrate the behavior of a real-life counterpart without reproducing all the details.[1] The principal concern of the gamer is to prepare a model that takes the important functional elements into account. Figure 5-1 shows graphically the principal functional elements in a very simple model of a battle situation.

War is an example of the complexity, the uncontrolled variability, and the impossibility of obtaining and recording desired data through manipulation and observation. Models make it possible to examine, manipulate, and analyze certain aspects of performance with greater precision and ease than is permitted by observation of the real-life process.[2] A model, serving as a substitute for and a simplification of the real-life process, brings the task to manageable dimensions. The model builder must, however, understand the entire process and the relation of the component systems. The model builder (one of the gaming specialists) builds models piece by piece as components of a larger, complex system.

---

[1] For further information on models see Charles D. Flagle, William H. Huggins, and Robert H. Roy (eds.), *Operations Research and Systems Engineering*, Baltimore, The Johns Hopkins Press, 1960; see in particular chap. 6 by Thornton Page and specifically p. 122.
[2] *Ibid.*, Naddor's section, p. 176.

Models: The Anatomy of Games    99

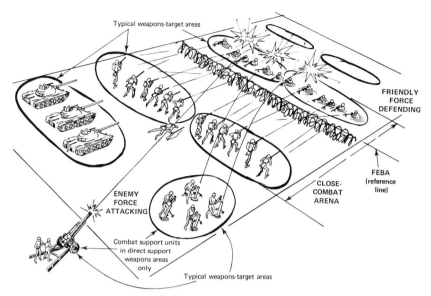

FIG. 5-1. Graphic model of a battle.

A model,[1] a representation of an object, a structure, a system, a process, or a series of related events,[2] is basic to the solution of many research problems. The term *model* has a family of common meanings as well as specialized meanings in mathematics, computer science, and gaming.

Macro- and Micro-models

In gaming the word model may refer (1) to a *whole* system, (2) to a *component* system, (3) to one *organic* part, or (4) to one *elementary* part. An example may be drawn from model railroading. The entire model includes all the trackage, switches, rolling stock, tunnels, bridges, terrain features, and everything else involved in the *whole* system. The trackage system (track and switches) represents a *component* system; a locomotive is one *organic* part; a tunnel or bridge, one *elementary* part. Yet each is a model. The larger, complex models may be called *macro-models*. These most often are composed of smaller, simple models which can be called *micro-models*. The

---

[1] Naddor, in Flagle et al., *op. cit.*, p. 175.
[2] McHugh, *Fundamentals of War Gaming*, 2d ed., p. 1-3.

macro-model (one whole system) may be a micro-model if and when it becomes a part of a larger system. For example, the railway system (a macro-model) of a given locality may become a micro-model in a representation (macro-model) of the entire transportation system of a country.

## Modeling of Combat Operations

Few activities of mankind are more complex than combat operations, and few have been studied as assiduously. The knowledge of warfare accumulated over the period of recorded history is a formidable mass of information. Fortunately much of this knowledge has been thoroughly evaluated and organized into military doctrine. This information is a rich heritage for war gamers. Data from past experience, available for use with current studies, enable the war gamer to turn his attention to game model design.

In war gaming the entire process of combat operations is divided into major components. These components or divisions of the overall macro-model are called submodels, or even micro-models, and often are given specific names such as the "air model" or the "SAM model" (surface-to-air missile model). To approach the problem in simple form a start is made with a model of a relatively localized phase of combat action—one that involves a *limited* number of elements. Each submodel or micro-model must include all needed elements without duplicating any elements of another submodel. For example, if the game is to simulate the employment of nuclear weapons, the elements of information pertaining to such weapons could be built into a *combat* model or into a separate *atomic* model but should not be in both.

In land warfare a combat model is required whether or not atomic weapons are used. A separate atomic model to represent the employment of nuclear weapons, when nuclear warfare is represented, facilitates gaming operations. Consequently, model builders choose to build separate combat and atomic submodels. In combat today, atomic weapons may or may not be used. If used, they would dominate the battlefield situation. For these reasons there should be a model for both combat and nuclear fire.

The specialized meaning of model, as used in operations research and war gaming, refers to any representation "that gives insight or facilitates reasoning about the problem at hand."[1] Because the object

---
[1] Page, in Flagle et al., *op. cit.*, p. 125.

Models: The Anatomy of Games    101

is to study interactions and effects in a precise manner, a quantified method is sought. This objective leads to the use of mathematical models.[1]

## MATHEMATICAL MODELS

Mathematical models are extensively employed in electronic data processing, operations research, and war gaming. A mathematical model is defined by one international association of war gamers as one "in which properties of the things represented and their interactions are expressed symbolically by means of mathematical expressions."[2] War gamers attempt to develop and use mathematical models to represent the actions in a game.

Mathematical Models of Battle

The model is all important in a simulation. In a war game some factors not explicitly quantified in the model can be added by the Controller, specified in the rules or input data, or covered or ruled out in the game assumptions. This is not done in a true simulation. If, for analysis purposes, only those aspects that can be built into a model will be needed, then the model will suffice without augmentation. However, model builders have not been able to construct models of battle to yield data on all questions for which answers are sought by military decision makers and operations analysts.

In short, effective models of battle can be built for some of the important factors involved in combat, but not all. Too little is known about the geometry of battle. Even information on the rate of casualty production and the resulting effects is inadequate to build precise mathematical equations. Formulas and models from historical records have some validity but their products are only grossly similar and seem to have predictive value in a general way only, or in terms of aggregations of many battles. Although such models are useful, they

---

[1] Qualitative determinations are acceptable and sought for some situations, particularly where means of quantification have not been developed or cannot be used with the available data. Forecasts of weather are frequently given in qualitative terms: clear, cloudy, rain, hot, cold, high or low humidity, windy, calm; sometimes in quantitative terms, as temperature, "expected high of 82." Quantitative terms are preferred where such measurement is possible, as in recorded events—yesterday: temperature, high 84°, low 65°; precipitation, 0.35 inch of rain; humidity 56 percent at noon.
[2] Quadripartite Conference on Military Operations Research, Ad Hoc Working Group on War Gaming.

do not provide all the interactions and measures of effect needed for many analytical purposes. The models are acceptable if the assumption is made that whatever errors exist will have an equal effect on Blue and Red tabulations. The object of the game will be served by comparative rather than absolute casualty figures. Such a model is only as good as the capability of the model builder to build military sophistication and battle-logic expertise into the model.

Mathematical models are suited to systems with known characteristics (values) and interactions. Few real-life systems are composed of factors whose important characteristics and effects are all known. Almost always some provision must be made (in the models) for the elements about which little is known. Substitutes for reliable quantified deterministic and probabilistic values must be employed to fill the gap.

Available Data

Where acceptable quantified data are available, no problem exists. Logistics input data are particularly susceptible to quantification. Appropriate data on practically everything that must be procured, transported, and used exist in some tangible form. Unit and dimension designations (for items such as vehicles, guns, or ammunition); the quantity (in volume and weight, for items such as fuel); special requirements (food, refrigeration, etc.); personnel, medical evacuation, and treatment facilities are items for which data have been accumulated. Distances, means of transportation, volume and weight to be transported, times in transit, loading, unloading, and turnaround all are relatively well known. The model builder uses quantified data much as an industrial engineer or architect does.

Unavailable Data. Example:
Determination of Firepower Scores (5-1)[1]

Consider the firepower scores for a 90-millimeter tank gun and an M-14 rifle. If the M-14 rifle is taken as a standard and assigned a value of 1, what value should the 90-millimeter have? 30, 40, 60, 100? This is a tricky question. If lethal areas, rate of fire, ammunition load, employment of weapon, type of target (e.g., personnel or armor), and such factors are considered, some basis of comparison exists and a

---

[1] These numbers refer to the Chapter Notes listed at the end of the chapter.

value for the 90-millimeter may be derived.[1] But is the value correct? Who knows? Military experience and expertise and familiarity with the game may engender a professional judgment, "such a value seems reasonable," but that may be as far as one can go.

Suppose a tank is operating alone (1) against infantry on foot or (2) against another tank without accompanying infantry, other artillery, or air support. Is this tank the equivalent of 30 infantrymen? If so, could 5 tanks replace an infantry company of 150 riflemen on the field of battle? Obviously, no. What then is the justification for this ratio? Basically the firepower score must be considered in terms of contribution to a force structure tailored to a particular purpose and must represent an effective mix of weapon types against specific targets, soft or hard. These kinds of input data the model builder and the game planner must consider. The process is involved and must be sound both in a technical and a military sense.

Other Input Data

Other input data relate to the performance of weapons, equipment, and men. Some doctrine, based on experimental field testing, is available, such as the normal number of rounds of ammunition that may be fired per hour (to avoid premature burning out of the bore) and the maximum rate of fire permissible. For other data, records are available, for example, the miles per day a tank company may advance over different types of terrain and what different trafficability conditions of surface or weather may be tolerated. Input data must provide answers to such questions as, What is the usable operational life of an item before overhaul or replacement? What percentage of tanks in a unit are "deadlined" (laid up for repairs) at any given time and for how many days? On an average, how many helicopters in a squadron will be operable? For how many hours per day and days per month will these helicopters remain operable, and how many hours per day and month may chopper pilots fly without a serious drop in effectiveness or without experiencing excessive loss rates? What are the kill ratios of Sidewinder missiles fired from fixed-wing aircraft against

---

[1] The firepower ratio between the M-14 rifle and the 90-millimeter gun used in the M-48 tank has varied from about 1:2 to 1:500, under different gaming situations. For example, against "dug-in" personnel, lethal tank fire would be less effective than against personnel in the open. Likewise, M-14 rifle fire against armor would be highly ineffective, except possibly to encourage the tanks to stay "buttoned up" with consequent reduced visibility and maneuverability.

other aircraft, or antitank missiles against tanks; and what is the kill ratio of tank fire per missile or round fired when attacking helicopters? To the data listed add those complications introduced by human factors, i.e., tank and chopper crews exposed to fire under combat rather than safe test conditions. Here the test data are not fully adequate.

Consider also the problem of developing input data for weapons never used in combat, such as tactical nuclear weapons. Not only are lethal areas significant, but troop density in the area at the time and the protective cover or exposure experienced are significant. Even these kinds of input data yield to some reasonable approximation of values, but no one really knows how close these values may be to real effects in actual combat.

Next, consider the never-never land of intangibles such as the numerical values that represent the effects of shock and fatigue on personnel; the relative resourcefulness, initiative, and leadership of the opposing commanders; the results of communication failures; and control mix-ups. It is understandable why war gamers usually omit these factors or "sweep them under the rug" by assuming equality of Blue and Red in these matters.

Unavailable data must be deduced or extrapolated from related data, or assumed, with some expectation that the values are within reasonable limits. An example: The human behavior, determination, morale, or courage of members of a unit in a particular situation cannot be foretold but, within limits, normal expectations can be supplied. Here past experience and knowledge of troops, troop training, discipline, and combat readiness furnish insights. Analogical models, relating new concepts or assumptions regarding future weapons, can be used to extrapolate from known effects to expected values for the new factors. Extreme caution must be exercised in drawing conclusions from extrapolated or assumed data.

Qualitative Data

Judgments of unit discipline, courage, and morale of fighting units can be assigned quantified values that represent different levels or degrees of these qualities. This process is called *scaling*. Thus, qualitative standards as "outstanding," "superior," "good," etc., may be assigned values, e.g., 1.10, 1.00, .80, . . . , to be used as multipliers of normal performance to upgrade or degrade expected performance. Most often such postulated values are avoided, and qualifying factors for morale are omitted. War gamers try to avoid mixing qualitative values with established and reliable quantitative data, but when a problem requires a mix a means exists for so doing. Mathematical models used in war gaming can be either *deterministic* or *probabilistic* in type, and individual models may contain combinations of the two types.

Deterministic Models

Deterministic models involve definite and constant relations which do not change from one application to the next or under varying conditions. The equation for calculating distance traveled when speed (rate of movement) and elapsed time are known is a deterministic model. To illustrate,

$$D = s \cdot t \quad (5\text{-}1)$$

where $D$ (distance traveled) equals $s$ (speed) multiplied by $t$ (elapsed time). The interrelation of the elements or factors is constant and furnishes a like result each time the equation or model is used.

The deterministic models of combat losses most widely recognized among war gamers and operations research practitioners are the Lanchester equations.[1] These, Eqs. 5-2 and 5-3, have become a standard for "reasoning about requirements for military forces and the effectiveness of certain improvements."[2] In war two forces ($M$ and $N$) "interact," that is, they fight. Each shoots at the other, and the military effectiveness ($A$ and $B$) changes with time ($t$) at a rate proportional to the strength of the opposing force. Both forces suffer casualties which decrease the respective force strength and military effectiveness. Two differential equations express these relations:

$$\frac{dN}{dt} = -AM \quad (5\text{-}2)$$

$$\frac{dM}{dt} = -BN \quad (5\text{-}3)$$

Where $dN/dt$ and $dM/dt$ represent the time rate of change of the two forces, each force's rate of change (rate of loss) is proportional to the strength of the opposing force. The number of casualties per day (or other unit of time) in force $N$ depends on the size of the opposing force $M$, and vice versa. (5-3) The differential equations above have been

---

[1] F. W. Lanchester, *Aircraft in Warfare, The Dawn of the Fourth Arm*, London, 1916, a great pioneering work (5-2).
[2] Page, in Flagle et al., *op. cit.*, p. 123.

checked against actual battle data by many investigators and have been found quite representative of simple battles where duel tactics were employed, that is, where each opposing force is within the line of vision of the other.

Lanchester's $N^2$ law epitomizes the mathematical means of expressing what happens when armies (or fleets) clash in battle. This law postulates that: "The *fighting strength* of a force may be broadly defined as proportional to the *square of its numerical strength multiplied by the fighting value of its individual units.*"[1] In such a relation the fighting effectiveness of each of two opposing forces would be equal when the squares of the numerical strength multiplied by the fighting value (rate of kill per soldier) of the individual units of each force are equal. Thus

$$Nr^2 = Mb^2 \qquad (5\text{-}4)$$

The graphs shown as Figs. 5-2 and 5-3 illustrate this relation.

Figure 5-2 shows a Blue force of 1,400 meeting a Red force of 1,000 in combat. As the battle progresses, casualties or losses reduce the fighting effectiveness of the two forces, each at a different rate. If the battle continued until one force was eliminated, Red would have lost 1,000 soldiers, Blue only 400. In this example the fighting *values* of individual units of the Blue and Red forces are considered equal.

Many variations of the $N^2$ law may be applied to correspond to the variations in forces represented. Another example illustrates the military principle of "divide and conquer." Suppose a Red force of 1,000, by some stratagem, could be split into two forces of 500 each, while a Blue force of 1,000 remained intact. Assume that the Blue force of 1,000 could fight one Red force of 500. Assume further that one Red force does not assist the second Red force. The effect would be that the Blue force of 1,000 could wipe out one Red force of 500. The loss to the Blue force would approximate 134. If the remaining 866 of the Blue force then could meet the other Red force of 500, the Blue force could annihilate the second Red force. The total force loss to Blue resulting from the two engagements would be only 293. See Fig. 5-3. However, the graphs of Figs. 5-2 and 5-3 are representative of the Lanchester time-dependent pair of equations that take into account successive casualties as the battle progresses in relation to the time variable.

The Lanchester equations, when applied to the great naval battle of Trafalgar, indicate that Nelson chose the optimum points to divide his

---

[1] Lanchester, *op. cit.*, p. 48.

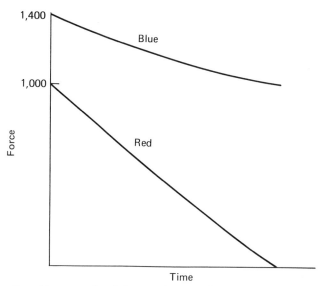

FIG. 5-2. Graphic example of the Lanchester $N^2$ law. Blue force of 1,400 opposing Red force of 1,000. (*Source: Lanchester, Aircraft in Warfare, The Dawn of the Fourth Arm, 1916, p. 45.*)

FIG. 5-3. Successive application of the Lanchester $N^2$ law. Blue force of 1,000 opposing two Red forces of 500 each, in succession. (*Source: Lanchester, Aircraft in Warfare, The Dawn of the Fourth Arm, 1916, p. 40.*)

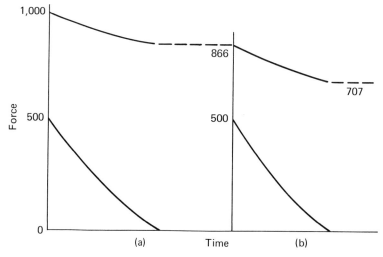

enemy's fleet and annihilate each of the divided parts.[1] (5-4) Although there is no indication that Nelson worked out his tactic by mathematics, the mathematical analog fits. Without detracting from the importance of the Lanchester laws as a foundation for analytical war gaming, the reader is cautioned to recognize their inherent limitations. Among these are the assumptions that each unit has a fixed ability to destroy an opposing unit—as in a duel. This assumption fits old-fashioned sea and land warfare in which two opposing forces advance toward each other ship by ship at sea, or on land in an open formation with infantry and artillery (and/or tanks) each fully exposed to enemy fire, as in the pattern of a Trafalgar or a Waterloo.

For hundreds of years the problem of naval commanders had been how to bring maximum firepower against an enemy fleet with minimum exposure of one's own ships to the enemy's firepower. John Clerk of Eldin, a civilian of the Scottish gentry, studied the great naval battles of history in order to learn how naval tactics could be improved. He reenacted those battles, and others he devised, with little wooden models as in war games. In these experiments, he made extensive drawings, worked out the geometry and mathematics of the wind, ship movements, firepower, and damages inflicted and received. By 1779 Clerk had recorded the concepts he had developed in a preliminary manuscript which he circulated among a few experienced ship captains and naval officers. Encouraged by their reactions, he published his findings in a more complete discourse in 1782[2] and circulated them privately to British Naval Authorities, including Admiral (Sir George) Rodney. Rodney put Clerk's tactics to the test of battle in a meeting with the French fleet in the West Indies later in 1782. The battle resulted in the destruction of the French fleet and the capture of its commander, Admiral de Grasse. Rodney credited Clerk with having devised the new, unorthodox tactic which he employed for "breaking the line" and which Nelson used so effectively in 1797 at Cape St. Vincent and again in the classic and decisive battle at Trafalgar in 1805.

After discussion of the Clerk, Nelson, and Lanchester models of warfare, the Fiske model also deserves mention. Commander (later

---

[1] Rear Admiral S. S. Robison and Mary L. Robison, *A History of Naval Tactics from 1530 to 1930*, pp. 478, 479, Annapolis, Md., United States Naval Institute.
[2] John Clerk, Esq. of Eldin, *An Essay on Naval Tactics, Systematical and Historical*, originally published January 1, 1782, with later expanded editions in 1790, 1804, and 1827, republished in 1964 under the same title as a reprint by University Microfilms, Inc., Ann Arbor, Michigan.

Models: The Anatomy of Games    109

Rear Admiral) Bradley A. Fiske, U.S. Navy, described a naval-battle model in a paper "American Naval Policy" published in 1905, 10 years prior to publication of the Lanchester equations. Fiske's model provided tables with arbitrary numerical values ranging from 1,000 down to 100, which values represent the aggregate comparative strengths (effectiveness) of the fleets, including all factors: ships, guns, personnel, and other factors which affect the effectiveness of a fleet. He formulated a logical system whereby, by time periods, a stronger fleet reduced a weaker fleet in a geometric progression, thus widening the difference in strength between the opposing fleets as time advanced. These tables, and the concept, without being reduced to equations—as done by Lanchester—nevertheless paralleled the $N^2$ law Lanchester enunciated later.[1] (5-5) The Fiske tables assume a battle ends only after one force is eliminated. In current forms of warfare, with potential employment of area (nuclear) weapons, or with defending forces entrenched or concealed, when forces are distributed in low density, or where the situation can be altered quickly, as by guerrillas disappearing, the basic Lanchester assumptions do not hold.[2]

Deterministic mathematical models do not represent real warfare where probabilities or chance factors are introduced. These circumstances require probabilistic models.

Probabilistic Models

A probabilistic model is one in which chance affects the result. A classic example is tossing coins and noting the chances (frequency or sequence) of heads or tails. Because chance factors are common in military situations, probabilistic models are often used.

War games require many models and many determinate inputs, such as the lethal area of an artillery shell as ascertained by proving-ground test firing. Also needed are many probabilistic inputs such as the casualty rate to be expected from artillery shells of a particular type and density falling on a particular target area. The data concerning exposure of enemy soldiers under varying postures such as attack or defense, if "dug in" in open foxholes or protected in part by terrain

---

[1] Bradley A. Fiske, "American Naval Policy," *United States Naval Institute Proceedings*, 31(1):1–80, March 1905. Also reported in Robison and Robison, *op. cit.*
[2] The Lanchester equations assume that when two forces meet they fight to annihilation without any other change in the situation.

features, constitute probabilistic input information. Each of the elements introduces a separate chance or probabilistic factor that must be incorporated if the model is to operate in a sound statistical pattern.

Probability and Random Numbers

The element of chance is provided in a game or simulation by use of random numbers. A random number may be selected by several different methods: by the throw of dice, the spin of a roulette-type wheel, a draw of a card from a deck, or a draw of a number from a table of random numbers. Any of these methods may be simulated and performed by a computer but, first, the frequency of occurrence of the real events must be determined from records or experiment. An example is weather. Various weather conditions—fog, cloud overcast, rain, snow, temperature, etc., each in a range of degrees of intensity or duration—occur with different frequencies and at different periods of time in each geographic area. Records of past occurrences, and frequency of occurrence, exist in most civilized areas. Such data have been or can be reduced to frequency distribution tables, related to calendar dates and time of day for each locale. Data of this type can be related to daylight and darkness, to forenoon and afternoon, or to clock hours. For any particular day of the year and hour of the day a probability distribution can be prepared,[1] and this distribution can be matched to one of the random-number generators, i.e., dice, roulette, cards, etc. If there should be a 50 percent probability of occurrence of an event, a red card drawn from a deck could mean rain, and a black card, clear. For a 25 percent probability of rain, a card from the club suit could mean rain; all others, clear. The probabilities of the conditions simulated can be matched to a set of the same normal frequencies of a scale or set of rules for the random-number generator.

The impact of a particular occurrence of a weather condition upon military operations can be pivotal, even critical—such as mud at the Battle of Waterloo—and that is part of the fortunes (or misfortunes) of war.

Where chance is involved, there is no way of being certain that a particular condition will result. The outcome or result of *one chance event* cannot be anticipated with certainty. This uncertainty applies to the results of war games. The fewer cases involved, the greater

---

[1] Almanacs base "predictions" on such data.

will be the uncertainty of results; conversely, the larger the number of cases involved, the greater the probability of a normal or expected outcome, if the variables and probability distributions are known. (5-6)

The results of games are expressed in quantitative terms such as number of casualties, kilometers advanced, tons of supplies consumed or destroyed—not in qualitative terms such as Who won? Each action involves chance events. For each chance event a random number is drawn and used in the computation. Quantitative results are obtained by the aggregation or accumulation of values of many separate actions. Thus the quantitative results of war games and simulations are sums or aggregations of many other aggregations or accumulations of a whole series of actions. (5-7)

One of the critical assertions about games is that one can tell nothing about results from a single game, because the results may be "way out," i.e., "a fluke." This judgment misses the point that a single game, or a single cycle of a game, involves hundreds of chance *events*, not just one. For example, in THEATERSPIEL it is estimated that in a typical cycle 1,000 random numbers are used in the computed results, and in TACSPIEL, up to 400. In effect, a single cycle of play may represent only 15 or 30 minutes of military operations in the case of TACSPIEL, or a 24-hour period in the case of THEATERSPIEL. Such a game represents a composite of very many chance events—1,000 or 400 per cycle—for each of which a random number is drawn. This phenomenon materially reduces the spread or deviation of results from one interval or one play, and a game may be composed of some 20 to 40 cycles of play (involving a total of 20,000 to 40,000 random values in THEATERSPIEL, and 8,000 to 16,000 in TACSPIEL). The effect of the large number of separate events, each determined by one value within the range of chance variations, reduces the opportunity for extreme or "fluke" results. Occasionally extreme outcomes can occur, particularly in some event such as the destruction of a bridge across a river which could seriously impair troop movements and therefore have a marked influence on the other events or results of the day's military operations. These are the chance events the Chief Controller or Game Director must be on the alert to detect and assess in terms of effect on the game objective(s).

The purpose of the foregoing exposition is not to explain the details of the statistical concepts involved, but to alert the reader to the fact that systematic statistical techniques are employed in sophisticated analytical games, and the results therefrom are indicative if not precise and infallible.

## GAME RULES AS PART OF GAME MODELS

The variability that could occur in judging each move in a war game must be avoided or reduced. This objective is achieved by making the game subject to specific rules and by rigid application of the rules. Without rules or guidelines, inconsistency is inevitable when individual players make a series of independent and free-lance decisions and issue orders to units and when controllers pass judgment and assess effects based on their own views of feasibility and effects of the Commanders' orders and moves. A complete computer simulation avoids inconsistencies of human judgment. One purpose of war games, as contrasted with completely automated computer simulations, is to provide the opportunity for use of the Commanders' initiative, resourcefulness, and skill to exploit opportunities; to make the most effective use of resources; and to conserve forces while maximizing the achievement of the mission or objectives. Consequently it is necessary to take steps to ensure consistency in game evaluation and assessment.

Rules may number in the hundreds—even thousands—for a particular game. Rules cover such matters as the distance a particular kind of unit may move in a given period over different types of terrain and under different weather conditions, and against different types of opposition, and the casualties a unit may suffer under different combat situations, with different force ratios and different types of units in different positions or postures. The rules of play are published in a manual, the Rules Manual, for each particular game and in large measure "round out" the game models.

Game Consistency

In effect, the complete set of rules should reflect the lessons of history and the results of tests and field experiments and should include mathematical and analogical models, deterministic and probabilistic data arrays, and the qualitative or experiential judgments of the military experts. All these are built into the design of the game and are determined in accordance with scientific techniques of operations analyses. Together, these are recorded and constitute the master model of the game. The overall system of game models, submodels, tables of data, and rules, when properly coordinated, represents a balanced unity which provides for reasonable consistency in the game.

When the term "model" is applied to war games, it should be remembered that the entire game consists of a master model that portrays the entire system. Separate models or submodels operate certain parts of the game. Many land war games have separate models for air actions, support weapons (artillery, mortars, etc.), and ground combat (movement, contact, firepower, units, etc.). These are, however, only some of the models (or submodels) employed in a simple war game.

A Conceptual Model

To illustrate the nature of component interrelations among submodels, the data employed, and the effect of interactions among these elements, an example is drawn from a logistics situation. The Logistics Model may be employed to study logistics problems, or it may be used as a submodel in a tactical or strategic game. In either case, if logistic problems in warfare are to be studied, the Logistic Model must be exercised in the context of a complete military operation—a whole war game.

In TACSPIEL,[1] for example, Logistics is one submodel. The game includes the following family of submodels (stated here to indicate the framework of models and submodels that make up the TACSPIEL game model).

Movement and Contact
Ground Operations
   Ground Combat
   Combat Support (artillery and missiles)
   Engineer Operations
Air Operations
   Ground-to-air
   Air-to-ground
   Air transport
   Reconnaissance and Surveillance
Logistics Operations

In THEATERSPIEL,[2] a theater-level war game, the Logistics Model is also employed as a submodel. The THEATERSPIEL sub-

---

[1] TACSPIEL is a tactical war game designed to study elements of land warfare in units from platoon or company level to division or corps level.

[2] THEATERSPIEL also is a tactical war game. The scope includes the employment of all the military forces in a theater of operations, such as Europe, Korea, or Vietnam. As indicated, the game includes an area large enough to encompass a self-contained ground war.

models, in the order processed in computer calculations, are:

Pre-nuclear Air
Nuclear
Air (conventional)
  Ground-to-air
  Air-to-air
  Air-to-ground
Combat Support
Ground Combat
Logistics
Combat Recovery (replacements)
Intelligence
Micro-battle
Command, Control, and Communications

Logistics and Mathematical Models

Logistics are amenable to quantitative treatment of the data and to representation in mathematical (or arithmetic) models. The items are the supplies that must be delivered to fighting units, the evacuation of casualties, the replacement of material losses, personnel replacements, distances traversed, and the means of transport involved plus the capacities of transport facilities. The logistics model input can be reduced to a set of forms on which data may be recorded.

A logistics model, essentially a series of detailed bookkeeping operations to trace and record the movement of forces and supplies to fighting units, was used in the SANDWAR series of games. The model was designed to provide a means whereby the players and controllers could determine rates of force buildup, supply expenditure, resupply, and the status of logistics (supplies) at any time during the game.[1] The model was only one of the many SANDWAR submodels. The design of the game was reduced to minimum essentials. Model design is a major undertaking and often consumes considerably more time than the play of the game.

---

[1] R. E. Zimmerman, *CARMONETTE, A Concept of Tactical War Games*, an internal staff paper in the Operations Research Office, Jan. 1959.

## TERRAIN ANALYSIS

Military personnel require detailed knowledge of the terrain on which military operations are to be conducted. The importance of terrain features varies in proportion to the effect of the features on military operations. The effects of terrain on troop cover, concealment, and movement are of particular importance. The Commander attempts to acquire knowledge of terrain with which he is concerned from standard maps and other available source data and by terrain reconnaissance and personal inspection. Military commanders always have sought and endeavored to hold observation points affording an extensive view of enemy-held ground. Today "taking the high ground" is augmented by aerial observation, and war-game input data are obtained through terrain analysis. Information must be described and recorded for use in a war game. These data could, in actual warfare, be acquired by physical terrain inspection and reconnaissance. For analytical war games precise input data of this type are derived from terrain analysis.

The characteristics of terrain that military geographers have isolated for consideration in war gaming include relief, micro-relief, slope and grades, soils, drainage, and vegetation. Other environmental elements to be considered are the vagaries of weather and climate: rain, snow, wind, temperature, etc., and the frequency, amount or degree, and probability of occurrence of each element on given dates or at times of day.

### Terrain Features and Military Operations

The effects of terrain on military operations were grouped into the following categories for use in the TACSPIEL models. These guidelines are typical.[1]

1. *Cross-country movement.* The TACSPIEL game models include terrain restraints on cross-country movement for wheeled vehicles and tracked vehicles, moving in column as well as tactical dispersion. Obstacles and hindrances both are played. In general, obstacles are depicted as stopping movement altogether but are regarded as a condition that can be alleviated by engineer efforts.

---

[1] Memorandum to author by David B. Doan, Feb. 4, 1965.

Hindrances are considered to slow troop movement over broad areas and are not considered a cause for movement stoppage.
2. *Line of sight.* At present a line-of-sight condition in military operations is simulated by lines representing ridges. Future games are being modeled to use a computer program that will simulate a closer approximation to the actual effect of topography.
3. *Streams and rivers.* Streams and rivers can have a variety of effects on military operations. In TACSPIEL, streams and rivers are given one of six designations and are categorized as being: fordable without delay, fordable, non-fordable 0 to 18 meters wide, non-fordable 18 to 40 meters wide, non-fordable 40 to 150 meters wide, and non-fordable more than 150 meters wide.
4. *Concealment.* The protection afforded by natural terrain features, such as vegetation, against air and ground observation.
5. *Observation.* The facility offered by natural features for seeing surrounding terrain. Observation is restricted by vegetation and irregular ground surface but is aided by the presence of topographic crests.
6. *Cover.* The protection from small-arms fire and fragmentation afforded by natural features such as rocks and boulders, gulleys, ravines mounds, large tree trunks, and any other natural feature large enough to protect at least one individual.
7. *Fields of fire.* Obstacle-free areas afforded by natural features through which flat-trajectory, high-velocity small-arms and company-component weapons can be fired. Fields of fire are degraded by vegetation but are aided or restricted by topography, depending upon the situation.
8. *Foot movement.* Movement on foot is expressed as speeds that dismounted troops can maintain while traversing an area.

The Grid System

Grid overlays are prepared for maps of the terrain to be studied. The dimensions of the grid may be of any size consistent with the nature of the problem under study, the forces engaged, and the area considered. Terrain features are designated for each square represented in the grid system; there is one or more designations for each square. The designation considered to depict the terrain features of the grid is that selected as the most extreme or advantageous for the military operations (cover and concealment, and movement of forces).

Terrain features affecting the amount of cover and concealment are grouped in eight terrain classes. The designations follow:

*Class 0.* Open country, no cover or concealment
*Class 1.* Open country with some scrub vegetation
*Class 2.* Open country interspersed with patches of woods
*Class 3.* Terrain cover of scrub growth, more or less complete
*Class 4.* Scrub cover interspersed with patches of woods
*Class 5.* Wooded land with little or infrequent open places
*Class 6.* Built-up areas such as towns or cities
*Class 7.* Concealed from ground visibility, but visible from the air

As is the case with the cover and concealment effects, the mobility features are related to the way in which they facilitate or hamper military operations. Terrain features affecting speed of movement are also grouped into eight classes:

*Class 0.* Movement in column only (as on roads); tracked and wheeled vehicles, both moving at half speed; deployment fully stopped
*Class 1.* Movement in column only; tracked vehicles at half speed; wheeled vehicles and all deployment fully stopped
*Class 2.* Movement in column or deployed; tracked vehicles at full speed; wheeled vehicles stopped
*Class 3.* Movement in column or deployed; tracked vehicles at half speed; wheeled vehicles stopped
*Class 4.* All movement at normal speeds
*Class 5.* Movement on foot only; normal speed
*Class 6.* Movement on foot only; half speed
*Class 7.* No movement by any means; all stopped

## A MODEL OF A BATTLE

Both TIN SOLDIER and the early CARMONETTE simulation set out to provide a model of a relatively localized phase of a combat action involving a minimum number of elements. The basic task was to build a model which would approximate the structure of a battle and which would duplicate the processes that occur in the natural sequence of a battle. The elements of a simple battle are shown in Fig. 5-1. The events which occur and their sequence are often thought of by gamers as "the logic of the battle." A simulation or game is without merit if it does not conform to a logic analogous to the real-life events it attempts to represent.

Zimmerman has described in considerable detail the step-by-step sequence of events, the Commanders' alternatives at each step and their decisions, and the execution of missions assigned to each of four platoons involved in a localized combat action.[1] That account is worthy of study by anyone who desires to look into the logic of a simple battle as an indicator of the elements which must be built into the model of a combat simulation or game.

Development of the models for a simulation or a game requires meticulous analysis of each of the very large number of factors and elements that are functionally important in the process being represented. The factors mentioned in the earlier discussion of terrain analysis is a simplification of the extensive study and research required for only one pertinent parameter or game feature.

Underlying the details of each factor is the major framework of the master model, the whole model which represents the interactions among the submodels and the variations which occur in each of the parameters or sets of data—some introduced as inputs, some generated by interactions within the simulation. The master model must truly represent the way the whole operation functions in real life. This means that any simulation must begin with a clear analysis of what occurs in the particular kind of military operation, business enterprise, management process, or whatever is being simulated. Thus, all the basic knowledge of the field, organized into known working relationships or principles, is called into play. The military refers to these principles and procedures as doctrine and standing operating procedures (SOP). In like fashion, there are many common practices and standard procedures in the business world, as in accounting, finance, law, and tax considerations.

It follows that the basic design of the model needs to conform to the principles and practices of its real-life counterpart. Next, each component must function in the model as it does in the situation it por-

---

[1] Richard E. Zimmerman's chapter "Simulation of Tactical War Games," pp. 711–762, in Charles D. Flagle, William H. Huggins, and Robert H. Roy (eds.), *Operations Research and Systems Engineering*, Baltimore, The Johns Hopkins Press, 1960.

trays. Thus the real-life activity must be analyzed searchingly and organized into a working model which performs in the same way. Much of the accumulated technical knowledge, and the experience of successful practitioners, must be built into the model. Yet the model should not be encumbered with unnecessary, nonfunctional, or useless data or relationships, or it will be hopelessly complex and extravagantly wasteful of time and energy in construction and use.

Some simulations and games have bogged down in the ambitions of their designers. It is wasteful to include in a model true-to-life details and trivia that have little real influence on the primary purposes for which the game or simulation is intended. The art of the simulation designer is to build a simple model of a complex real-life counterpart but one which functions like its counterpart. Model building of simulations is a highly specialized activity calling for expertise both in the enterprise to be modeled and in the model-building process. The latter specialization involves the zooming field of computer science and technology. The dual specialization, infrequently found in the same person, usually calls for a team approach.

## Chapter Notes

5-1. All military commanders attempt to bring their combat units to a high state of combat readiness and to maintain that level. In war gaming some measure of unit combat effectiveness, generally referred to as the Index of Combat Effectiveness (ICE), is required. ICE's are the result of factors that cannot be measured directly. For example, the intangibles of training, morale, teamwork, command, control, organization, and coordination contribute to maximizing employment and effective use of unit weapons. The effects are tangible and are represented by aggregated firepower scores (or values) of the weapons in the unit. For each weapon employed against various targets under different conditions, a certain value represents the kill or casualty-producing effect. Operations research, gaming, and military efforts have struggled to find scientific bases for firepower scores, but those accepted and employed usually have been the result of a consensus of military judgments, underwritten by experimental and test data. United States experimental and test data sources include the Army's Ballistic Research Laboratory at Aberdeen Proving Ground and the Naval Proving Ground at Dahlgren, Virginia. Army Manual, FM 101-5, provides tables of firepower scores. The Army War College, the Army Combat Developments Command, and the Research Analysis Corporation have made combat-effectiveness studies.

5-2. Frederick William Lanchester (1868–1946) was an engineer of wide interests and great talent. He was a pioneer designer and manufacturer of automobiles: the Lanchester car, first produced in 1895 in England. By 1907 Lanchester's serious interest had encompassed aeronautics, his two-volume work on *Aerial Flight* appearing in that year. Relating flight to warfare, he espoused his mathematical models of combat losses in his 1916 book, *Aircraft in Warfare*, in which his $N^2$ law is presented in chap. 6 and its limitations discussed in chap. 17. His chap. 6 is reproduced in Morse and Kimball, *Methods of Operations Research*, pp. 63–77, and in Newman, *The World of Mathematics*, vol. IV, pp. 2138–2157, New York, Simon & Schuster, Inc., 1956. Lanchester was an early proponent of aircraft in warfare and regarded it as the fourth combat arm, along with infantry, cavalry, and artillery. Lanchester, familiar with Clerk's work a century earlier, formulated mathematical equations to express principles of combat losses in warfare. His mathematical models included the principle of concentration of force (firepower): the $N^2$ law. Morse and Kimball reported both the linear law and the $N^2$ law. Since Lanchester's time, many others have studied and constructed mathematical models of conflict, including many different types of combat or warfare. Herbert K. Weiss has written a number of articles on applications and modifications of the Lanchester equations. See also P. W. Kingsford, *F. W. Lanchester, A Life of an Engineer*, 1960, pp. 148–149. See also Joseph F. McCloskey, "Of Horseless Carriages, Flying Machines, and Operations Research," *Operations Research*, April 1956, pp. 141–147.

5-3. Page, in Flagle et al., *Operations Research and Systems Engineering*, p. 123. Also, Edward M. Lee, in "A Model of Combat with Both Space and Time Variables," ORO-S-453, pp. 14–16, worked out modifications of Lanchester's equations and applied them to a World War I battle at St. Lo, where the results of the model and the battle correspond. J. H. Engel, in "A Verification of Lanchester's Law," *Operations Research*, May 1954, pp. 163–171, reported the equations were found applicable in the capture of Iwo Jima in World War II.

5-4. Lanchester, *Aircraft in Warfare*, summarizes the conclusions of Clerk/Nelson/Lanchester analogs on p. 66. The graphs and contexts reported by Lanchester are widely reproduced in other works and hence are quite familiar to those who read the literature of models and analytical studies of warfare. Material reproduced here was drawn directly from its original source in *Aircraft in Warfare*. Most of the chapters in that book, however, were previously published by Lanchester in a series of articles in *Engineering* during the latter part of 1914.

5-5. Herbert K. Weiss, quoting from Robison and Robison's *A History of Naval Tactics from 1530 to 1930*, discusses the "Fiske Model of Warfare," in *Operations Research*, 10:569–571, July–Aug. 1962. This article drew a comment from Joseph H. Engel, in the Jan.-Feb. 1963 issue of *Operations Research*, 11:147–150, that "It may be of interest to observe that the Fiske example quoted by Weiss is indeed a close analogue of the situation treated by Lanchester" (p. 147). Further, Engel stated, "The Fiske Square Law is still sufficiently powerful to permit a simple determination of which side will be victorious in a [naval] engagement fought until one side is completely destroyed" (p. 149), and "Thus we see that the Fiske Square Law yields the same result as the Lanchester Square Law with respect to predicting the winner of the battle" (p. 150).

5-6. The spread or scatter of the quantitative values expressed as means or averages may be measured (in units of so-called standard deviations) as distances or differences from a "true" average ("true mean"). The degree of such spread varies as the square root of the number of random events. Because of this phenomenon, the greater the number of random numbers drawn from one population (the same batch of data) for one event included in one summation, the smaller will be its standard deviation. Example: If the measure obtained is the result of drawing 100 random numbers (each drawing of a random number representing one chance occurrence), the standard deviation as a measure of probable difference of that average from a "true" value would be regarded as one-tenth as much as if the average had been derived from a single random number. That is,

(a) $\dfrac{1}{\sqrt{100}} = \dfrac{1}{10}$

(b) $\dfrac{1}{\sqrt{1}} = \dfrac{1}{1}$ or 1

Thus (a) is 1/10 of (b). This relationship is discussed in books on elementary statistics. If a normal distribution of random numbers is assumed, the result in (b), although based on only one random number, would represent a value that probably would fall within the same area with 68 out of 100 results if 100 cases were drawn.

5-7. The Lanchester equations have been subjected to many critical, analytical studies in recent years. These studies took extensive historical data and treated them statistically. The gist of these studies were: (1) For large battles over extended periods of time the equations have little if any predictive value; (2) Lanchester's $N^2$ law—not his linear law—seemed to have some relevance in studies involving small samples of recent battles; (3) in short duration, small-scale battles, the $N^2$ law has greater applicability; (4) in air and surface-fleet sea battles, the $N^2$ law has some relevance; (5) force ratios have shown no significant or dominant influence and seemed to have appreciable influence only when they exceeded 1:4 in the samples (in land warfare) studied; and (6) land warfare is too complex, with its great dependence on command decisions, to be represented by Lanchester's relatively simple equations. Lanchester himself made many of these same points.

Although these studies indicated that the analyses were insensitive to errors of 10 percent in the casualty data in certain cases, a modified version of the Lanchester equations "predicted" the outcome of eight out of ten battles. Sheer chance would have given correct results in five out of ten cases.

The great utility of the Lanchester equations was to stimulate a search for more reliable mathematical models and to challenge students of warfare to develop theories having greater validity of battle and battle outcomes.

Among the analytical studies are:

R. H. Brown, *A Stochastic Analysis of Lanchester's Theory of Combat*, ORO-T-323, 1955.

R. L. Helmbold, *Lanchester Parameters for Some Battles of the Last Two Hundred Years*, CORG-SP-122, 1961, and *Historical Data and Lanchester's Theory of Combat*, CORG-SP-128, 1961.

R. N. Snow, *Contributions to Lanchester Attrition Theory*, RAND RA-15078, 1948.

H. K. Weiss, "Lanchester-type Models of Warfare," *Proceedings of First International Conference on Operational Research*, 1957. This paper relates the Lanchester laws to battle statistics of many World War II battles in the Pacific campaign.

D. Willard, *Lanchester as Force in History: An Analysis of Land Battles of the Years 1618–1905*, RAC-TP-74, 1962.

Quincy Wright, *A Study of War*, The University of Chicago Press, 1942.

Among the historical studies are:

Gaston Bodart (ed.), *Militar-historisches Kriegs-Lexikon* (1618–1905), Wein und Leipzig, C. W. Stern, 1908.

Lewis F. Richardson, *Statistics of Deadly Quarrels*, 1960.

# Types and Characteristics of Games

## TYPES OF GAMES CLASSIFIED BY TECHNIQUES

Present-day war games stem from the elder and younger Von Reisswitz game development in the early nineteenth century. The next turning point in war gaming was marked by the Meckel and Von Verdy departure from *rigid* KRIEGSSPIEL and the initiation of *free* KRIEGSSPIEL. It is therefore appropriate to consider the characteristics of these two types of games, the subsequent directions of development, and the conduct of each type of game in updated form.

Rigid Games

Rigid games are conducted in accordance with detailed, nondeviating rules. The game rules usually provide for chance events; the role and the range of chance results are not left to the judgment of an umpire or controller. Chance effects are provided for by such means as the throw of dice, the spin of a wheel, the draw of a card or random number, or, in the simplest case, by the flip of a coin. All possible outcomes are predetermined and built into the rules. For each possible result of chance one possible outcome is decided. For example, if there is a 50:50 chance of either of two particular outcomes, such as a "hit" or "miss" of a missile fired at an aircraft, the rules indicate how the result will be determined. If a toss of a coin is employed, heads will mean hit; tails, miss. If cards are used, a red card may mean a hit;

the black, a miss. If a wheel is spun, any number falling in the lower half will mean a hit; in the upper half, a miss.

Rigid games usually have voluminous rules. The game controllers or umpires must be familiar with these rules to make assessments in a reasonable time so that play of the game may proceed unhampered.

Rigid games are believed to yield more consistent or reliable data. The rules provide for use of identical procedures, methods, and source data in each application. Rigid games are laborious to manage, and the results are difficult to record. Nonetheless rigid games are better suited to research and analytical uses than are free games.

Free Games

A free game is conducted without a mass of fixed rules, and in accordance with the judgment of an umpire or controller. The players, i.e., opposing commanders, may use identical tactics and plans in either a rigid or a free game. Only the basis on which the Controller, Umpire, or Game Director makes decisions and assessments is different. A free game is less tied to the mechanics of play and assessment. Various shortcuts can be taken. Fewer detailed records are needed, and although less reliable than a rigid game, and less suitable for research and analytical purposes, the free game may be more desirable for training use.

The most significant differing characteristic between rigid and free games relates to the manner in which the Controller assesses the actions. If the Controller makes assessments without recourse to fixed rules, tables, and charts and relies upon judgment (based on military experience and insight) the game is called a *free* game. If, instead, the Controller bases assessments on detailed and comprehensive rules, rather than on judgment, the game is called a *rigid* war game. As one experienced gamer expressed it, the quality of a free game depends on the competence of the Controller; the quality of a rigid game depends on the competence of the people who formulated the rules.

Open and Closed Games

An open game may be either rigid or free. In an open game the opposing players have full knowledge of the positions and actions of their own and enemy units. In a closed game each player has firsthand knowledge only of his own situation. Partial knowledge of the situation of opposing units is transmitted by Control. The closed game is complex and formal; the open game, simple and informal. From these conditions a number of characteristics follow. In an open game there is no need for, or play of, the intelligence function. The game may be played in one room on a single game board. Both teams and the Controller are present at all times. See Fig. 6-1. The game is characterized by informal procedures, and its chief usefulness is for training purposes. Free-type games require minimum preparation, little or no record keeping, and the least time to play to completion. Evaluation of open-game moves and results may take the form of a critique at the end of the game, when the major lessons demonstrated or learned are pointed out. Or a continuous critique, interspersed with moves as the situation develops in the game, may be conducted. In this type of critique, situations can be interrupted as they arise, and alternatives pointed out or even played through at that time for training purposes.

A closed game approximates the reality of actual warfare. Each player has incomplete knowledge of the situation as it develops in the conflict. The Commander issues orders based upon his mission, his estimate of the situation, and his own operational or campaign plan. Each Commander must await the feedback of information to know what resistance. if any, his units have encountered; the degree to which his orders have been carried out with effect; his casualties; and

FIG. 6-1. A free game in operation, Naval War College, circa 1914. (*Navy Department Photo.*)

other such information about his own and his opponent's forces and movements.

From the closed-game characteristics a number of requirements arise. A minimum of two game rooms, Blue and Red, must be provided and each kept under security so that Blue players have access only to the Blue Room and Red players only to the Red Room. The Controller or Control Staff has access to both the Blue and Red rooms. More often a closed game is played in three rooms, the third room serving as the Control room. Each of the three rooms is equipped with a game board, maps, and markers. The Blue and Red rooms each have an identical and complete layout, knowledge of deployment of its own units, and such information about enemy units and positions as may be obtained through intelligence sources. Each Commander must include intelligence operations as a part of his play. Reconnaissance flights over prescribed paths in enemy territory can be flown in an attempt to detect units and other targets. What Commanders learn from such overflights and reconnaissance missions is determined by the Control Staff who attempt to report to the respective Commander only that data the reconnaissance mission would be likely to observe had it actually flown the route at the assumed time, with the equipment indicated, and under instructions the Commander had issued. Such factors as weather, cloud cover, natural cover (trees, etc.), and the position and activity of enemy units and installations are considered.

It follows that a closed game is suitable where a sophisticated or detailed type of data-generating game is required. Detailed preparation and a staff larger than that required for an open game are necessary. There are a great volume of message exchange, detailed recording of assessments, and use of sophisticated analytical techniques. Formality is required in the game, the rules, and the procedures.

The objections to "the vogue of military mathematics" in the early nineteenth century and the modifications of rigid KRIEGSSPIEL, advocated by Meckle and Von Verdy, in the 1870s were directed at a simplification of game procedures and a freeing of the game from extreme formalism, complexity, length of time required, and the difficulties of learning game conduct. All these efforts were directed at improving games for training purposes.

Hand-played and Computer-assisted Games

Until quite recently all war games were the so-called hand-played or "manual" type. Simply stated, a manual or hand-played game means that all the pieces are moved and all the work is done by human players and controllers  Not until the advent of the electronic computer was an alternative to hand-played games practical. As rigid games are prone to complexity and involve much computation in the assessment process, they are admirably adapted to computer support. Hence, in modern war gaming another type of game has been added to the spectrum of war games. (The spectrum of games extends from the manual game to a computer-assisted game in which human decisions are made only at critical intervals.  There is a further extension of this spectrum to a complete computer procedure, more appropriately called a simulation, where there is no element of human decision involved in the complete run of the conflict situation.)

In manual games forces are represented by miniature models, pieces, pins, or symbols, and the participants move them about by hand on a board, map, chart, or terrain model that depicts the area of operations.  Contacts and interactions between forces are evaluated in accordance with the professional judgment of the umpires or with the rules, measuring devices, tables, graphs, and formulas specified in the game manual or handbook.  Manual games represent the earliest attempts to simulate, realistically, conflict between armed forces. Because of the great adaptability, ease of use, and low equipment cost of manual games, they enjoy wide use and are of great value.

Map Maneuvers

Map maneuvers may be considered a rudimentary form of war game and a forerunner of today's game. The map-maneuver technique was used by famous military leaders such as Nelson, Napoleon, and Montgomery. Also, map maneuvers have been used extensively to train military officers in most of the Western nations since the Napoleonic era.  Almost always hand-played, map maneuvers have wide adaptability and ease of use and will undoubtedly continue to furnish practice important in the military art.

The simplest type of map maneuver is the so-called "one-sided" maneuver.  In the one-sided game the opposition is played by the Game Director or by the principal player, who alternately represents one side of the game and then the other, in successive moves.  In effect, this is the process that almost every commander goes through in developing an operational plan or the details of an estimate of the situation.  Specific mention was made of this technique in connection with the Allied plans of World War II.

One-sided map maneuvers, like one-sided games, sometimes are used for special purposes. One-sided map maneuvers may be used for the introduction of tactical concepts and presentation of principles of positions, firepower, and maneuver. In addition, problems in field-force logistic support, intertheater movement of supplies, or air defense against a missile attack on the North American continent can be resolved on a one-sided basis. Assumptions are made as to the nature and extent of the problems imposed by enemy deterrent action. From the definition used in this book, a game is always a two-sided or multi-sided situation; that is, conflict is present and, to have conflict, military action is involved between opposing forces. The one-sided map maneuver is included here among types of games because it actually represents a two-sided situation of conflict. Although only one side of the game or maneuver is played in detail, the actions and alternatives available to the enemy always are considered.

Two-sided maneuvers present a realistic and challenging type of activity. As in real war, these maneuvers give each commander an opportunity to gauge the opponent and test his knowledge and capability in a competitive situation (against the skills of the opponent). Both commanders are given freedom of initiative, constrained only by the missions assigned, the requirements of the military situation, and the resources available.

Two-sided maneuvers may be open or closed. In the open two-sided map maneuver both commanders play on the same map or terrain board in full view of each other and in complete knowledge of each other's disposition of forces and moves. The only feature in which each Commander does not have full and open knowledge of his opponent's course of action is his opponent's plan of action not yet revealed by his moves.

The two-sided open map maneuver is one step beyond the one-sided maneuver in the learning process. It is useful in making each Commander aware of the advantages, limitations, and disadvantages of each tactic employed, and it stresses his ability to react to opponent moves, whether or not such moves are anticipated.

The purpose of an open two-sided maneuver may go beyond the training experience furnished participants and observers. Since each Commander is stimulated to employ any means he can devise to gain an advantage over his adversary, such maneuvers provide an opportunity for the injection of new ideas. These new ideas may open fields worthy of further investigation and study.[1] Map maneuvers often are presented in service schools in the form of map problems.

---

[1] Farrand Sayre, *Map Maneuvers and Tactical Rides*, 3d ed., pp. 65–66, Army Service Schools Press, 1908–1910.

In the open two-sided map maneuver the Umpire or Game Director has a somewhat different function. Like the one-sided maneuver, the two-sided maneuver is initiated by the Director or Umpire who specifies the forces involved, the disposition of units, and the general situation. From that point the initiative in moves is with the Commanders. The Director or Umpire serves only to adjudicate and rule upon moves, to assess the effects of the moves, and to see that both sides conform to the rules of the game.

A map maneuver frequently is used as a learning device. The problem may be presented to a group of officers or addressed to each individual officer. In either case the officer considers himself the Commander. When one officer is selected to present his plan, other members of the group may imagine themselves in the Commander's position, even though each will observe and critique the play of the other.

The map problem, or maneuver, may be presented in a written statement or in an oral briefing. There are two typical parts: the general situation and the special situation. The general situation refers to the whole problem and is intended as information to all commanders and participants. The special situation refers to one of the opposing sides, and certain information is made available only to the members (or specific members) of that side. Thus, in a two-sided map maneuver or game there is a separate special situation for Blue, revealed only to the Blue team and to the Control group, and another special situation for the Red team, made available only to the members of the Red team and the Control group. These special situations represent a case where "game security" is involved and information must be restricted to certain members of the group.

The general situation usually covers such information as the disposition of forces on each side; the relative number of units of each type expected to be involved; and how the action is initiated, that is, who is the aggressor and what he does to initiate hostilities. Very often the general situation may describe a series of background events leading to hostilities. Normally it contains such other details as the date, time of day, weather conditions, the locale, and similar items which each Commander must know in order to carry out the intended purposes of the problem or maneuver.

The special situation for each Commander usually specifies the list of troops he is to command, troop location, the exact hour operations are to begin, or the circumstances to which he is to react when the initiative is taken by his opponent. Also given is any special information about the enemy that may have been acquired prior to hostilities through normal intelligence activities.

In an open game there may be little or no need for "special situations" or "game security" types of information. In a closed game, i.e., a game in which each Commander knows only what he can learn about his opponent through contact or clash of units and through intelligence activities, special situations are prepared and made available to each Commander.

In artificial warfare it is desirable to restrict the number of nonessential details to the lowest level consistent with keeping the problem or maneuver realistic. The special situation should make clear the mission or objectives assigned to the Commander. Instructions from higher political authority, including constraints on the Commander, if any, in the employment of his forces should be specified. Directed political constraints include such things as the specification of weapons not to be used, e.g., chemicals, nuclear, etc., and sanctuaries, such as cities above a certain population, buildings and works of historic significance, art treasures, and the like.[1]

If a set of prescribed rules apply to the game, these should be made known to all participants well in advance. Usually a manual or handbook that gives the rules, procedures, and data is provided. The maneuver problem or game is conducted in conformity with these rules. Assessments are made on outcomes resulting from adherence to the rules. If, instead of a handbook of rules, "sound military practice" is to be employed, each Commander is allowed initiative within his own judgment, but the Game Director or Umpire will make all final judgments and assessments on the basis of his own (presumed superior) knowledge and experience.

## TYPES OF GAMES CLASSIFIED BY PURPOSE

Three uses of games have been discussed: (1) gaming to train, (2) gaming to test plans, and (3) gaming for research. These purposes can be bracketed within two general types of games: training games and analytical games. A training game provides a learning situation and experience in the application of military principles and command decisions. The outcome and the details of the game are of secondary importance, perhaps even inconsequential. Such games are valuable only in terms of what they contribute to learning by the participants and observers.

---

[1] *Ibid.*, pp. 67–68.

Training Games

Training games are an important part of the advanced service school curriculum intended to prepare officers for high command and staff-position responsibilities. Training games also are useful in active commands; e.g., games may be employed to simulate and emphasize details of operational problems and command requirements. A *NATO war game* is an example of this use of games.

One of the most difficult problems concerning NATO military forces relates to the possible use of atomic weapons. This problem is of extreme political significance to NATO members and poses some of the most fundamental problems in the planning and conduct of combined operations (operations of military forces of the member nations). Legal and security restrictions are imposed on the military forces of the members of the nuclear club. Within this framework of constraints and restrictions it is necessary to share certain information about weapon employment, delivery means, yields, effects, and deployment. To serve this purpose, a NATO Special Weapons School was provided by the U.S. Army, as a component of SHAPE Headquarters, Europe. The school was established at Oberammergau, Germany.

To assist in training of NATO military personnel, an Atomic Air-Ground War Game was developed in 1953. The game was to demonstrate some basic principles of tactical air operations in land warfare keyed to the tactical use of atomic weapons delivered by piloted aircraft in a limited or localized situation. Close coordination of air and ground forces, an essential for employment of this type of weapon, was stressed in the game.

The game design was flexible. Two "Joint Staffs" could participate. The Joint Staffs could be represented by an individual player or by a considerable number of participants representing many of or all the Joint Staff functions. The objective of the game was to stress the factors of timing and the situation at each stage and the impact of these factors on the allocation of air sorties and weapons. A number of special problems faced the participants. Blue and Red, to keep from being annihilated, each used air forces and atomic weapons in direct ground support and directly against enemy army units. To prevent the game from becoming unmanageable in scope, interdiction was limited to the time and zones of active enemy assault. Concurrently, failure to reduce the opponent's air force through attacks on airfields might result in an unfavorable outcome. An official report describing the purposes and nature of the game included the statement: "No significance is to be attached to the outcome of this game."

The game, conducted for a variety of echelons of military personnel, was designed to be played in a half day following prior study of rules and briefings. One series of the game was conducted to augment the General Officers' course.[1] Another series was offered for NATO Staff Officers, mainly those in field grade (Major through full Colonel).[2] Lieutenant Colonel Jesse C. Peaslee, USAF, was the Officer in charge of developing and managing the courses.[3]

Analytical Games

Analytical games may be used for a number of different purposes although generally not for the training of officers. Analytical games are used most frequently for research, to test concepts and doctrine, and to test or evaluate war plans.

Research-type analytical games usually are constructed and conducted as controlled experiments so that comparisons can be made. A controlled experiment means the situation is designed so that the concept to be studied will have an opportunity to come into play in a normal manner and the particular data resulting from such actions and interactions can be recorded in a form permitting systematic analysis.[4]

Analytical games seek to generate meaningful quantitative data, preferably data in a form that can be treated statistically and mathematically. An analytical game calls for detailed record keeping and, for the most part, follows a rigid set of rules and assessment procedures. Analytical games require more time for preparation, more time in the playing period, and considerably more time for analysis and interpretation of the resulting data than does a training game.

Analytical Games and Their Uses: INDIGO; TACSPIEL

INDIGO, an acronym for Intelligence Division Gaming Operation, was developed to produce data needed by the Intelligence Division

---

[1] Members of the General Officers Course #3 (June 16 to June 19, 1953) included fourteen general officers from Brigadier to Lieutenant General, with most in the Major General rank, from the armed forces of Britain, France, United States, Belgium, Germany, and the Netherlands.
[2] Such a course in Class #3 (July 14 to July 24, 1953) included some thirty officers representing the countries already named and, in addition, Portugal, Turkey, Greece, Norway, Italy, and Denmark.
[3] The information reported was obtained from Colonel Peaslee.
[4] The reader is cautioned that "control" as used above in no sense implies that the game is rigged to provide a favorable or wished-for result. Quite the opposite is meant. The experiment is designed to avoid, as far as possible, any bias or distortion.

## Types and Characteristics of Games    133

of the Operations Research Office of The Johns Hopkins University. INDIGO was used to implement a study undertaken to produce data on Army requirements and future capabilities for acquiring combat intelligence information. The problem dictated a study of Army communication requirements projected to a future time frame. A means of generating the kinds of messages involved in a combat situation was needed. Combat target-acquisition information involving either conventional or nuclear weapons systems was desired.

To meet the specific requirements, INDIGO was designed as a two-sided, free-play, rigidly assessed, manual war game. The rules permitted play of individual surveillance devices, patrols, platoons, companies, and battalions. All actions developed in the game were the direct result of orders issued by Commanders, or players, to respective forces. The Control Group maneuvered the forces in accordance with these orders and, in accordance with the rules, conducted and assessed the resulting interactions. The game was played as INDIGO I in the spring and summer of 1958. See Fig. 6-2.

A major undertaking was the analysis of data produced in the play period. Analysis was not completed until the spring of 1960. A report covering the design of the game and the data generated (but not analyzed) in the first (1958) series of plays was issued (five volumes). The report covered a description of the game, information on target acquisition, combat-intelligence production, communications requirements, and electronic-warfare problems. Although game details were classified, much of the basic data produced, both in the pre-

FIG. 6-2. INDIGO terrain model and game in process. *Left to right:* Dr. Harry N. Hantzes, Col. Lawrence S. Simcox, Edward W. Girard.

analysis and post-analysis stages, and some relevant information not included in the report were submitted to various Army research agencies who could make use of the information generated.

Many by-products of analytical games can be put to significant use. Some by-products can be anticipated; others are recognized later. As an example, one by-product from INDIGO I could be used in a subsequent study of tank-crew radiation exposure in a tactical situation employing nuclear weapons.

Game cycles or intervals of 30 minutes were used throughout the play of INDIGO I. Commanders had an opportunity to issue new orders after getting reports of each 30 minutes of combat time. All data and assessments for each of these periods were recorded and became a data bank or reference library for future analytical studies.

INDIGO, designed as a war game because a particular study needed the kind of information such a game could produce, had utility as a war game for study of other tactical problems beyond that of intelligence requirements. Thus INDIGO, modified, became the basis of TACSPIEL, a game of wide applicability to the study of tactical problems. TACSPIEL continues in operation as of this writing. Further, the detailed set of rules constructed and retained, and the comprehensive description of methodology, permit reactivation of INDIGO should a need arise. Some of the principal war gamers in ORO, Research Analysis Corporation (RAC), and the Naval War College are shown in Figs. 6-3 and 6-4.

Problems of warfare and of command of military forces vary widely in accordance with the echelon of force. A patrol or a squad on a reconnaissance mission faces a different set of problems than a company or a battalion in perimeter defense or an attack. In like manner, the commanders of a battalion, brigade, division, corps, and army all deal with problems differing in scope, mix of elements, mobility, firepower, security, and resupply. A war game that provides sufficient detail for adequate study of a lower tactical unit would be unsuited for the study of certain higher-echelon problems. As indicated in IN-

FIG. 6-3. Some principal war gamers who developed recognized games in ORO and RAC. *Left to right, top to bottom:* H. E. Adams, ORO; M. W. Brossman, RAC; J. G. Christiansen, ORO, RAC; W. Eckhardt, ORO, RAC; E. W. Girard, ORO, RAC; J. O. Harrison, Jr., ORO, RAC; J. G. Hill, ORO, RAC; B. H. Himes, Sr., ORO, RAC; J. W. Johnson, ORO, RAC; M. N. Little, ORO, RAC; W. L. Pierce, ORO, RAC; R. B. Ryan, ORO, RAC.

Types and Characteristics of Games 135

FIG. 6-4. Some principal war gamers who developed recognized games in ORO, RAC, and the Naval War College. Left to right, top to bottom: L. S. Simcox, ORO, RAC; A. D. Tholen, RAC; E. P. Visco, RAC; R. G. Williams, RAC; F. J. McHugh, NWC; E. J. Murphy, NWC.

DIGO, the mass of data accumulated can be overwhelming if the details of concern to lower units are multiplied by the number of such lower units involved in a larger force. To minimize this difficulty, war games usually are designed to bracket a limited spread of command echelons. War games may be grouped according to the level of command at which each is best suited. A number of these different echelons of war games, and some examples of games intended for each, follow.

Games for Small-unit Operations

The smallest tactical units appropriate for a gaming situation would represent a fire team of the Blue forces opposed to a fire team of the Red forces.[1] The unit might take the form of a patrol, a fire team

---

[1] The lowest possible conflict level is between one soldier of each side, but gaming is not needed on this level. Actual practice in field training exercises covers this type of problem.

within a squad, a squad, or a platoon. Special cases might involve a fire team such as one tank crew; one crew-served weapon, such as a recoilless rifle or mortar; or one artillery piece on each side. An engagement, for example, between one Blue and one Red tank might result in a duel between the opposing tanks. Various simulations and games have been designed to deal with tactical problems at this level of combat. Examples of a simulation and a game designed for this level of gaming follow.

## CARMONETTE III

Through continuing development CARMONETTE has served as a simulation of considerable value in research and analytical studies. The latest version is CARMONETTE III.

CARMONETTE III is a computerized simulation or war game used to simulate combat between small units (up to combat battalion size) during a brief, intense engagement phase lasting from a few minutes to approximately one hour. It is a two-sided, closed, rigidly assessed, analytical simulation not requiring human intervention; it includes three types of military activities: move, shoot, and communicate. Four types of units—infantry, armor, artillery, and Army aviation—can be incorporated in the action.

CARMONETTE III can simulate fighting units from one individual soldier or vehicle up to approximately a platoon-sized unit. If the smallest unit considered is of platoon size, the simulation can be used to describe combat team action to battalion level. The objective is to provide a realistic representation of close combat during a brief, intensive engagement. The activities simulated relate principally to movement and firepower.

CARMONETTE III simulates battlefield interactions that occur between terrain and other environmental aspects and the very complex interactions of weapons systems assigned to the four combat elements involved. The simulation has the capacity to deal with forces organized in no more than thirty-six units (on each side). Each side of the conflict may have up to 2,000 killable elements (men and/or pieces of equipment). The killable elements can be in the form of a mix of individual personnel casualties, individual tanks and aircraft incapacitated or destroyed, or platoons of infantry or armor. This simulation can play the development of the combat episode, create an event history, report the number of elements killed by each weapon type, and furnish various operational statistics, engagement ranges, multiple firing numbers, casualties per round, duration of intervisibility, and similar data. Reliability of results can be studied by repeated computer runs and by comparison between alternative weapons systems and tactics, real or hypothetical. Comparative combat potentials of low-echelon forces may be evaluated, and data for higher-level studies or games and for analytical purposes can be generated. These uses are made possible by a fully computerized simulation, without human intervention as in human-decision war games.

Army use of CARMONETTE has included games in the Army Tactical Mobility Requirements study and in exploration of the airmobile concept. Other potential uses are to evaluate helicopters and their operational requirements and to evaluate missiles of the antitank, forward-area air-defense, and surface-to-surface varieties.

CARMONETTE also has been used in international studies. Canada, the United Kingdom, and the United States have cooperated in running comparison games, conducted in the United Kingdom in 1964, in Washington, D.C., in 1965, and in Canada in 1966. The British game, developed by the Royal Armament Research and Development Establishment (RARDE), is hand-played. The Canadian contribution, from the Directorate of Land/Air Operational Research (DLAOR), is a computerized armored-fighting-vehicle model.

Although CARMONETTE III has specific capabilities, it also has specific limitations. This simulation can play only short intervals of combat and cannot be used to study questions such as those related to resupply, casualty evacuation, maintenance of equipment, force reorganization, retreat, and exploitation. Most of the latter aspects have to be studied by other games or other analytic means.

MINIGAME

MINIGAME was developed to fulfill a need for a flexible game to study problems at the small-unit operations level, e.g., fire teams made up of a single tank and a supporting infantry squad or less. The game was an outgrowth of requirements of a 1964 study directed by Major General Lipscomb, Director, Special Studies Group,[1] U.S. Army Combat Developments Command.

The Strategy and Tactics Analysis Group (STAG) of the U.S. Army had attempted to improvise a small-unit operations game by adaptations from a larger computer game. The result of this modification ef-

---

[1] Later renamed Advance Tactics Group; still later, Directorate of Special Studies; most recently, Institute of Special Studies.

fort was called "the micro-game."[1] As the name implies, micro-game was intended to examine in greater detail aspects of a larger-unit game. The process of conversion proved too complicated and slow. Rapid development of a full-blown analytical game to meet the Combat Developments Command time requirements for the study was not possible.

Accordingly, General Lipscomb directed his military staff to develop a free-play semi-rigidly assessed game, with technical support from CORG, suitable for gaming critical events or selected episodes of certain battle situations involving new tactical concepts. (6-1)[2] General Lipscomb described the effort as a "miniature game."[3] The team that developed and played the game found it convenient to shorten the name to "minigame."

---

[1] A similar development preceding this is RAC's THEATERSPIEL, where a micromodel was developed and used to play smaller units and individual weapons as part of the play of larger forces.
[2] These numbers refer to Chapter Notes listed at the end of the chapter.
[3] Miniature game(s) is a term also used in reference to war games of the toy or hobby type, like miniature railroading. That meaning is not intended here. H. G. Wells and Donald F. Featherstone played these types of recreational or parlor war games and wrote books on them. See Fig. 6-5.

---

FIG. 6-5. H. G. Wells and friends playing the parlor war game he developed and published, with detailed rules, as *Floor Games* in 1912, and as *Little Wars* in 1913.

MINIGAME drew upon rules and experience developed in the SYNTAC game. MINIGAME has the advantages of being less complicated, requires less preparatory work, and is adaptable in manner of play, time intervals, skip gaming, or critical-event play. Record keeping is pared to minimum essentials.

MINIGAME was used in three games in 1964 to evaluate certain aspects of small-unit tactics involved in new concepts explored at that time. Two more MINIGAMES were played in 1965, and another in 1966; all were in support of CDC studies. (6-2)

Separate models are employed in the game for each of the following activities:

Intelligence
Command, control, and communication
Movement
Resupply
Engineer activity
Air and air defense
Artillery
Tank, antitank
Infantry engagements

A special feature of MINIGAME is event sequencing. The detailed actions in the game are played in a true time sequence, i.e., as they occur, rather than as aggregate actions that occurred within fixed time intervals (such as 15 or 30 minutes of combat). The flexibility gained by this process permits skip gaming or critical-event gaming. The latter terms refer to the play of only those periods of combat during which something under study is happening. The intervening time periods when little or no action occurs are skipped.

In one project the objective of MINIGAME was to provide data and insights for analysis of offensive and defensive operations in a specified type of warfare. The game was intended to simulate a battle with an initial one-sided introduction of selected munitions followed by transition to employment of similar munitions by both forces.

The action portrayed a reinforced Blue mechanized infantry battalion attacked by a Red motorized rifle regimental task force in a West European area. Time, September 1972. Units considered in the game were resolved at the level of individual vehicles and dismounted infantry squads. Each fire support mission was assessed separately to determine casualties and materiel losses. The game was conducted as a sequence of critical events with a time resolution of 1 minute or less.

The dynamic gaming covered 4 hours and 12 minutes of play. In this period, over 600 messages were prepared and 280 fire missions were conducted. The data collected included casualty numbers, materiel losses, ammunition consumption, mix of rounds expended, and the weapons systems utilized. The report included data, game findings, answers to the study questions, and a number of insights developed by the participants.

MINIGAME is similar to most two-sided, closed, stochastic,[1] hand-played, rigidly assessed, detailed war games in its general form, use of rules for assessment, and facility requirements (three rooms, display panels or screens, etc.). See Fig. 6-6.

The principal unique features of MINIGAME are centered in wide flexibility, capability to simulate and study the dynamics of combat actions at the lowest tactical levels (or at higher levels, if desired), and

---

[1] Actions and effects played on the basis of the probability of occurrence.

FIG. 6-6. MINIGAME gaming facility. U.S. Army Combat Developments Command.

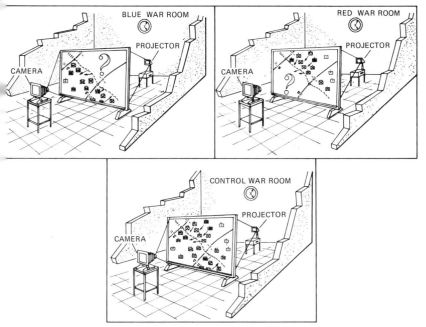

142     Venture Simulation in War, Business, and Politics

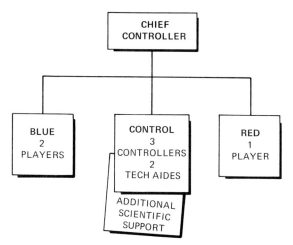

FIG. 6-7. MINIGAME organization chart. U.S. Army Combat Developments Command.

selectivity. By selectivity is meant that the game can generate and isolate specific kinds' of detailed data needed for analytical study without involvement in other related data not under study.

MINIGAME is modest in overall requirements of personnel and calendar time. Personnel usually consists of a team of one chief controller, three assistant controllers with two technical aides, two players on the Blue team, and one player on the Red team. See Fig. 6-7.

One MINIGAME can be completed in 3 months with the time allocated as follows: game preparation, 1 month; playing time, ½ to 1½ months;[1] analysis and reporting, ½ to 1 month. Analysis conducted concurrently with the playing period can shorten the third period. (6-3)

## TACTICAL-LEVEL GAMES

Most so-called "company- to division-level games" are tactical games. Organized combat operations on the battlefield normally involve units of company and battalion size, and most tactical problems involved in the study of combat operations involve units of this size. The early

---

[1] Experience indicates that about one hour of combat time can be played in one work week.

KRIEGSSPIEL game is one of this group. Primary attention in present-day war gaming is focused on this level.

Significant development in analytical war gaming followed World War II. Among the few surviving military powers, the severely war-weary United Kingdom was almost alone in having the remnants of an empire to defend. With large geographic dispersion, bled resources, and limited military forces, perhaps it was natural that the United Kingdom spawned a resurgence of interest in tactical war gaming in search of a means of maximizing military capabilities.

The United Kingdom had pioneered operational research as a valuable asset in studying military problems. This asset was retained and the Army Operational Research Establishment in Surrey, England, undertook the development of a research-oriented war-gaming program. War-gaming models were designed to study operations ranging from battalion to corps and army level. By 1957 the AORE game had reached a stage of development that permitted game application to the study of a number of problems, such as vulnerability of communication lines, personnel casualties to be anticipated in nuclear war, logistics, and a number of tactical problems encountered under threat of nuclear warfare. The newborn nuclear age had become an era of acute and urgent concern to the military forces that might be called upon to fight in Europe.

Army Operations Research practitioners in the English-speaking countries developed a community of interest during this period and began an exchange of experience. The United Kingdom, the United States, Canada, and Australia organized various means of sharing knowledge and laying groundwork for combined military operations such as might be required by developments in the world situation. All four countries recognized that war gaming, research, and analytical techniques provided an opportunity to study the many new problems confronting them under the new and expanded requirements of a rapidly changing world situation. The British and the Canadians, with military forces of moderate size, focused interest in gaming at the tactical level. Canada organized the Canadian Operational Research Establishment, patterned on that of the United Kingdom, and proceeded to adapt gaming to Canadian needs.

The United Kingdom and Canadian War Games

The United Kingdom and Canadian war games are referred to as the AORE and CAORE games. Each is a hand-played, tactical game used as a method of research in problems of tactics, combat organization, and weapons and equipment systems at operational levels ranging from company to corps. Between 1957 and 1961 emphasis was placed on corps-level games played to explore general problems arising in the combat operations of a brigade group in nuclear warfare in Northwest Europe (in the 1960 time frame). In 1962 a more detailed series of games was initiated to study combat operations at the company level.[1] A start also was made to adapt the models to computer support. (6-4) Some war gamers of CAORE are shown in Fig. 6-8.

U.S. War Games

Operations research units serving the land forces of the United States also set out to develop analytical war games. Both the U.S. Army and the U.S. Marine Corps supported this kind of effort to serve United States needs. The Army had its Operations Research Office, conducted under contract with The Johns Hopkins University. Wargame development for the Army's operational-research requirements began in earnest at this facility in 1950. At this same time, RAND, which provided similar research support for the Air Force, was actively pioneering in the field of gaming for analytical and research purposes. A number of research organizations for the most part under the sponsorship and support of various elements of the U.S. Armed Forces soon followed in the development of gaming programs. Concurrently, a number of "in-house" components of the military services also began gaming activities. Some of these groups drew upon operations research organizations to provide the scientific background and support for the games while the military service provided officers to furnish military expertise.[2]

Soon advanced schools of the U.S. military services introduced rigid-type war games, suitable for analytical studies as well as for training purposes. The Army War College, at Carlisle Barracks, Pennsylvania, was a leader in this movement and developed the *Army War College War Games Manual*, a book of rules that became a guide for many games involving land-force units. Among the war games

---

[1] See *Directory of Organizations and Activities Engaged or Interested in War Gaming*, p. 42, Bethesda, Md., U.S. Army Strategy and Tactics Analysis Group (STAG), 1962.
[2] Among the scientific or operations research organizations active in war gaming in this era were Technical Operations, Inc. (1954); Booz, Allen Applied Research, Inc. (1960); the Applied Physics Laboratory of The Johns Hopkins University; The Mitre Corp. (Massachusetts Institute of Technology); as well as ORO (1950) and RAND (1953) and, somewhat later, the Raytheon Co. Source: STAG *Directory of Organizations and Activities Engaged or Interested in War Gaming.*

Types and Characteristics of Games 145

FIG. 6-8. Some war gamers of the Canadian Army Operational Research Establishment. *Left to right, top to bottom:* C. E. Law, E. W. Rae, J. Stafford, R. J. Sutherland.

that drew upon the aforementioned experience and the Army War College manual[1] were SYNTAC, INDIGO, TACSPIEL, the MARINE CORPS LANDING FORCE GAME, and the CAORE game.

SYNTAC as an Example

SYNTAC was started in 1953 under CORG (Combat Operations Research Group) of the U.S. Continental Army Command at Fort Monroe, Virginia. The Operations Research Office (ORO) of The Johns Hopkins University was asked to establish a field office at Fort Monroe to provide scientific support to CONARC. One early activity was the initiation of work on the CONARC game later called SYN-

---

[1] At the time the Army war-game manual was under development, the Army War College had a representative of the Operations Research Office as a faculty member. Through this liaison there was a close working relationship and a mutual exchange of information. The ORO gaming experience and thinking were used along with those of Army sources in the development of the *Army War College War Games Manual*.

TAC. This game was intended as a research-type, two-sided, closed, manually played and assessed, and rigidly controlled game. Computer support was added later.

SYNTAC, intended to bracket a wide spectrum of command levels, can be used to study problems involving units from squad to field army in size. Early applications were largely centered on the lower tactical levels. The game was in operation in 1954 and was used in the study of certain tactical problems. In 1955 ORO gave up its CORG contract, and Technical Operations, Inc., has continued the effort since that time. In 1956 SYNTAC was shifted from CORG into the command structure of CONARC, with CORG continuing to provide scientific assistance on a supporting basis. Currently the game is operated by a war-games division of the Combat Developments Command. SYNTAC is one of the U.S. Army's earliest in-house, computer-supported gaming facilities for research purposes.[1]

The MARINE CORPS LANDING FORCE WAR GAME, Another Example

The LANDING FORCE WAR GAME was developed at the Landing Force Development Center, U.S. Marine Corps, Quantico, Virginia, for the primary purpose of development and analysis of new tactics, techniques, and weapons systems. The secondary purpose was to develop an operational gaming capability that could be employed to study those operational problems for which the Marine Corps is characteristically employed: amphibious operations, expeditionary forces supplied and supported by the Navy, and tactical operations in which the Corps engages after a landing.

The game is two-sided, closed, free-played, rigidly assessed, and manually operated. The game has been in use and in the process of continual refinement and development since 1960. Games are played on maps or terrain models, normally of large scale, i.e., 1:25,000 or 1:50,000. Hydrographic charts are used in addition to terrain models and land maps, where necessary, to extend the gaming area seaward. The game was designed to be used in various forms, including an open one-sided or free-play type of situation, and to be computer-supported if required. In addition to the basic forces involved, provision is made to include terrain and weather, command

---

[1] E. A. Johnson, "History and Future of War Gaming in Operations Research," unpublished paper, p. 17, presented at Third War Games Symposium, University of Michigan, Oct. 6, 1960.

and control, movement, disposition of units, intelligence, ground combat, tank-antitank actions, artillery, naval gun fire, air operations, air defense, NBC warfare,[1] communications, logistics, engineer operations, electronic warfare, nuclear-weapons effects, conventional-weapon effects, special operations, cover and deception, and unit effectiveness. A manual, including detailed rules, is available. Units of company size or smaller may be played as components of a larger force that may exceed a division level. The game intervals may be varied but most commonly are ½ to 1 hour each. If details of play at battalion or company level are significant in a game, the game interval may be lowered to 10 or 15 minutes.

## DIVISION TO THEATER-LEVEL GAMES

Although some of the games discussed were designed to extend beyond a single division-size force, a number of games were designed specifically to study problems at the theater level. The models to deal with these larger forces become hopelessly complex and produce mountains of data if details of operations at lower command echelons are included in the game. Personnel requirements also multiply, and the time required to run through a desired amount of game combat time is stretched beyond practical limits.

For reasons of economy and to make the research aspects manageable, theater-level games must be divested of details quite essential for problems studied in company and battalion tactical situations. Theater games can meet this requirement in two ways. The simpler of the two, and perhaps more suitable, is the game designed to deal only with the larger units, such as divisions and special units. The second, and more involved, method, suited only for a complete computer-supported game or simulation, is to design the models so that details of the lower-level units (companies and battalions) can be bypassed. The latter is more laborious and perhaps a more error-prone method. It requires that all details be built into a game framework (and computer models), a whole mass of complex submodels being required but not used. It is then necessary to design computer routines by which these submodels can be bypassed in the operation of a theater game.

Some theater-level games developed over the years include the Theater Battle Model (TBM-63), a manual game, and TARTARUS, a

---

[1] NBC warfare, as used by the Marine Corps, refers to nuclear, biological, chemical.

computer-assisted game, both developed at STAG; VALOR, a manual game, developed by Technical Operations, Inc.; and THEATERSPIEL, a computer-assisted game, developed by the Operations Research Office and Research Analysis Corporation.

TBM-63 may be played as an open or closed manual game. The tactical combat operations involve all types of theater forces (ground, sea and air) and conventional or nuclear weapons employed in any geographic area except polar regions.

TARTARUS is a two-sided, free-play, closed, computer game that assesses an engagement involving up to a total of 300 brigade- or division-size units for movement, fire, casualties, weapons attrition, and ammunition and fuel expenditure. A machine-run simulation, TARTARUS can game at about twenty times real time.

VALOR is a two-sided, closed model that simulates limited war situations in any geographical area, with any reasonable force mix, and with conventional or nuclear weapons. VALOR includes land, sea, air, logistic, and intelligence operations.

THEATERSPIEL is a two-sided, closed, computer-assisted game that can play conflict situations in any geographical area, with any reasonable force and weapons mix (conventional and nuclear) in land and air warfare. Logistics and intelligence operations are included. Some comparative features are shown in Table 6-1.

Table 6-1. Some Comparative Features of Selected Types of Maneuvers, Simulations, and Games

| Type (Examples) | Courses of action | Rules | Outcomes |
|---|---|---|---|
| Field maneuvers, CPX's (SWIFT STRIKE; LOGEX) | Free play; freely governed (some constraints) | Specified but incomplete; assessments judged, scored | Mostly scored and judged; some computed, e.g., logistics |
| Map maneuvers (free KRIEGSSPIEL) | Free play; freely governed | Variable; incomplete | Some computed, mostly judged |
| Mathematical formulas (Lanchester equations) | Specified, fixed, limited | Rigidly circumscribed | Formula solution; deterministic; not probabilistic |
| Computer simulation (CARMONETTE III) | Specified, fixed | Rigidly circumscribed | Computed, probabilistic; data analyzed |
| Hand-played analytical game (INDIGO, MINIGAME) | Free play; rigidly governed by rules | Rigidly assessed by rules | Computed, probabilistic; data analyzed |
| Man-machine analytical game (SYNTAC, TACSPIEL, THEATERSPIEL) | Free play; rigidly governed by rules | Rigidly assessed by rules | Computed, machine bookkeeping, probabilistic; data analyzed |

Types and Characteristics of Games    149

## OTHER TYPES OF WAR GAMES

There are many types of war games and simulations in use beyond those mentioned in the preceding sections. In general the additional types are of two general classes: Supra-theater Level Games and Special-purpose Games. They will be discussed in the next chapter.
*Supra-theater Games* or Multi-theater Games are in the strategic category. Such games usually give attention to political, economic, and other factors that reach beyond purely military considerations.
*Special-purpose Games* and simulations are numerous. These games are designed to serve rather specific needs, such as:

1. ZIGSPIEL,[1] the Air Defense of the United States, at RAC.
2. NAR (M-2), Nuclear Assessment Routine, by U.S. Army Strategy and Tactics Analysis Group for U.S. Army. The game assesses casualties and damage to military units.
3. MABS, Mixed Air Battle Simulation, by Stanford Research Institute.
4. NUCAT, Nuclear Cost Assessment Technique, by CORG for USACDC.
5. War-at-sea. Protection of sea lanes from enemy submarines, by WSEG/IDA,[2] for DOD.
6. SWIM, Sea Warfare Integrated Model, by APL/JHU[3] for U.S. Navy. Naval task force versus missile-equipped submarines.
7. NEWS, Navy Electronic Warfare Simulator, by Naval War College for U.S. Navy. Evaluates tactical concepts and operations in naval warfare. See Fig. 6-9.
8. FAST-VAL, Forward Air Strike Evaluation, by RAND for U.S. Air Force. To assess indirect fire weapons against ground targets.
9. PENTAC, Tactical Penetration Model, by Boeing for U.S. Air Force. Simulation of aircraft penetrating an air-defense system of SAM (Surface-to-air Missiles) and fighter interceptors.
10. COBRA, Comprehensive Blast and Radiation Assessment System by CEIR for NMCSSC.[4] The game determines damage levels resulting from nuclear attacks.

---

[1] ZI of ZIGSPIEL for Zone of the Interior, or Continental United States.
[2] Weapons Systems Evaluation Group, Institute for Defense Analysis.
[3] Applied Physics Laboratory, The Johns Hopkins University, under contract to USN Bureau of Weapons.
[4] CEIR, Council for Economic and Industrial Research. NMCSSC, National Military Command System Support Center.

FIG. 6-9. Electronic display screen in the control room of the Navy Electronic Warfare Simulator (NEWS) at the Naval War College. This system has been in operation since 1958. (Official photo: U.S. Navy.)

## EXTENT OF WAR GAMING

In late 1961 STAG[1] surveyed U.S. Army agencies to identify agencies that had developed, or were conducting, war games for the Army. Sixty organizations were reported, thirty-six "in-house" (within military organizations) and twenty-four civilian universities and corporations, all engaged in military research. The list included gaming to serve the Navy, Air Force, and Joint Service Agencies (Joint Commands and Staffs).

In a 1963 study E. W. Paxson of RAND estimated that 200 organizations in the United States were doing studies of the operational-research type for the military; that these organizations produce close to 3,000 reports per year; and that "gaming or simulation techniques

---

[1] *Directory of Organizations and Activities Engaged in or Interested in War Gaming*, U.S. Army Strategy and Tactics Analysis Group, June 1962.

or results are used to some degree in about one-quarter of all projects undertaken."[1]

Subsequently, the Joint Chiefs of Staff created the Joint War Games Agency (JWGA) which in 1966 prepared a *Catalog of War Gaming Models*. Some 103 war games and simulations were listed and described. These activities were conducted by fifty organizations, twenty of which were civilian. The largest number of games and simulations listed were for the Army, followed in order by the JWGA, the Navy, the Air Force, the Marine Corps, and other parts of the military establishment.

## Chapter Notes

6-1. Among the personnel directly involved in the development and play of MINIGAME in early 1964 were Lt. Col. William C. Trefz and Sgt. Roger W. Gregory of AVTAC, with technical support by Edward R. Williams, Donald K. Fogelsanger, Harvey Hawley, and Jesse N. Moore of CORG, and a number of retired officers engaged as players and controllers: Lt. Col. Richard Duckwall, Col. B. M. Davis, Lt. Col. Guy C. Meiss, and Lt. Col. Wayne Williamson.

6-2. D. K. Fogelsanger, "Minigaming," p. 1, paper presented to the East Coast War Games Council's Fifth Symposium at Miami Beach, Fla., May 6, 1966.

6-3. MINIGAME information, for the most part, was obtained by the author's personal observation of the game in operation, briefings, and contacts with the individuals concerned, including General Lipscomb, Colonel Farrell, Director, Institute of Special Studies, CDC, and Dr. Edward R. Williams, Chief, Tactical Analysis Division, CORG, Technical Operations, Inc.

6-4. Common interests of the U.K.-U.S. Army Operations Research Groups led to an exchange of analysts between the two countries. Graham Komlosy, a United Kingdom operations analyst, who had served with AORE in England and in Malaya, was designated as the United Kingdom guest analyst assigned to RAC, 1962–1964. Komlosy first worked with the TACSPIEL war-game group and then devoted a major portion of his time to development of a computer model for the AORE tactical-level game.

---

[1] *War Gaming*, p. 4, Santa Monica, Calif., The RAND Corporation, RAND Memo RM-3489-PR, February 1963.

# Short-cut and Special-purpose Games

## TIME-SAVING GAMING TECHNIQUES

A principal limitation on the use of analytical games is the time required to plan and conduct a game and analyze the results. Six months to a year may be required to complete the process. Although modern computers have expanded the amount of data that can be managed in a game, they have not materially lessened overall game time.

Many analytical-game applications are so important and far-reaching that a complete and time-consuming gaming process is justified. Contingency-plan testing, evaluation of proposed equipment and weapons systems, and the evaluation of new concepts of deployment or logistic supply that may reach a time frame of from 5 to 10 years hence are examples of research or analytical studies that merit slow, detailed gaming. Other problems requiring study by gaming methods are in urgent need of early answers.

There are less urgent problems that could be solved best by gaming but do not warrant a long, expensive, answer-getting process. These statements refer to human-decision games and do not apply to computer simulations not requiring human intervention. The longest computer simulation can be processed in a few hours.

A practical maxim of real life—in military and nonmilitary matters—is that a reasonable course of action taken promptly often is worth more than a better course of action that involves substantial delay.

Consequently, shortcuts, long sought for educational and training games, also were sought for analytical games.

The CARMONETTE simulation, a quick-running analytical war game, has been mentioned. That game is useful at the small-unit end of the conflict spectrum. Equally important or perhaps more important is short-cut gaming at the larger-unit end of that spectrum.

Quick Gaming

Quick Gaming is a technique for a two-sided, open, semirigidly assessed, grossly aggregated,[1] manually operated war game for feasibility testing of plans and exercises, and for gross evaluation of force capabilities. The game has little or no value for training purposes and at first consisted only of a "ground combat model."

Tactical decisions are made by the players and deal with the allocation of military resources to different battle-area sectors. These decisions provide for determination of which units and respective firepower will (can) be brought to bear on enemy units in contact or within reach.

The models provide for reduction of each unit's Index of Combat Effectiveness (ICE) in proportion to the casualties suffered. The rate of advance depends on force ratios and posture. The casualties depend on posture alone and are specified by standardized rates drawn from historical examples. Quick Gaming requires a basic assumption that there is a fixed number or level of casualties per day of combat, based upon certain percentages of the force becoming casualties as a function of exposure.[2]

Although QUICK GAME is designed to have great flexibility, a set of rules with reference tables is used. The rule book for this game consists of about 100 pages. Players having some reasonable under-

---

[1] *Grossly aggregated* means that all weapons, men, and equipment in a larger unit are lumped together and dealt with as a whole, in much the same way that the guns, firepower, and capability of a battle cruiser are dealt with as one unit, rather than as separate elements of so many 8- or 6-inch guns in main batteries along with 5- and 3-inch antiaircraft guns. In a business game, e.g., a supermarket game, such aggregation may be total daily sales. Resolution means that all data are summarized into "pieces" which represent the size of the unit, such as divisions. A business game may use $1,000 as a unit. The Quick-gaming unit of resolution is normally one (Army) division. Other-size units such as brigades or corps could be used. At this point it is necessary only to understand that a single numerical value, based upon known data and systematic analyses, represents the aggregated or full combat effectiveness of the unit.

[2] Research has given some credence to this assumption.

standing of military operations can learn to play the game through participation and become quite proficient, "on their own," if given 2 to 4 weeks' experience.

The real payoff in Quick Gaming comes from the small number of participants required and the rapidity of play. One analyst can play the game, and two analysts per game is common; teams of up to five analysts can be used if desired. Several games can be conducted concurrently, each with from one to five participants, according to the need and the availability of personnel. A high order of military experience is required to control the game as any QUICK GAME substitutes sophisticated military judgment for some of the usual body of rigid rules.

For the early short version of the QUICK GAME the rate of play averages 2 or 3 days of combat time for each day of gaming. About 30 game-days have been sufficient for those analytical studies that have been undertaken. A 30-day period of combat has been played in 2 to 3 weeks of working gaming time. Analysis of the data generated in the play then requires another 2 to 4 weeks.

Thus, with a team of analysts familiar with the game, a study could be completed within the following time schedule: game preparation time, 2 to 4 weeks; playing time, 2 to 3 weeks; and analysis time, 2 to 4 weeks. A total of 6 to 11 weeks, or roughly 2 to 3 months, is required for a typical study having a series of from two to five games. On a crash basis this process has been completed in as little as 4 weeks.

Super-Quick Gaming

In 1962 a study undertaken by RAC for the Department of the Army and the Department of Defense required the simultaneous gaming of situations in each of four theaters of operation. The game results were included in a larger study, as is often the case with war games. The due date for this entire study required a greater compression of game time than the Quick-gaming technique provided. With this challenge, the originators of the Quick-gaming technique sought ways to speed up the play.[1] The requirement was met, and the new system was named Aggregated Quick-Gaming, a title soon shortened to Super-Quick Gaming, sometimes referred to as Quick-Quick Gaming.

The scheme and methodology for Super-Quick Gaming were devised in 1 week. It was immediately applied to the problem which

---

[1] There was insufficient time to program the manual assessments and computations for the computer; hence that method of speedup was not used.

was the mother of its invention. In each SUPER-QUICK GAME 30 to 60 days of combat first were played at the rate of 3 days' working time. At the end of the game series, 30 to 60 days of combat were played in about ½ day. These games were manual. The players made all assessments and compiled records without computer support.

The speedup was accomplished by aggregating the advance of the forces in a sector from phase line to phase line, rather than on a daily basis. Graphs were prepared that integrated the data on daily rates of advance and the net effect of casualty rates and replacement policies.[1] The sector system uses a schematic of the theater battlefield based upon actual map and terrain features. The schematic shows principal lines of communication and important centers. The graphs were designed from precalculated force ratios and movement rates as a function of time. An example of such a schematic is presented as Fig. 7-1.

Critical elements in Quick and Super-Quick Gaming include typical movement rates from a number of actual campaigns, including the Korean War; rates of advance as related to force ratios of attacker to defender; type of terrain; posture of defender[2] and type of force; firepower scores; casualty assessments; and replacement rates. Professional military judgment plays an important role in the conduct of the game and must be involved in the play of the game.

Computerized Quick Gaming

Although the hand-played QUICK GAME represents a "great leap forward"[3] over slow-moving, analytical games, as does the SUPER-QUICK GAME over Quick Gaming, the next "great leap," computerized Quick Gaming, promises to be of equal significance. In late 1965 work began at RAC to computerize a revised, late-model QUICK GAME. In a computerized QUICK GAME, one expects to get (1) a fast game and (2) a game better suited to replication. Replication raises the level of confidence in the results obtained.

The computerized QUICK GAME has taken two major strides

---

[1] The graphs were prepared from the more voluminous data in the tables. Once constructed, the graphs provided a quicker, albeit rougher, way to use the data.
[2] Posture refers to the standard categories of exposure of the defender and normally ranges from a fortified position to a disorganized retreat. Seven postures are used.
[3] A term used by Mao Tse-tung; the quick-gaming effort proved more successful than did Mao Tse-tung's application of the concept in geopolitics.

156  Venture Simulation in War, Business, and Politics

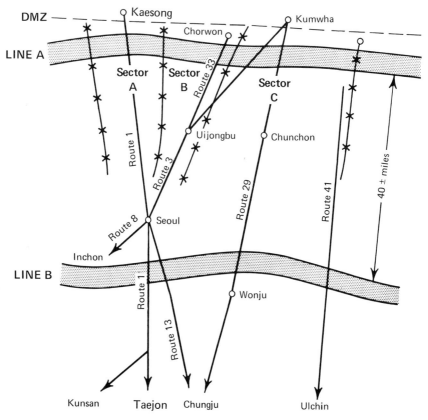

FIG. 7-1. Theater-battlefield schematic, based on the sector system used in a SUPER-QUICK Game. Korea.

forward. First, the game was improved to furnish an updated representation of ground combat. Three important models were used:

1. A tactical air operations model
2. An intratheater, logistics model
3. A simple tactical decision model to make allocation decisions formerly made by players

Moreover, some of the tenuous and questionable basic numbers or values for firepower scores were revised. These values now are based upon recent, reliable casualty records and casualty potential data.

The second stride came from the computerized version[1] of a four-model, improved game that speeds up the play by a factor of 1,000. With this version, a 60-day theater war can be played in 10 to 20 minutes. Game preparation and data analysis have not been speeded up. But, most important, the sensitivity testing of results to crucial assumptions is permitted, as is extensive variation of the tactical scenario and choice of theater. Computerized models were completed in late 1966, and the whole version became operable in mid-1967.

SCHNELLSPIEL: A Seventh Army Planning Game in Europe

This game was a 1963 adaptation of Quick Gaming developed by members of the RAC Field Office, Europe, to meet the needs of the U.S. Seventh Army Headquarters, Germany. Planning officers needed a game to evaluate defense plans and field-exercise plans prior to conduct of an exercise with troops and equipment in the field. The game provided for optional play of tactical nuclear weapons in defense situations.

## GAMING FOR COMBAT READINESS: AN EIGHTH ARMY OPERATIONAL GAME IN KOREA

When Gen. Carter B. Magruder was assigned, in 1959, as Commander of the United Nations Command in Korea, he made a joint war-gaming section a part of his Staff.[2] Many of his Staff Officers were rotated through this section where games were conducted throughout the 2 years of General Magruder's direction. The exercise at any given time in this period usually involved twelve officers, many of whom had had previous war-gaming experience.[3]

General Magruder made this heavy commitment of senior officers to provide training and furnish a basis for certain critical decisions. More specifically, he felt gaming stimulated intelligence operations. In these war games the Red side always was given the resources and capability the enemy was believed to have (as reported by the In-

---

[1] Programmed for the IBM 7040/7090 computer.
[2] Reported by Gen. Carter B. Magruder, USA (Ret.), to the author, and in "Logistic Gaming as Applied to Army Problems," in Martin W. Brossman (ed.), *Proceedings, Fourth Symposium on War Gaming*, East Coast War Games Council, pp. 9–14, Research Analysis Corporation, Aug. 1965.
[3] At the Army War College or other training schools.

telligence Staff of the UN Command). The Red Commander in the game would press the UN Intelligence Staff for information about capabilities of the Communist side. This questioning stimulated and guided the Intelligence Staff in information-gathering and interpretative activities.

Further, these war games challenged and trained the Operations Staff. In a static situation such as existed in Korea in 1959, General Magruder points out that an Operations Staff can easily saturate its daily schedule with problems of training or the planning of operations in a routine manner much as a football team may run through a signal drill. Only if "opposition is provided that has plenty of resources and is just as smart as we are," does an exercise provide real value. The gaming stimulated planning and training in the context of imminent conflict.

In addition, the war games were used to train the Logistics Staff and keep it on its mettle. General Magruder knew from his experience in World War II that most military officers are oriented primarily to tactics and strategy and inclined to neglect consideration of logistics. Yet the hard realities of logistics are usually the limiting factor in whatever campaigns can be undertaken.

Finally, General Magruder emphasized, there were many questions to which he wanted answers. He wanted the best Staff answers possible to such questions as:

What strength could the enemy bring to bear in the middle of Korea?

At what rate could UN buildup be accomplished?

What enemy measures would give the earliest warning of an impending attack?

What were the enemy's most promising methods and directions of attack?

How could UN forces best counter enemy actions? For example, could the UN materially increase deployment depth to be better deployed against a possible nuclear attack and retain defense capabilities against a conventional attack?

Could UN operations in the campaigns visualized be given adequate logistics support?

What were the UN logistics weak points? What actions should the Logistics Staff take to reduce vulnerability?

Although this gaming was undertaken partially for training purposes, a concomitant goal of the UN war-gaming section was to assure for the Eighth Army and Republic of Korea Army the highest possible state of combat readiness. The units that comprised the United Nations Command always were under threat of hostilities. That UN units be well prepared to meet all types of enemy attack, and whatever else might face the Command, was imperative.

## A LARGE-SCALE COORDINATED LOGISTIC STUDY

Never before has a nation undertaken a logistic load as large as that assumed by the United States to deploy and support armed forces around the world. The dollar cost is in billions. With sizable forces on three continents and the contiguous seas, logistic requirements are tremendous. The challenge is to provide adequately for the mobilization, equipage, transportation, deployment, supply, maintenance, and support of these forces under the various contingencies to which they must be ready to react. The careful planning of optimum levels, sustained production, and the husbanding of resources are prime requirements.

In early 1967 the Research Analysis Corporation was asked to assist the Department of the Army in solving logistics problems which have plagued military planners for years. RAC did so by designing and employing a coordinated program of war gaming, logistic simulation, and cost-effectiveness studies. The problem was tackled as described here.[1]

Study Objectives

The official statement of purpose was to "adapt several existing models into a working prototype system for analyzing force requirements (with associated costs)" and apply it to "alternatives for two scenarios (Korea and SEA)."

Planners contended with the study problems and formulated work-

---

[1] The substance of this account is derived from an unclassified, staff seminar briefing given at RAC, Oct. 12, 1967. The presentation was made by Brig. Gen. Paul D. Phillips, USA (Ret.), Project Chairman, and Albert D. Tholen. The study reported was the product of several RAC departments: Military Gaming, Logistics, Combat Analysis, and Economics and Costing.

able solutions. Time was at a premium, and economy in costs was an additional requirement. The effort was focused on creating a planning tool or system to deal with a sophisticated and complex concept while retaining simplicity. The object was to employ existing procedures and models, as these were usable; develop a workable system; and test that system. The solutions were applied and tested in two theaters: Northeast Asia (Korea and associated areas) and Southeast Asia (Vietnam and associated areas).

Component Models and Relationships

A framework for the system, built on specific-purpose models, took shape. Five models were incorporated. The models and the output of each model are as listed below:

| | |
|---|---|
| QUICK GAME Model | Time-phased major-combat-unit requirements |
| SIGMALOG Model | Time-phased deployment and resupply tonnage requirements, and time-phased troop lists for all units required |
| Strategic Deployment Model | Least-cost fleet requirement |
| Dual-state Cost Model | Restructuring, redeployment, and peacetime operating costs for all Army forces as affected by alternatives |
| Force Costing Model | Initial investment and operating cost of the force in a given theater |

A description of the salient contributions and relationships of each model follows.

The QUICK GAME

The QUICK GAME generates the major combat-unit requirements—by divisions and types of divisions—to accomplish the objectives within a time-phased schedule. To achieve this the QUICK GAME simulates theater-level warfare. The theater is divided into sectors which may be regarded as corridors of advance, each sector being the width required for a U.S. Corps. See Fig. 7-2. The game plays both Blue and Red forces. Movement is, primarily, a function of three factors: firepower, terrain, and posture. Posture, here, refers to

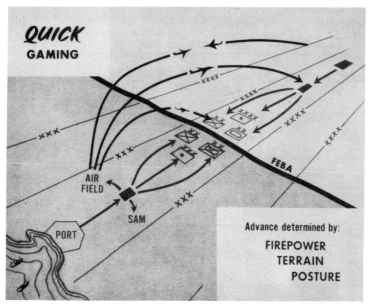

FIG. 7-2. QUICK GAME. Factors affecting advance.

the military activity of each major combat unit at a given time. Logistic and air operations also are simulated.

The entire process performed by the QUICK GAME Model employs four submodels. These are the Tactical Decision Model, the Logistics Model, the Tactical Air Model, and the Ground Combat Model. See Fig. 7-3. The Tactical Decision Model is used to determine the allocation of resources (men and material) within the theater according to the sector needs. The Logistics Model simulates the flow of units and supplies through each sector from ports of entry to points of employment or use.

The Tactical Air Model simulates the use of combat aircraft and the means of destroying these aircraft. Four missions are represented: air defense, counter air defense, lines-of-communication and supply-point interdiction, and close air support. The Ground Combat Model, using inputs from the other models, simulates the two-sided combat by computing force ratios, considering the posture of engaged troops, and determining the rate of advance of the attacker. A daily determination is made thereby of the change in location of the Forward Edge of Battle Area (FEBA)[1] for each section. Play of the game ends when one of three events occurs:

---

[1] The "front" at which opposed forces are in contact.

162  Venture Simulation in War, Business, and Politics

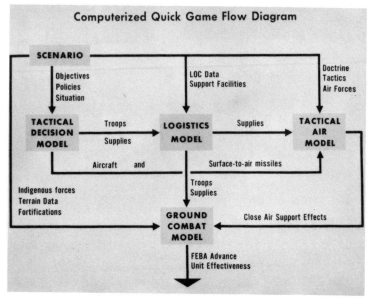

FIG. 7-3.  QUICK GAME.  Flow diagram of game model.

1. A prescribed number of days has elapsed.
2. Friendly forces have driven the enemy back to the border of the defended country.
3. Enemy forces have reached a friendly port of debarkation.

The SIGMALOG Model

The purposes of the SIGMALOG Model were threefold. First a detailed troop list was to be produced. Second, the gross tonnage of unit equipment of the forces deployed to the theater of operations was to be ascertained. Third, the gross tonnage of resupply material needed and shipped to the theater was to be determined. See Fig. 7-4. To accomplish these purposes input data were derived from the QUICK GAME. The nature of these inputs is shown in Fig. 7-5.

A "building-block" structure was employed in the SIGMALOG Model to provide flexibility in modification of force structure and to facilitate computation, analysis, costing, and planning. The building blocks are prestructured groupings of troops used to designate and represent force types and units which may be assigned to a theater. The types of units represented include infantry, armored, airborne, airmobile, and mechanized divisions; an infantry brigade; and several types of army and corps support forces.

## Short-cut and Special-purpose Games 163

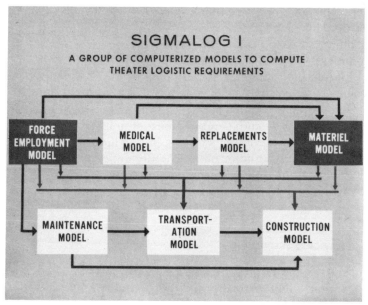

FIG. 7-4. SIGMALOG. Flow diagram of models.

FIG. 7-5. Inputs to SIGMALOG from QUICK GAME.

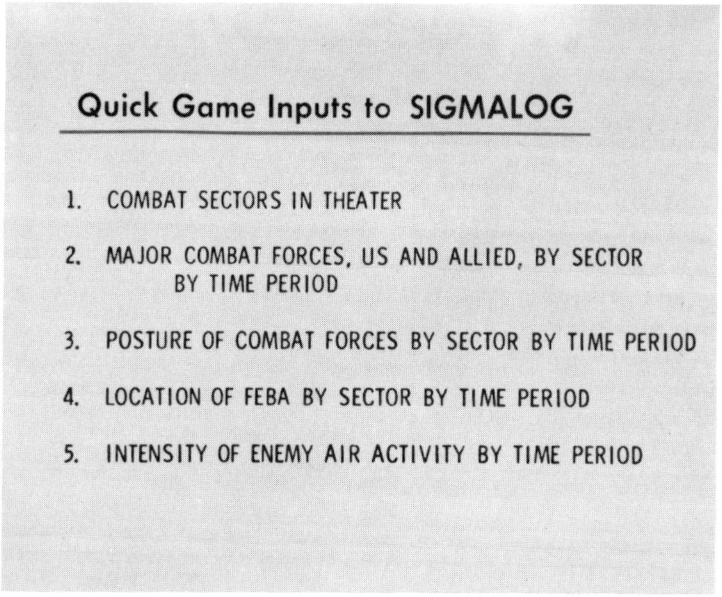

### Quick Game Inputs to SIGMALOG

1. COMBAT SECTORS IN THEATER
2. MAJOR COMBAT FORCES, US AND ALLIED, BY SECTOR BY TIME PERIOD
3. POSTURE OF COMBAT FORCES BY SECTOR BY TIME PERIOD
4. LOCATION OF FEBA BY SECTOR BY TIME PERIOD
5. INTENSITY OF ENEMY AIR ACTIVITY BY TIME PERIOD

Increments of a building-block structure may be treated separately as required for each force type represented, and as components of several different-size forces, be this force a division, a corps, an army, or communication forces (supply and support) in forward and rear zones. The building-block scheme is portrayed in Fig. 7-6. By this means each building block may be used to represent a unique set of requirements that, when combined, may represent a force made up of any of a variety of numbers and combinations of units. The blocks also provide convenient groupings for computing and comparing costs for initial supply, transportation, and resupply under dual states. For example, one state may be a peacetime environment, as in the United States, whereas the other state is one of a wartime environment, as in Vietnam.

The Coordinated System

The system, in prototype form, is shown in Fig. 7-7. The interrelation of various functional parts and the flexibility of the system are indicated. The acronym ADROIT used in the figure refers to Analysis

FIG. 7-6. Building-block scheme for requirements of units, as used in SIGMALOG. SRC: Standard Requirement Code of Army Unit. ISI: Initial Supporting Increment. SSI: Sustaining Supporting Increment.

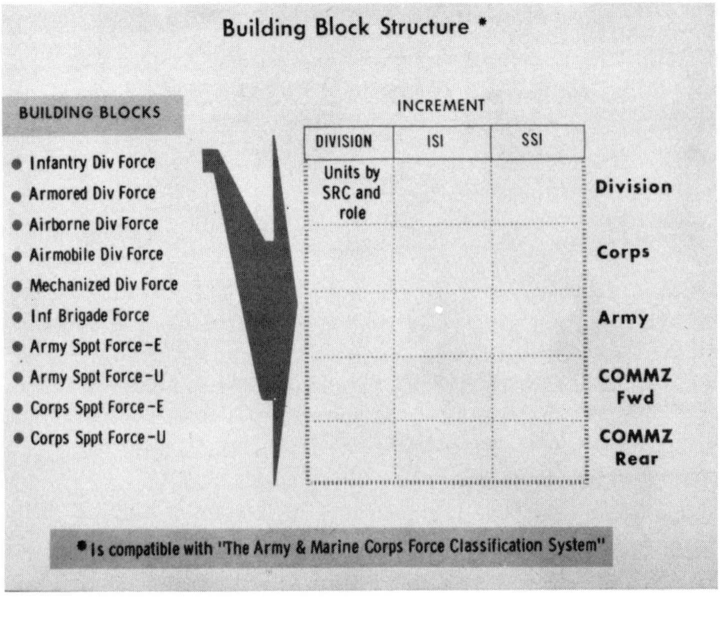

**Building Block Structure** *

| BUILDING BLOCKS | INCREMENT | | |
| --- | --- | --- | --- |
| | DIVISION | ISI | SSI |
| • Infantry Div Force<br>• Armored Div Force<br>• Airborne Div Force | Units by SRC and role | | Division |
| • Airmobile Div Force | | | Corps |
| • Mechanized Div Force<br>• Inf Brigade Force<br>• Army Sppt Force-E | | | Army |
| • Army Sppt Force-U<br>• Corps Sppt Force-E | | | COMMZ Fwd |
| • Corps Sppt Force-U | | | COMMZ Rear |

* Is compatible with "The Army & Marine Corps Force Classification System"

Short-cut and Special-purpose Games    165

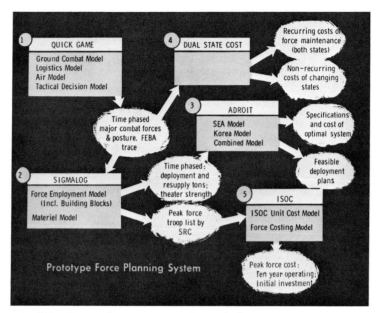

FIG. 7-7. Force planning system interrelations.

of Deployment and Resupply Operations—Inter-Theater. Delivery to the theater (intratheater) is treated separately. The acronym ISOC means Individual System or Organization Cost.

The prototype here described is of direct interest to a major operating part of the Department of the Army: The Force Planning Analysis Directorate. The system here described relates to operations involving the annual expenditure of billions of dollars. In addition to the complexities of planning and providing the support required, the uncertainties of changing contingencies on the worldwide scene put a premium on the time required to formulate new or modified plans (including alternative actions), procurement contracts, and operational orders. The gaming, logistic simulations, and planning techniques involved in this coordinated system illustrate a highly significant contribution to national security and to the national economy. Experience gained in the development of an earlier simulation, LOG-SIM, is carried over into SIGMALOG (Fig. 7-8).

## THEATER STAFF GAMES

Theater Staff games are important because gaming has become a significant adjunct to operational planning at the command levels. The

FIG. 7-8. LOGSIM. Early stage in the development of a logistic simulation based on a planning system used in the U.S. Army. Roles of Technical Services represented by analysts serving as planners, ORO, 1960–1961. *Left to right:* Stanford Hoffman, QM; Wanda Porterfield, Medical; Stephen Coffin, Signal; Martin Brossman, Coordinator and planner for Air Force, Navy, and other agencies involved; Paul Gowen, Engineers; Robert Busacker, Chemical; Ralph Hafner, Ordnance.

hope is that through such planning—and testing by war gaming—theater forces may be ready to meet requirements, and big and little potential "Pearl Harbor" situations of the future may be prevented or quashed at inception.

Theater Staff games as reported here also are germane to illustrate other special requirements of decentralized planning staffs. First, the games must be suitable for use at the Headquarters, without requiring elaborate facilities, specialized equipment such as computers, and special gaming expertness on the part of the staff; second, the games must provide a range of applicability spreading over most of the spectrum of conflict to which that Command must be ready to respond. Thus the gaming system must be flexible and suitable to use in whole or in any reasonable combination of parts. These requirements are met by a family of games designed to be used in concert or separately.[1]

---

[1] The theater battle models here described provide an operational example of the concept of a family of games, often enunciated but rarely achieved.

When deployment of forces extends to remote areas, command is decentralized to those areas. In this process Theater Commanders are designated and each assumes responsibility for preparation and continuous refinement and revision of his theater's plans. This responsibility creates a requirement for operational games which can be used in the process of testing, evaluating, and updating normal contingency and operational plans. An accompanying requirement is for games which can be played in the theater headquarters, using officers of the Headquarters Staff. Games have been developed to serve some of these specialized purposes in a number of headquarters. Examples already discussed include an Eighth Army (Korea) Logistics Game and SCHNELLSPIEL, a Seventh Army (Europe) Tactical Game.

In 1962 the U.S. Joint Chiefs of Staff initiated a drive to develop hand-played games suited to operational use in the theater commands. The Department of the Army, through the Deputy Chief of Staff for Operations, directed the Army's in-house Strategy and Tactics Analysis Group (STAG) to develop suitable Theater Battle Models. This action resulted (in 1963) in a Master Battle Model, the TBM-63, a manual war game intended for use by the Joint Staff or Joint War Gaming Groups, the individual Services, and the Specified and Unified Commands.[1] TBM-63 is used for war-gaming joint operations.[2] Subsequent experience with the TBM-63 proved it too complex and demanding in Staff time. Also, this model was too limited in flexibility to meet requirements arising from the increase in guerrilla-type "wars of liberation" as in Malaya and Vietnam.

In 1967 RAC was asked to develop a new manual which would simplify and update TBM-63, in a new version of the game, to be called TBM-68, after the year in which it was scheduled for dissemination. A description of the new version follows.[3]

---

[1] Specified and Unified Commands—established by and designated by the President through the Secretary of Defense with the advice of the Joint Chiefs of Staff: Specified—normally composed of forces of one Service; Unified—under a single Commander but composed of significant, assigned components of two or more Services. Army Regulations AR 320-5. Joint refers to two or more Services of the same nation—in this case the U.S. Army, Navy, and Air Force.

[2] The TBM-63 was fully documented with complete instructions for Control and players, including the necessary reference data in the form of tables and charts for assessment.

[3] Reported in an unclassified briefing for Rear Adm. E. R. Zumwalt, Director, Systems Analysis Division, Center for Naval Analysis, U.S. Navy Department, presented by Richard E. Zimmerman, RAC, Sept. 21, 1967.

## TBM-68, A FAMILY OF MANUAL WAR GAMES

Two objectives of the TBM-68 models were:

1. To provide flexibility for gaming at different levels of conflict
2. To provide a range of choice among levels of aggregation and resolution to adapt to requirements and available staff resources at various headquarters

Another objective in RAC's effort was to advance the state of the art of war gaming. As a part of this last objective research was conducted on firepower scores and Indices of Combat Effectiveness (ICEs).

The first two objectives were achieved by the design and development of five models or games which might be used separately or in combination as required. The TBM-68 models, or games, are:

The Theater War Game (TWGM)
The Theater QUICK GAME (TQGM)
Division Operations (DOM)
Amphibious Warfare (AWM)
Counter Guerrilla Warfare (CGWM)

The Theater War Game Model

The Theater War Game Model (TWGM) is based in part upon the RAC THEATERSPIEL game, aggregated to division level for ground combat forces (with supporting artillery at the battalion or group level) and to the wing level for air units; however, results and attrition for air operations are assessed on the capabilities of individual aircraft. Although twenty to thirty people are required to play the game, a normal complement would be twelve controllers, six players for Blue and six players for Red. A playing cycle is 24 hours of conventional operations (non-nuclear). Eight, or even four, hours may be played in a cycle when tactical nuclear weapons are involved. Two cycles can be played in one working day. The model may be played in open or closed gaming.

Daily results are measured in terms of success or failure of operations, the extent of advance, losses incurred, and the capability that remains to continue combat. Results are derived from the interactions among seven submodels:

Ground Combat
Combat Support (including Naval gunfire)
Intelligence
Logistics (intratheater)
Air Support and Air Defense (including Naval and Marine Air)
Nuclear
Chemical and Biological

The last two submodels are optional; i.e., they are employed only in war games which (are to) employ such weapons.

The Theater QUICK GAME

The Theater QUICK GAME Model (TQGM) conforms to the hand-played RAC QUICK GAME. The TQGM may be played either as an open or a closed game. A computerized version also is available. This QUICK GAME has the same advantages and uses already discussed earlier in relation to the RAC QUICK GAME work, that is, small requirements for staff and work time, simplicity of play and of assessments, and use to test the results of differing operational concepts or to identify specially important issues to be submitted to more elaborate gaming work. The QUICK GAME may be used as a substitute, or as a preliminary, for the more complete and sophisticated Theater War Game. Only Ground Combat, Air (including Naval and Marine Air), and Logistics (intratheater) submodels are played in the QUICK GAME.

The Division Operations Model

The Division Operations Model (DOM) is based upon experience with RAC's TACSPIEL game and provides a capability of war-game tactical operations up to an army corps. In this game, details of operations at the battalion or company level may be identified, isolated, and aggregated for each battalion or company involved. Three to eight hours of combat can be played per working day with six to nine participants. Two to ten days of combat can be gamed in 15 to 30 working days. The intensity and complexity of actions, and resolution of detail, affect the working time and staff requirements. The model requires the play of at least the first four submodels shown below.

The remaining six submodels are optional and may be used as the purposes of the game require. The required submodels are:

Movement and Contact
Ground Combat/Combat Support
Intelligence
Command and Communications

The optional submodels are:
Air Support/Air Defense
Logistics
Air Mobility
Engineer Support
Nuclear
Chemical, Biological

The Amphibious Warfare Model

The Amphibious Warfare Model (AWM), not having a precedent, was constructed from scratch. The AWM was designed to cover the landing operations only—of a force up to corps—for the period of H hour + 30 minutes, i.e., for the 30 minutes of the actual assault phase when an amphibious force is put ashore. Actions start with the arrival of the advance force and continue through the actual assault, including any subsequent landing and offshore combat support. This model can be played through in about 2 working days by a team of six to eight participants. Once the forces are ashore, the continuing combat is gamed in the Theater War Game and/or Division Operations Models. The AWM involves seven submodels, which are played in the order listed:

Intelligence Submodel
Fire Support Submodel
Air Submodel (attack and defense)
Ship to Shore Submodel
Mine Warfare Submodel
Small Craft Operations Submodel
Nuclear Submodel

Advance Force Operations are gamed starting at 0000 hours (midnight) D-day and continue until 30 minutes after the landing force is put ashore, H hour + 30 minutes.

### The Counter Guerrilla Warfare Model

The Counter Guerrilla Warfare Model (CGWM) is based upon RAC's TACSPIEL Counter Guerrilla Model. Only overt ground combat between guerrilla and counter-guerrilla forces is considered, i.e., the strictly military actions of smaller units, mostly platoons and companies. The model can deal, however, with forces up to division size. It is not capable of evaluating political, psychological, and other quasi- and non-military factors involved in insurgency situations. The model includes eight submodels which are normally played in the following order:

The Command and Communications Submodel
The Movement and Contact Submodel
The Air Support and Air Defense Submodel—concurrently with Submodel 4 when it is played
The Engineer Support Submodel—concurrently with Submodel 2 when used
The Air Mobility Submodel—concurrently with movement and contact when played
The Ground Combat Submodel
The Intelligence Submodel—continuous and concurrently with all
The Logistics Submodel—continuous and concurrently with all

### Summary

The TBM-68 family of war games consists of five models or games and thirty-five submodels—an unusually large repertoire of interrelated models for a flexible and coordinated system. Furthermore, the system provides the documentation, directions, and data needed by a theater staff to operate these games and the methodology to adapt data to local conditions. A summary of TBM-68 capabilities, requirements, and characteristics is shown in Fig. 7-9.

The TBM-68 models could be useful to military staffs of other nations with which the United States has security pacts. Although the

### Theater Battle Model (TBM-68)

| Games and models | Game time (days) | Playing time (days) | Game staff (people) | Level of aggregation |
|---|---|---|---|---|
| Theater War Game | 60–180 | 60–90 | 20–30 | Division (for combat troops) Battalion (for combat support) |
| Division Operations | 2–10 | 15–30 | 6–9 | Battalion and Bn task force (combat and combat support units) |
| Amphibious Warfare | 2 | 5 | 6–8 | Battalion landing team and individual ships |
| Counter Guerrilla Warfare | 20–30 | 30–40 | 6–9 | Battalion-size or smaller, normally at platoon level |
| Theater QUICK GAME | 60–180 | Several days to a few weeks | 2–4 | Highly aggregated |

FIG. 7-9. The TBM Model and the five games which compose the family of models, and their characteristics, requirements, and capabilities are indicated.

size of a nation is no measure of its military sophistication, many small countries have lacked the resources necessary for development of highly sophisticated war games. Manual games of the TBM type are well suited to the needs of these nations. Perhaps they, in turn, can contribute to and advance the art of gaming insurgency and counter-guerrilla warfare beyond the very rudimentary level to which these types of games have been developed by the major military powers. In this last area significant advances remain to be made.

### 1. TACSPIEL

A Division-level Computer-assisted Analytical War Game

TACSPIEL is a flexible, two-sided, free-play, manually operated, rigidly assessed, computer-assisted, division-level war game, operated with retired military officers on the Blue and Red teams and on the Control staff. Control assessors are principally physical scientists, mathematicians, engineers, and military officers. The support staff includes computer specialists and others with specialized skills. Together, the whole group makes up a functioning operations research unit.

FIG. 7-10. Insignia of TACSPIEL war game. Stars in outer circle represent the four members of the Senior Review Board staff who overwatch and critique the game from the earliest planning stage through the completed report of the analysis.

The TACSPIEL game is designed as a flexible instrument for use in examination of problems of ground combat at and below Army division level. It provides for the study of interactions of combined arms in battle and for the operational performance (as contrasted with the purely physical performance) of weapons, equipment, organization, and doctrine. For each game the locale, combat organizations, equipments, tactical doctrine, situations, force missions, and resolution of details of actions are built into the game design to represent the problems involved in the purpose of that game. TACSPIEL provides a small-scale facsimile of battle, somewhat analogous to an experimental laboratory in the physical sciences. In this laboratory it is possible to subject to precombat study and analysis such matters as force structures not yet in being and projected concepts of mobility, firepower, communications, logistics, combat surveillance and target acquisition means, and air attack and defense. Measures of the relative merit of postulated mixes of forces and systems can be obtained by careful game design in play against any given enemy as a control. Figures 7-10 to 7-19 show details of the physical features of TACSPIEL games in operation.

Analysts in charge of gaming in TACSPIEL and of its forerunner, INDIGO, first under the Operations Research Office of The Johns Hopkins University until late 1961 and subsequently under the Research Analysis Corporation, were: INDIGO, Maj. Gen. James G. Christiansen 1958–1961; TACSPIEL, Edward W. Girard 1961–1964, Lawrence J. Dondero 1964–1965, Lawrence S. Simcox 1965–1968.

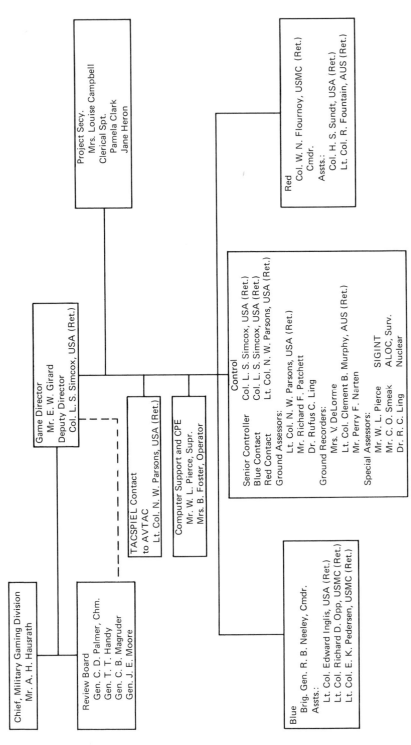

FIG. 7.11 Staff organization for a TACSPIEL Game in the OREGON Trail Study, at RAC, July 1, 1964.

Short-cut and Special-purpose Games    175

FIG. 7-12. Game Controller, Colonel Simcox, conducts a pre-game check of the player display of Organization for Combat. At each cycle of play this display is photographed and retained as a permanent record for use in post-game analysis of game results, and for future research. (1963)

FIG. 7-13. Blue game room in TACSPIEL game. Blue team of three players plans all operations, including artillery, engineer, and air support in this series. These plans are then translated into orders and submitted to Control. On receipt of such orders from the Blue and Red teams, Control assesses the interactions and submits the results to each team. *Left to right:* Lt. Col. John R. Boatwright, Lt. Col. I. H. Baker, Lt. Col. L. W. Williams. (1962)

FIG. 7-14. Part of the Control group in an early TACSPIEL game. Technical aides assist controllers in preparation of punch cards, printouts, and game records. Blue and Red rooms communicate directly with Control but not with each other. *Participants, left to right:* Miss Joyce Sangston, Mrs. Betty Foster, Miss Margaret Mix, Dr. Dorothy Clark, W. L. Pierce, C. O. Smeak. (1961)

FIG. 7-15. Operations analysts assess logistic interactions relating to the resupply of Class III (fuel) and Class V (ammunition) in a division-level war game. *Participants, left to right:* W. L. Pierce, C. O. Smeak. (1964)

Short-cut and Special-purpose Games    177

FIG. 7-16.   General view of the TACSPIEL Control room.   Two game tables are in operation, each working on a different game.   The table in the foreground shows a group assessing and recording the disposition of units and their status. The fan-shaped templet, visible in the center foreground, represents the scope and range of radar equipment and is used in assessing possible detections. Reference maps, tables, game-time clock, and random-number generator are visible on the rear wall.   *Participants, left to right, foreground:* C. J. Lynch, R. F. Patchett, Col. C. B. Murphy; *background:* Col. W. N. Flournoy, Dr. R. C. Ling, C. O. Smeak, Gen. S. Whipple.   Col. L. S. Simcox standing at right.   (February 1966)

FIG. 7-17.   Two members of the Control team in process of assessing results of interaction between opposing forces.   These assessments are recorded in coded form and copies sent to Blue or Red.   Interactions assessed include the results of battle, movement, artillery fires, contacts, aviation, engineer operations, communications, logistics, and many more.   Note the ridge lines on the game-board map.   *Participants, left to right:* Dr. R. C. Ling, Col. W. N. Flournoy, Col. N. W. Parsons, P. F. Narten.

178  Venture Simulation in War, Business, and Politics

FIG. 7-18. A portion of an earlier Control terrain board used in TACSPIEL. Pieces locate and identify both Red and Blue units. Tags indicate unit orders, and chained loops portray units in battle at close range. Map scale is 1:25,000. (1962)

FIG. 7-19. Order-of-battle and status-of-units display board. Game date and clock give date and hour of game time shown by this posting. Col. R. D. Opp placing markers. Note lower center portion of board showing Ineffective Units, i.e., units which have received casualties to the point of being ineffective for combat. Different standards are used for attack and defense capability. (1965)

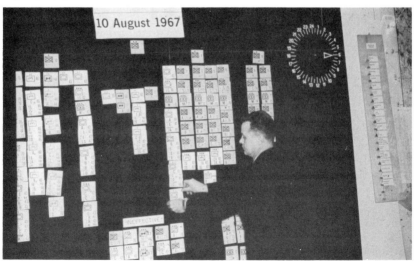

## 2. THEATERSPIEL

A Theater-level Computer-assisted Analytical War Game

THEATERSPIEL is an analytical game having application to the testing and analysis of proposed military operations and proposed military systems. It has settings in Western Europe, the Middle East, and Southeast Asia. The game is played by two competing player teams. Each player team has full knowledge of the location and condition of its own forces

FIG. 7-20.  View of the THEATERSPIEL Control room setup in ORO, with a game in process.  Shown are opaque maps on movable tracks, game-time clock in front of ceiling track, random-number generator on table in right foreground, interroom communication system at extreme right, flexowriter recording and communicating system behind tracks.  Opaque maps, both flat and three-dimensional, mounted on tracks for movement between Control and player rooms.  This system was superseded by rear-view projection, with opaque maps used only for reference and basic planning.  *Analysts at work, left to right:* Adm. M. N. Little, USN (Ret.), Brig. Gen. J. G. Hill, USA (Ret.), R. M. Bentz, W. H. Sutherland, B. L. Himes, secretary, Capt. J. D. F. Dorsett, USN (Ret.). (October 1960)

but has only a limited knowledge of the disposition of enemy forces such as may be acquired through intelligence activities.

The Control Group reviews the mission orders of the Blue and Red Commanders to make sure that they are realistic and that they conform to the requirements of the game being played. The orders are punched into card decks, which become inputs to a computer run.

The computer assesses the individual actions, determining casualties and equipment losses, and modifies the status-of-forces records to reflect these losses. After all such assessments have been made, the computer, taking logistical limitations into consideration, attempts to resupply each side's forces. Finally, the computer prints out force disposition and intelligence reports for each player team, and the next cycle is ready to start. Figures 7-20 to 7-32 show details of the physical features and activities of THEATERSPIEL games in operation.

An analysis and reporting group studies the results of each cycle of play in the light of what the player teams attempted and generates the reporting records required for each game.

FIG. 7-21. Details of the earlier system of handling game maps on tracks between game rooms.

FIG. 7-22. THEATERSPIEL Red players working on three-dimensional opaque maps. On the left, Maj. Gen. T. S. Timberman is checking his plans for movement of Red units in attack on Blue positions. To the right Brig. Gen. C. C. Sibley is checking on details of logistics involved in Red operations. Game set in Southeast Asia (SEA-1). (February 1962)

FIG. 7-23. THEATERSPIEL rear-view projection system. Projector in right foreground; center background shows the 8- by 8-foot mirror which reflects the projected image on a translucent screen, shown at the left. (February 1962)

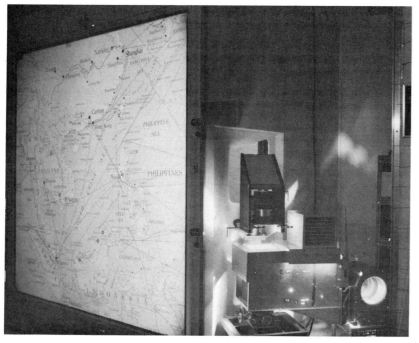

182  Venture Simulation in War, Business, and Politics

FIG. 7-24. Rear-view transparency projector, designed and built for RAC's THEATERSPIEL game is used to project situation maps on Control- and player-room screens. (February 1961)

FIG. 7-25. THEATERSPIEL map supervisor mounting and placing a transparency on the rear-view projector. Lt. Col. G. B. Murphy.

FIG. 7-26. Schematic sketch of later three-room game facility, with the computer used in support of THEATERSPIEL at RAC, McLean, Virginia. All bookkeeping and most of the assessments are programmed for and processed by the computer. Cutaway diagram showing general arrangement of three-room setup, Blue and Red player rooms at extreme left and right, Control room in center. Each player room has one rear-view projection screen with overlays. Control room has two screens with overlays; cutaway on right projection screen booth shows projector; random-number generator for use by Controllers and Assessors is visible on upper right wall of Control room. Message Center and clerks shown in foreground at left and right of Control room. (June 1965)

FIG. 7-27. THEATERSPIEL Control room. At the left and right in the background are the two rear-projection display screens, one for Air and Ground Operations, the other for Logistics and Intelligence Operations. Control party at work. Overlays, which can be placed over the rear-projection screens as needed, are shown in a rack between the two rear-projection screens. *Participants, left foreground:* Maj. Gen. P. C. Wehle; *right background:* Maj. Gen. R. J. Butchers and Mrs. Ruth F. Voigt checking and posting Intelligence data; *right foreground:* Brig. Gen. J. M. Worthington. (February 1966)

FIG. 7-28. Group of Controllers at work in THEATERSPIEL Control room. Wall map for reference purposes in left background, and in the right background the rear-projection screen with overlay is visible. *Participants (left to right):* Maj. Gen. D. Dunford, Col. N. C. Bonawitz, Maj. Gen. P. C. Wehle, Maj. Gen. H. J. Vander Heide. Col. H. H. Figuers, Input Coordinator, may be seen at the projection screen checking markers representing deployed units. (February 1966)

FIG. 7-29. THEATERSPIEL Blue team checking out moves. *Participants (left to right):* Maj. Gen. R. W. Zwicker, Blue Commander; Colonel J. E. Harper, Assistant. (February 1966)

## Short-cut and Special-purpose Games 185

FIG. 7-30. THEATERSPIEL operations analyst, Edward P. Kerlin, checking printout from the computer and summarizing data for future analysis. (February 1966)

FIG. 7-31. THEATERSPIEL staff briefing in process. Purpose: to acquaint Controllers with details which have developed in the situation during the past period of play. *Participants:* Brig. Gen. J. A. Elmore, briefing. *Other personnel left to right:* Maj. Gen. H. Vander Heide, Intelligence Controller; Maj. Gen. R. W. Zwicker, Ground Combat Controller; Col. J. E. Harper, Logistics Controller; Maj. Gen. O. C. Troxel, Command and Control Controller; A. H. Hausrath, Chief, Military Gaming Division; Col. L. R. Seibert, Air Controller. (1965)

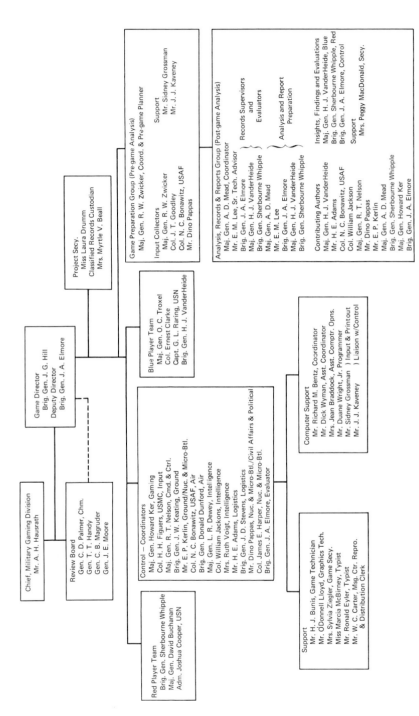

FIG. 7-32. Staff Organization for a THEATERSPIEL Game in the OREGON TRAIL Study at RAC, July 1, 1964. All officers not otherwise indicated are retired from the U.S. Army.

Brigadier General John G. Hill, USA (Ret.), an operations analyst, was Game Director of THEATERSPIEL during most of its development and was Project Chairman until his retirement in late 1965. During this period THEATERSPIEL was involved in extensive gaming of new Army concepts. Richard E. Zimmerman preceded General Hill as Project Chairman. Major General Armistead D. Mead, USA (Ret.), and later Maj. Gen. Howard Ker, USA (Ret.), served as Chairmen until their respective retirements from RAC.

# Business Games and Management Simulations

It was probably inevitable that the techniques of simulation and gaming would be applied to business and industrial management.[1] Having been proved useful for military purposes, the employment of these techniques in business applications presented no insurmountable problems. Also, as in the military, the development of operations research and the advent of electronic computers extended the potential for quantitative analytical methods and contributed to the rapid spread of simulations and games in the business world. Together, these developments supported the advance of management science.[2]

### EVOLUTION OF MANAGEMENT SCIENCE

*Management science* is the term used in recent years to represent the field of science applied to the problems of management. In earlier days it was called by such names as shop management, factory management, efficiency engineering, and scientific management. The

---

[1] The terms business game, management simulation, business simulation, and management game are used interchangeably in the literature of this field. The more precise meanings attached to the terms—models, simulations, games—generally employed in war gaming and in this book were not observed in the earlier literature on business games.

[2] *Management Science* is the name of the journal of The Institute of Management Sciences (widely referred to as "TIMS"), the masthead of which carries the statement, "An international society to identify, extend and unify scientific knowledge pertaining to management."

188

last term is attributed to Frederick W. Taylor, "the father of scientific management," who used it in the title of a book he published in 1911.[1] For three decades straddling the turn of the century, Towne, Taylor, Gantt, Gilbreth, and their cohorts pioneered the concept that underlying industrial management was a body of fundamental truths or principles which could be discovered by the methods of science.[2] Taylor and others set themselves at the task of discovering these principles. Some of the early significant studies were reported in papers presented at meetings of the American Society of Mechanical Engineers (ASME) in 1895 and 1903.[3] From those beginnings "scientific management" became a field of intense and sustained effort in the United States and abroad.[4] Through such efforts there emerged principles of management which could be used as the foundation blocks for the management sciences and for simulations and games in which these principles could be applied.

Professionalizing of Business Education

Coincidently with the emphasis on scientific management in industrial and engineering circles, the field of business administration was

---

[1] Frederick W. Taylor, *The Principles of Scientific Management*, New York, Harper & Brothers, 1911.
[2] Henry R. Towne, "The Engineer as an Economist," *Transactions of the American Society of Mechanical Engineers*, 7:425–432, 1886. For an account of the work of Frank Gilbreth, see Edna Yost, *Frank and Lillian Gilbreth, Partners for Life*, New Brunswick, N.J., Rutgers University Press, 1949. Henry L. Gantt (1861–1949), best known for the widely applied Gantt charts, worked closely with Taylor between 1887 and 1893 and intermittently until 1904 to improve management techniques. Gantt and Gilbreth stressed the human side of management and always tried to advance the welfare of the worker as well as to increase production efficiency and the profit to ownership. Taylor's interest focused on efficient production and greater return to the entrepreneur. Wallace Clark, *The Gantt Chart*, 3d ed., London, Sir Isaac Pitman & Sons, Ltd., 1952, gives examples of the use of Gantt charts in diverse industries in many countries. The first American edition of Clark's book was published by The Ronald Press Company, New York, 1922. An informative biography and record of Gantt's professional activities, prepared by L. P. Alford under the sponsorship of the ASME, was published by Harper & Brothers, New York, in 1934 under the title *Henry Laurence Gantt, Leader in Industry*.
[3] Frederick W. Taylor, "A Piece Rate System," *Transactions ASME*, vol. 16, 1895, and "Shop Management," *Transactions ASME*, vol. 24, 1903. Subsequently, these two papers were reprinted in book form together with Taylor's testimony before a Special House Committee of the U.S. Congress in 1912 set up to "investigate the Taylor and other systems of shop management." Frederick W. Taylor, *Scientific Management*, New York, Harper & Brothers, 1911, 1939, 1947.
[4] Taylor's works were translated into French, German, Italian, Japanese, Dutch, Russian, and other languages, and were widely circulated. Clark's book *The Gantt Chart* was also published in all the aforementioned languages, and in Polish, Czech, Greek, Spanish, and Bulgarian.

becoming professionalized. The creation of the Wharton School of Finance and Commerce at the University of Pennsylvania in Philadelphia in 1881 marked the beginning of the era of formalized business education in the universities. Several other leading universities added schools of business administration prior to World War I. A great surge in formalized business education followed, until almost every major university had a business school or department by the time of World War II. This period also was marked by much experimentation and increasing specialization in business education.

The teaching of management techniques spread in universities both in the engineering and business schools. The field of business education expanded to encompass planning, forecasting, and management-employee relations, in addition to the established fields of accounting, banking, finance, transportation, and marketing. The field of management was developing its own body of organized knowledge, and with the advent of operations research, the newer term *management science* came into more general use in centers of business education and leadership.

Techniques in Business Education

The techniques of teaching also were undergoing changes in professional schools during this period. Harvard University had introduced the case method in its Law School in 1870. A half-century later Harvard's Graduate School of Business Administration, which had been established in 1908, embarked upon a systematic program of continuing research on the case-study method in business education and has been a principal center for such research since that time. The case-study method has been widely used as a vehicle to teach the principles involved in business operations. But case studies stop short of operational decision making by the student. Neither do case studies involve the pragmatic, cause-and-effect assessment of the results of individual business decisions made by the students; nor do they show the cumulative consequences of a student's series of decisions made in a step-by-step sequence in a given business venture. Furthermore, case studies do not provide the student with the opportunity and necessity of making his decisions in terms of the subsequent impacts ("feedback") of his competitors' decisions, all of which have to be made under the pressures of time which are characteristic of the real business and management world. Simulations and games are unique in offering these opportunities in the learning situation.

It has been reasoned that simulation partakes of many of the values attributed to some of the varieties of the case-study and role-playing methods. A business game can be considered as an elaboration of a case study in which the data are subjected to the actions of the players as the game progresses (interactive) rather than consisting only of a review of the record of an earlier completed example.[1] One summarizing statement put it: "Business Games represent an advance in business simulation techniques, because they combine the principles illustrated by case study, the dynamic character of role-playing, and the new element of feed-back. . . . Business gaming is the logical expansion of case study and role-playing into a new form of controlled participation."[2] William R. Dill of Carnegie Institute of Technology commented as early as 1958 that as a library of business games is built up, and a variety of specialized games are developed, they should be able to present nearly the same repertoire of problems that were then available in casebooks and work books.[3]

## MANAGEMENT SIMULATIONS GET STARTED

Meanwhile the military establishment was making extensive use of the newer advances in science, technology, and management techniques. Among them were the new technique of operations research, the use of management consulting organizations ("think tank" companies) and specialists in the universities, simulation, and computer developments. In fact, the military establishment was a prime factor in supporting the development of these fields. Some of the early military simulations, designed and developed in the "think tank" companies and in the aircraft industry, demonstrated the utility of the process in the broader area of management decisions. Many of these simulation studies have never been reported in the open literature because of proprietary rights in industry and security considerations in the military. By the time such restraints were no longer needed, much of both the substance of the simulations and the simulation

---

[1] Stanley C. Vance, *Management Decision Simulation*, pp. 1–3, New York, McGraw-Hill Book Company, 1960.
[2] John Whedon Acer, *Business Games: A Simulation Technique*, p. 8, Iowa City, Iowa, State University of Iowa, College of Business Administration, Bureau of Labor and Management, Information Series 3, Nov. 1960.
[3] Dill presented this view at a National Symposium on Management Games held at the University of Kansas, Dec. 12–13, 1958.

techniques had been overtaken by newer developments in the state of the art.

One of the unpublicized early simulations was initiated in 1948. It was designed by the Operations Research Office (ORO) of The Johns Hopkins University to provide a functional model for the study of antiaircraft weapons systems for the military establishment.[1] After several years of use in hundreds of analyses and progressive refinements, this experience was incorporated in what is believed to be the first computer simulation in operations research. It was used to study the air defense of the continental United States in 1953 and 1954. Thousands of computer runs and analyses were made, assessing the effect of hundreds of combinations of simulated bomber and missile attacks. One form of these simulations has been mentioned in the literature under the name ZIGSPIEL (Zone of the Interior Ground Defense Game). Some of the problems encountered in the development of the mathematical model and the computer simulation in an air-defense study are described in an article by Rich.[2] He describes a relatively simple physical device which was used in developing the mathematical model for a computer simulation for studying fleet air defense. The ORO air-defense simulation was a form of war game in which the computer played both the attack and the defense. As such it was a forerunner of some of the complex computer simulations used in later war games and in business games.

Another transition exercise between war games and business games was a simulation developed in 1954 by the U.S. Air Force Project RAND (later organized as the RAND Corporation). This effort resulted in an Inventory Management "Game," called MONOPO-LOGS. It is a one-player game which was designed to simulate the more complicated system in use in the U.S. Air Force at that time. The game centers on a hypothetical replacement part of relatively high value but low demand, which can be repaired at certain depots. The player is given data on costs and lead times for various supply actions. He is also supplied piecemeal a month-by-month set of requisitions for the part from the different bases—a simulated case history of demand for this aircraft part over a period of several years. The player assumes the role of an inventory manager for this aircraft part,

---

[1] This simulation was programmed for the first model of the Univac computer and was run on that computer.
[2] R. P. Rich, "Simulation as an Aid in Model Building," *Journal of the Operations Research Society of America*, 3(1):15–19, Feb. 1955. Rich was a staff member of the Applied Physics Laboratory, The Johns Hopkins University, Silver Spring, Md., where the fleet simulation was developed at that time.

and he must manage a simulated supply system of one depot and five bases. As the game progresses, the player initiates procurement, plans repair schedules, and sets inventory and distribution policies. The demand for the part is a random variable beyond the control of the player, but he must make predictions of future demands and establish inventory levels. He makes decisions as the game progresses and from time to time gets reports of the results of these decisions. At the end of the game the total costs are computed. These costs include normal and premium procurement, repair costs, storage, transportation, and inventory carrying charges. Penalties are levied for any intervals when the part is out of stock. The player's goal is to keep costs low, meet demands, and avoid excesses and shortages.[1] The game is really a problem-solving exercise in which the player uses work sheets. His decisions are scored on score sheets. The game serves as an example of a simple hand-scored simulation of a specialized function which is one segment of a much more complex system. Any number of players may play the game simultaneously, but each would play it *solitaire*. In some respects MONOPOLOGS resembles the In Basket exercise which will be discussed later.

A more complicated supply game is represented by the U.S. Army's Logistic Simulation (LOGSIM-W) developed cooperatively by the Operations Research Office (ORO) of The Johns Hopkins University and personnel of the U.S. Army Logistics Center at Fort Lee, Virginia, in 1958–1959.[2] It is non-interactive; i.e., it is not a game in which one company competes against other companies for the market. It simulates operation of a national inventory control center and involves procedures and practices of effective supply management. Two to five teams of six or more participants each are engaged in play of the game for a 3-day period near the end of a 12-week supply management course. Both computer and manual calculations are employed. Details of later and more sophisticated logistic simulations are given in Chap. 7 under the subhead SIGMALOG.

The U.S. Army Management School, established in 1954 at Fort Belvoir, Virginia, was an early user of simulations as instructional devices in courses in management science. Noncomputer simulations were developed and used first to provide financial management,

---

[1] Reported by William Hamburger as a Research Memorandum RM-1579 in U.S. Air Force Project RAND, Jan. 3, 1956, and issued by the RAND Corporation, Santa Monica, Calif.
[2] The reader will recall earlier mention of the large-scale logistics exercises, LOGEX, conducted on an annual basis at Fort Lee since the early 1950s. LOGEX exercises include features of field maneuvers and were conducted as war games.

logistics, and personnel management exercises: FORT IRWIN in 1958; FORT ROOT in 1959; FORT SIMULATION in 1960, computerized in 1961. A more recent simulation, FORT ARMS developed in 1966, incorporated judgment situations.[1] MONOPOLOGS, FORT SIMULATION, LOGSIM-W, and SIGMALOG represent games which were meeting grounds between the military, with its earlier war-games tradition, and business and industry with its newly developed interest in the gaming technique. Although these four simulations or games have been singled out for mention, there have been many others. The techniques of management and the problems of logistics are fields of common concern in both military and business circles.

Along with the management-science movement, systematic efforts arose to train practicing managers to higher levels of proficiency and to improve their skills in management decision making. The American Management Association (AMA) was one of the leading forces in this movement and helped it grow and spread in the world of big business. Management development programs began to appear within leading companies, and universities offered conferences and short courses to assist in the movement.

## BUSINESS GAMES BEGIN

In 1956 the American Management Association started to adapt war-gaming experience to the business world. A research team of AMA officials and operations analysts from management consulting firms had been formed.[2] This team visited the Naval War College at Newport, Rhode Island, to confer with the staff which had had long experience in war gaming, and the team enlisted the cooperation of the International Business Machines Corporation. The team set about the task of building a mathematical model of business and a group of cause-and-effect formulas which could be used to determine the results of each set of decisions made in the game.[3] The model was

---

[1] Information from Lt. Paul Nelson, in charge of research at the school.
[2] Franc M. Ricciardi, Clifford C. Craft, Joel M. Kibbee, and Richard H. Rawdon of the American Management Association; Donald G. Malcolm, Charles Clark, and Richard Bellman of Booz, Allen and Hamilton's Chicago office. Bellman also was associated with the RAND Corporation office at the same time. See Franc M. Ricciardi et al., in Elizabeth Marting (ed.), *Top Management Decision Simulation*, pp. 6, 59, New York, American Management Association, 1957.
[3] *Ibid.*, p. 59.

programmed for an IBM 650 computer. An IBM 650 computer was installed in the AMA headquarters in New York, and a demonstration game was played in May 1957 with twenty corporation presidents participating. The well-publicized demonstration of the game was hailed as "a major break-through in management education" and it touched off a wave of enthusiasm—and controversy.[1] One complication was that the AMA game was computer-linked; that is, it required a computer to apply the data and mathematical equations of its models in order to calculate the results of the players' decisions.

The game was constructed for and featured in the Executive Decision-making Program and the Management Development Seminars at the AMA Academy for Advanced Management at Saranac Lake, New York. Participants—some 400 by early 1958—were those whose enthusiasms and official positions served to stimulate action in their own organizations.[2] Interest was excited further and particularly in management circles of business concerns by a surge of articles in the business press: *Business Week, Fortune, Factory Management,* etc.[3] With such impetus the AMA Top Management Decision Simulation became the first widely known business game.[4] AMA reports helped.[5] But with the expanded market for business and management games the AMA Academy could reach only a fraction of those whose interests had been aroused. Even if the AMA Game could have been made generally accessible to schools and business firms, trained personnel and the computers required to operate the game were not available. Clearly, other approaches were needed.

Management simulations and business games were desired in a form that did not require a computer. The potential use for noncomputer games was far greater than for those which required computers. G. R. Andlinger was among those who recognized this need, and he set out to develop a management game of that type. Simultaneously

---

[1] *Ibid.*, p. 107. See also Lois Stewart, *A Survey of Business Games*, pp. 16–26 and specifically p. 16, AMA Management Report 55, 1961.
[2] *Fortune*, March 1958, p. 140.
[3] *Business Week*, "In Business Education, the Game's the Thing," July 25, 1959, pp. 56–58, 63–64.
*Fortune*, "The Business Decision Game," by John McDonald and Franc Ricciardi, March 1958, pp. 140–142, 208, 211, 213–214.
*Factory Management and Maintenance*, "'War Game' Will Train Managers," June 1957, pp. 132–133.
[4] Kalman J. Cohen and Eric Rhenman, "The Role of Management Games in Education and Research," *Management Science*, Jan. 1961, p. 135.
[5] Among these were the AMA Management Report 55 and Ricciardi et al. in Elizabeth Marting (ed.), *Top Management Decision Simulation*.

with the AMA game development in 1956, Andlinger started an experimental model of a noncomputer game; he then joined the management consulting firm of McKinsey and Company, where he and Jay R. Greene completed the Business Management Game. This game is also referred to as the Andlinger game and the McKinsey and Company game. Andlinger described this game in the first of two articles in the *Harvard Business Review* in 1958.[1]

Andlinger reported that his game "incorporated the key elements of a one-product capital goods company which is competing with one or two other companies for the same market."[2] The elements include the market, marketing, advertising, research and development, production, finance, and competition. Each of the competing companies is operated by a team of three or four participants, although it is possible for one person to operate a company. An Umpire, or Control Team of three or four people who know the game, is required. The Control Team has full information about the game and calculates the operating results of the decisions of each company team. Control also decides the trends of the market and injects the chance factors from tables of random numbers and probability distributions to simulate the types of variability which occur in real business competition. The game has considerable flexibility and can be adapted to a wide variety of situations.

The Andlinger articles in the *Harvard Business Review* were followed by featured articles in several other professional journals of graduate schools of business administration and in management and engineering periodicals.[3] These journals stimulated consideration of business games in the academic side of the business world and added to the surge of interest in the new applications of this old technique of the military profession.

Training directors in business and industrial concerns saw games as

---

[1] G. R. Andlinger, "Business Games—Play One!," *Harvard Business Review*, 36:115-125, March-April 1958, and "Looking Around: What Can Business Games Do?," 36:147-152, July-Aug. 1958.
[2] Andlinger's first article, p. 117. The game description was drawn from this source.
[3] See as examples: E. W. Martin, Jr., "Teaching Executives via Simulation," in the Indiana University journal *Business Horizons*, Summer 1959.
James R. Jackson, "Learning from Experience in Business Decision Games," in the University of California (Berkeley) journal *California Management Review*, Winter 1959.
Jack D. Steele, "Simulated Management Experience: Some Comments on Business Games," in the University of Kansas journal *Kansas Business Review*, Oct. 1958.
Hans B. Thorelli et al., "The International Operations Simulation at the University of Chicago," in the University of Chicago's *Journal of Business*, July 1962.

an advance in methodology, as a tool for use in conducting management development programs, and as an aid in winning management support for such programs. The *Journal of the American Society of Training Directors* carried a spate of articles between June 1957 and 1962, reporting such uses of games.[1]

Management simulations and business games of the types discussed provide practice in the essence of realistic decision making. The critique which follows play of the game, as well as the computed profit and loss results of decisions made by the players, presents lessons not quickly forgotten. Feedback of game results as a direct consequence of the participant's own decisions and those of his competitors in the game provides valuable learning experiences, and at much less cost to the decision maker's employer than would similar learning experiences in the real world of business. These advantages of management simulations and business games were promptly recognized, and a lively rush to employ this new approach resulted.

## BUSINESS GAMES FLOURISH

For these and other reasons business games caught on. They captured the imagination of business leaders and business educators who saw in them a device to train employees and students in the process of decision making in business, much as the military had viewed war games in their own professional development. All sorts of special situations were suggested in which games and simulations could make a contribution. Needs were broad, and the time and the state of the art were ripe for the development of business games and management simulations on a wide front. As was the case with war games, business games developed in a wide variety of forms, applications, and degrees of complexity.

Many organizations lost little time in developing games suited to their own purposes. Business and industrial enterprises, educational institutions, and government agencies constructed simulations and games and used them with their own personnel. Kibbee listed and described some ninety-five business and management games reported in use by August 1960; the data were obtained in a Remington Rand

---

[1] See Robert M. Robbins, "Decision-making Simulation Through Business Games," in the Sept. 1959 issue, and John W. Zimmerman, "Non-mathematical Simulation—Dynamic Development Guide," in the June 1962 issue of the *Journal of the American Society of Training Directors*, as examples.

Univac survey.[1] Another book, by Greenlaw et al., published in 1962, gave summaries of some eighty-nine games.[2] In 1969 the American Management Association published a handbook which described 182 business games.[3]

By mid-1959 the Boeing Airplane Company is reported to have used games with more than 2,000 of its management trainees.[4] In 1960 some 600 men had been trained by September with the United States Army's Logistic Simulation, LOGSIM, an Army supply system game;[5] some 250 had played the Pillsbury Company games by October;[6] another 264 men had been trained at the U.S. Army Management School in its game FORT SIMULATION, U.S.A.;[7] and at Minneapolis-Honeywell over 3,600 people had played its Top Brass game[8] by November. By mid-1961 more than 15,000 employees in the lower and middle management echelons of the Bell System had played the American Telephone and Telegraph Conpany's financial management game.[9] Inclusive estimates made by Lois Stewart in 1961 stated that "over 30,000 executives have participated in at least one of the more than 100 management games" that existed at that time and that the trend was continuing upward in number of participants and of games being developed.[10]

A survey of the 107 institutional members of the American Association of Collegiate Schools of Business showed that, of 90 responding, 64 were using business games in their regular curriculums in 1962, and 6 others expected to start using games in the ensuing year.[11] The rapid rate of acceptance of games in these schools is indicated by the number of institutions using games, year by year: four in 1957; twenty-nine in 1959; sixty in 1961.[12] In general, business and manage-

---

[1] Joel M. Kibbee, Clifford J. Craft, and Burt Nanus, *Management Games. A New Technique for Executive Development*, pp. 315–336, New York, Reinhold Publishing Corporation, 1961.
[2] Paul S. Greenlaw, L. W. Herron, and R. H. Rawdon, *Business Simulation in Industrial and University Education*, pp. 270–341, Englewood Cliffs, N.J., Prentice-Hall, Inc., 1962.
[3] Robert G. Graham and Clifford F. Gray, *Business Game Handbook*, New York, American Management Association, 1969.
[4] *Business Week*, July 25, 1959, p. 56; Cohen and Rhenman, *op. cit.*, p. 144.
[5] Greenlaw et al., *op. cit.*, pp. 335–336.
[6] *Ibid.*, p. 321.
[7] *Ibid.*, pp. 336–337.
[8] *Ibid.*, pp. 313–314.
[9] Kibbee et al., *op. cit.*, p. 171.
[10] *Ibid.*, p. 165; AMA Management Report 55, p. 16.
[11] Alfred G. Dale and Charles R. Klasson, "Business Gaming," *Survey of American Collegiate Schools of Business*, pp. 2, 4, 6, Austin, University of Texas, Bureau of Business Research, 1964.
[12] *Ibid.*, p. 4.

ment games appear to be most effective beyond the beginning instructional level.[1]

The same survey also revealed the extent to which computer and noncomputer games and general-management or total-enterprise and specialized functional games were being used in business schools.[2] See Tables 8-1 and 8-2.

Stimulated by the publication of Andlinger's papers in 1958, Fairhead and his associates[3] in the Department of Industrial Administration at Birmingham (England) College of Advanced Technology developed and used six management simulations, grouped into three types. These types and simulations were:

A. Functional
1. Production Planning
2. Marketing Management
3. Personnel Management
B. Control
4. Supervisory Management
5. Executive Management and Policy
6. Top Management

These authors, although acknowledging the widespread use of the broad term Business Games, particularly in the United States, express a British preference for the term Business Management Exercises.[4]

---

[1] Nicholas Radell, "Concepts of Management Gaming," *Systems and Procedures Journal*, 15:(2)24–29, issue no. 64, March-April 1964, particularly p. 27.
[2] Data in tables from Dale and Klasson, *op. cit.*, pp. 34–43.
[3] J. N. Fairhead, D. S. Pugh, and W. J. Williams, *Exercises in Business Decisions*, London, English Universities Press, Ltd., 1965.
[4] This preference is also attributed to the American Management Association and the International Business Machines Corporation in the games they sponsored. See Fairhead et al., *op. cit.*, p. 15.

Table 8-1. Numbers and Types of Games Reported in Use in Business Schools, 1962

| Type | Number of games | Types of games | |
|---|---|---|---|
| | | Gen. mgt. | Functional |
| Computer games | 28 | 16 | 12* |
| Noncomputer games | 29 | 11 | 18 |
| Totals | 57† | 27 | 30 |

* Only one game used in more than one school, and that game in two schools.
† Most of the fifty-seven games were used only in one, two, or three schools.

**Table 8-2. Most Frequently Used Games in Business Schools, 1962**

| Games (all management type) | Computer or noncomputer* | Number of schools using game |
|---|---|---|
| Management Decision-making Laboratory | | |
| Form for IBM 650 Computer | C | 16 |
| Form for IBM 1620 Computer | C | 8 |
| UCLA Executive Decision Game No. 2 | C/N-C | 14 |
| UCLA Executive Decision Game No. 3 | C | 10 |
| Dynamic Management Decision Games† (Greene and Sisson) | N-C | 11 |
| Executive Action Simulation (Herron)‡ | N-C | 9 |
| Carnegie Tech. (CIT) Management Game Mark 1 | C | 7 |

° C refers to computer type; N-C, noncomputer type.
† A series of seven games purporting to represent each level of management in a business organization. Described and with player instructions in Jay R. Greene and Roger L. Sisson, *Dynamic Management Decision Games*, New York, John Wiley & Sons, Inc., 1959.
‡ A manual for this game: Lowell W. Herron, *Executive Action Simulation*, Englewood Cliffs, N.J., Prentice-Hall, Inc., 1960.

These Business Exercises, they state, "have opened up a new dimension in the training of all levels of management, first-line supervisors, functional specialists, top executives."[1] They elaborate further: "So far [simulation exercises] have been found useful, and have been applied in three main spheres of activity: (1) research into management problems, (2) familiarization with, and training in, the techniques of operating specific business systems, (3) management education."[2]

Both in the United States and in Europe management simulations cover most of the functional areas of industrial, commercial, and public administration enterprises. Included are inventory control, purchasing, production control, transport, distribution, marketing, finance, and personnel.[3]

Exercises of the control type are usually directed toward the coordination and control of decisions arising in or related to rather specific functional areas of activity. These decisions are likely to differ at the

---

[1] *Ibid.*, p. xi.
[2] *Ibid.*, pp. 15–18.
[3] *Ibid.*, p. 22.

distinct echelons of management and normally involve different sets of variables or considerations. For example, the executive level would be concerned with finance; capital investment; production; marketing in terms of price, salesmen, and advertising; and research and development. At lower management levels, such as the department superintendent or manager, and in some respects at the foreman or supervisor level, the concerns would likely include the operating budget; material requisitioning and flow; work scheduling; work methods; inspection and quality control; industrial relations, labor, and supervision.[1] In some management simulations, at the upper levels of decision making, many of the above variables also would be considered. In addition these simulations might go into consideration of technological developments, their impacts on employees, and the general economic and financial health of the enterprise. The climate and trends in external factors affecting the enterprise in terms of the economy and social and political developments may be considered.[2]

The spectrum of business games and management simulations covers a broad range of types and applications. At one extreme are the one-person, one-sitting games; at the other extreme are the multi-team games with multi-person teams competing and interacting over an extended period of time. Some games use manually calculated results, others are computer-assisted, and still others are completely computerized and yield game results in computer printouts. Some games center about a single, specialized function of management such as production control; others encompass the complete management of all the functions normally involved in general overall management of a large industrial concern. Some games deal with real products such as detergents or tires; others involve hypothetical "widgets," "hackles," and "pockles" on the assumption that emphasis is thereby directed to the principles involved rather than to the product. Some games apply only to small-scale, retail merchandizing operations, whereas others may extend to international operations bridging multiple industries. Games also differ in the level of operations, ranging from the individual employee on the sales force or in the stock room or office, through the first-line supervisory level or shop foreman, to the department head, factory superintendent, sales manager, top executives, and the policy makers.

---

[1] *Ibid.*, p. 23.
[2] *Ibid.*, p. 24.

Games have spread into a wide diversity of business fields. One set of games developed at Purdue University includes a Farm Management Game, a Farm Supply Business Management Game, a Dairy Management Game, and a Supermarket Management Game.[1] Games have been constructed about the men's retail clothing industry,[2] underwriting aspects of a property and liability insurance company,[3] and banking and finance in relation to management of loans and the operations of the Federal Reserve System.[4] In fact, it is difficult to find a management field not yet touched by games or simulations.

Business games originated in the United States and spread to Europe and elsewhere, reversing the pattern established by their progenitors, war games. Business games and management simulations have been reported in use in Canada, the United Kingdom, France, Germany, Italy, Spain, Belgium, the Netherlands, Switzerland, Denmark, Sweden, Finland, Australia, Japan, Argentina, Brazil, Mexico, and several other countries in Latin America.[5]

## SIMILARITIES WITH WAR GAMES

Many game characteristics and techniques which evolved in war gaming have been carried over to business games. Some differences also have developed in business gaming usage. Similarities exist principally in many of the features pertaining to the game and simulation models, the facilities and equipment, and administrative details. Both kinds of games utilize simulations and models which are interacting, with feedback; are dynamic; aggregate data; and can use both deterministic and probabilistic data. Facilities usually provide for discussion, space suitable for briefings, and for separate rooms or work spaces for the protagonist teams and for the control team, to assure game security. Equipment may be simple or complex, ranging from paper and pencil at one extreme to sophisticated data processing

---

[1] These games are described in E. M. Babb and L. M. Eisgruber, *Management Games for Teaching and Research*, Chicago, Educational Methods, Inc., 1966.
[2] Robert Earl Schellenberger, *Development of a Computerized, Multipurpose Retail Management Game*, Chapel Hill, N.C., University of North Carolina, Graduate School of Business Administration, Research Paper 14, 1965.
[3] The Allstate Insurance Company's Managerial Game for Insurance Companies, reported in Greenlaw et al., *op. cit.*, p. 274.
[4] The Boeing Airplane Company's Operation Federal Reserve Game, reported in Greenlaw et al., *op. cit.*, pp. 282–283.
[5] Examples of games from most of these countries can be found in Kibbee et al., *op. cit.*, pp. 177, 315–343; in Greenlaw et al., *op. cit.*, pp. 270–341; and in Graham and Gray, *op. cit.*, pp. 107–456.

and computer installations at the other for either type of game. Supplies such as reference data sheets or manuals, work sheets, and report forms are common requirements.

Administrative details which have been found useful in both war and business games include a game director; a chief umpire, controller, or referee; a control staff (or computation and reporting staff); the division of the game into periods or cycles of play; the usefulness or necessity of pre-play orientation and post-play analysis, critique, and discussion; the provision for "briefings" before and during the game to explain situations and clear up misunderstandings; and expert leadership thoroughly familiar with the game, the game models, game data, the situation, and the subject matter of the game. This last requirement is extremely important and calls for the highest level of professional competence in the discipline(s) involved in the game. The game director must have a clear concept of the game purposes and objectives, and this information is shared with the control staff under normal situations. In some research games all but the director may be unaware of the research objectives, in order to avoid slanted decisions. The same applies in some teaching situations.

In both fields rigid, closed games are more generally preferred because quantified data are available and better suited to analysis, but free, open games can be used. The latter type is more likely to be employed when a game is improvised, more or less spontaneously, to illustrate some tactic or principle an instructor may wish to make more vivid. In business and management circles the unstructured, free, open game is more akin to role playing and is usually thought of in that category rather than as a game.

The principal differences in employment of military and business games are likely to stem from the objectives and uses the respective games are intended to serve. In the military, games are used extensively for training, testing of contingency plans, and for research and the development of new concepts. In the training situation, military games are more frequently confined to experienced and senior military officers, where the objectives of the game relate to command and control of a complete establishment, or force, dealing with the totality of the situation. This use requires full-scale war games. Business games and management simulations may be used in a similar way with top executives, and many games have been developed to serve this purpose. But business games are used more extensively than war games at the undergraduate level in professional schools. At this level of instruction less complex general-management games are required, and many instructional needs are well served by simple games which illustrate the limited range of details in a specialized function such as inventory control, production scheduling, etc.

Serious efforts are being made to use business games and management simulations for research purposes. This use requires highly quantified, rigidly assessed, fully recorded and documented games— features best served by computerized games. These requirements also make such games expensive to construct, maintain, operate, record, analyze, and publish. Only organizations with ample funds and well-subsidized universities can undertake extensive research of this type. A further difficulty in employment of the complex and elaborate games required for research is that the gaming staff requires long experience with the particular game and the game-related technology, after having attained a high degree of expertise in their own disciplines. And much of this game expertise is not readily transferable from one game model to another. The less complex games, intended for educational or training purposes, can be introduced and employed with relative ease.

## WHAT HAPPENS IN A BUSINESS GAME

The description which follows refers to the grandfather of business games, the American Management Association's Top Management Decision Simulation. This game simulates the problems of top management of manufacturing companies in allocating limited working capital to the factors of production in a competitive market. "As in real life, the objectives are (1) to produce a product efficiently by minimizing production costs, and (2) to achieve the maximum return on investment by improving the product and successfully presenting it in the market."[1] The game was designed for use with top operating executives in American business firms, the executives being organized into teams of three to five members, each team representing one of the competing companies.

The five teams or "companies" compete against each other in a hypothetical market over a period of 5 to 10 years of simulated time. Each game year is divided into quarters, and each set of decisions is made in order to direct (manage) the business for the ensuing quarter.

---

[1] Quoted from the American Management Association's official description as reported by Franc M. Ricciardi et al., *Top Management Decision Simulation: The AMA Approach*, p. 79. The description given above is drawn principally from that source, particularly pp. 79–93. This game also has been reported in many other publications too numerous to list here.

A single product is involved in the competition, each company manufacturing the identical item, but pricing of the product and management of its own company are independent decisions of each team.

At the start of the game each team is given the initial operating statement of its company. This statement indicates such information as the company's total assets, unit cost of production, price of product, and percentage of the market controlled by the company. The assets of the company are shown in amounts in cash funds, opening inventory, and plant capacity. All these initial figures are the same for each company.

Play of the game begins with each team, working independently and within a fixed time limit, making basic budget and price decisions for the first quarter of the year's operations. Report forms are the basic working documents for the players. Each team's decisions are recorded simply by circling the decision alternatives listed on the report form. This system saves time and facilitates feeding the teams' decisions into the computer.

The alternative decisions listed on the report form are shown as specific sums of money for production, marketing effort, research and development, and additional plant investment, and a range of prices which may be charged for the product. The team can also indicate how much money, if any, it will use for market information or other research and development, and whether it will sell part of its plant to raise cash. Rules of the game and controls limit the sum of the expenditures so that it must not exceed the total working capital available to the team. Drastic changes in allocation of resources in any one quarter are restricted, with specified limits on changes which can be made in one quarter. As in real life, the effects of changes do not show up until after appropriate time lags.

At the end of the first decision period the report forms showing each team's decisions are collected, and from these forms key-punch operators prepare punch cards to be fed into the computer. The computer is programmed to process these decisions and compute the results for each company, based upon the calculated effect of its own and its competitors' decisions. These calculations take less than 2 minutes of IBM 650 computer time. The output cards from the computer are then fed through an IBM 407 automatic printer which prints out a new quarterly operating report for each team. These reports are then distributed to the respective teams, and a new quarterly decision period begins. This cycle of play is repeated for as many quarters as the game is to run.

In the initial period of play, teams may be allowed about ½ hour to decide upon business strategy and to make and record decisions. As teams become more familiar with the operations, quarterly decision-making periods can be shortened to as little as 10 minutes toward the end of the game.

At the end of each year of simulated operations each of the competing teams receives a report on similar information for each of its competitors. These annual statements simulate the real-life situation in which such information would become available in the form of the company's annual reports. An extra 10 or 15 minutes are allowed each team in the decision-making period when annual statements are distributed to them.

Along with the company's own quarterly report, each team receives an income statement, reporting net income as in standard accounting procedure, i.e., gross sales income, with amounts deducted for cost of goods sold, operating expenses, and taxes. The quarterly report form also shows the total market changes for the previous quarter, indicating the size of the market and one's own company's share of that market. A company may acquire more information about the whole market and its competitors' shares of the market by buying "market research" (intelligence) out of its R&D funds in fixed amounts, e.g., $5,000 or $10,000 per report. Other kinds of information also can be purchased, such as what share of the market a company might have acquired if a different set of decisions had been made. Such reports are charged against the company at a specified rate such as ½ of 1 percent of gross sales income.

Each quarter a company must make eight decisions. Five of them are basic and required; three are optional. The basic decisions are the allocation of working capital funds to production, marketing, research and development, and new plant investment, and the establishment of the price for the product. Optional decisions relate to such matters as the purchase of market information, total or competitors' shares, and disposal of old plant to raise cash. The company may continue to operate at the same level, or at increased or reduced levels, as it chooses from quarter to quarter. New plant investment does not come into production or cash raised from the sale of old plant does not accrue until a lapse of one quarter.

Although the report forms and statements are the basic operating documents for the players, they may choose to keep a running record to show trends as on charts, graphs, or tables.

A Control room is operated by the game controller's staff, who plot a record of each company's operations, quarter by quarter, on cumulative charts. These charts are used in the critique session after the game is ended. The Control room is "off limits" to the players during the game, and individual teams operate in their own secure places of business. The critique session may be held in the Control room, and all game records are available for review and discussion. A whole game sequence of forty quarters may be played through and critiqued in about 2½ days of a management institute. Computer support makes it possible to compress 10 years of simulated operations, and forty sets of decisions made from some 2 million possible alternatives available to each company, into some 20 hours of actual time.

How Business Games Are Played

Business games are usually based on operations covering a period of years, with plans and decisions made for each 3-month period. To play such a game the participants may be organized into teams of three to nine persons, and there may be several such teams each representing a competing company. The game may specify the industry and the products involved in the competition. As much information as may be required to understand the competitive situation, the industry, the companies, and the products involved is given to the participants, along with general instructions about how the game is operated and the participants' objectives in the game. Usually the objective of each team is to improve the competitive position of its company and increase profits.

When the teams have been designated and the play of the game is to begin, participants first must study and analyze the game information made available to them. Then they must organize their teams. This involves assessing their experience and/or simply assigning each participant to a particular role or responsibility in the team, as may contribute effectively in conducting their business, and determining how these responsibilities can be coordinated for team action. Then they are ready to tackle their planning and decisions for the first quarter's business. For each quarter the team is given a statement of the business at the start of that quarter. This statement attempts to give only such details as may be necessary for the team to make the decisions pertinent to the purpose of the game. Unnecessary detail only complicates the operation of the game and increases the time required for each period of decision making for the quarter's operations. The quarterly statement is accompanied by a number of data sheets and a number of work sheets. The data sheets contain reference data pertaining to the market, the industry, the company, actions already taken by other companies which may affect each company's planning and decisions for the ensuing period, and in more detail the current status of the team's own company.

The work sheets are usually forms on which each team's plans, calculations, and decisions are recorded. They may call for dollar amounts allocated to advertising and sales promotion; procurement orders; production, distribution, and transport schedules; inventory changes; price adjustments; labor modifications including new hiring, layoffs, or overtime, wage or fringe-benefits negotiations; capital expansion plans and financing; expansion or contraction of research and development investment; introduction of new products; and other factors which impinge on the business. Plans and decisions on such matters are called for in enough detail and in such combination as may be required according to the nature and purpose of the game.

The teams are given a fixed and limited amount of time to make their decisions, record them, and submit them to the game Controller. In a simple game, the team may have as little as 30 minutes for this process; in a complex game, several hours, overnight, or even 2 days (as between class meetings in a business school course). Work sessions for teams may vary, often being longer for early periods in the game, or for more involved "assignments" such as the annual reporting period. All teams are required to turn in their work-session decision reports at the same time for each period of operations.

When the team reports for a period are submitted to the game Controller, the Control staff—or the computer, if the game is programmed for a computer—works up the data which will result from each team's decisions and the interactions of other teams' decisions upon each of the competing teams, or companies. These results are calculated by predetermined formulas, specified in the game rules (or in the computer programs). When the Control staff has completed computation of the results for decisions in one period, the results for each company are submitted to the team operating that company. Thus the ball is passed back to the company team, and the team proceeds with its deliberations and decisions for the next period (quarter) of the game. Along with the Controller's reports of results of the previous period's decisions, the teams receive any new data and work sheets needed for the next period of decision making. The play of the game proceeds in this manner through the whole sequence of periods until the game is completed. In the American Management Association Top Management Decision Simulation the game usually runs through a series of twenty or forty quarters, simulating 5 or 10 years of operations.

Steps in Building a Business Game

Ready-made games may serve the needs of organizations other than those for which they were developed.[1] Use for educational purposes, as in business schools, and use in management development institutes are examples. For the more specialized operations of particular business enterprises and management applications, tailor-made games may be preferred. This is the kind of problem which confronts training directors in many companies. John D. Herder, Supervisor of Education for the Southern New England Telephone Company at New Haven, Connecticut, has addressed himself to this problem and reported it in the *Journal of the American Society of Training Directors*.[2]

Herder outlines six steps as he sees the process of preparing a manually played business game. They are (1) select the level of decision making, (2) analyze the decision making (at that level), (3) collect the data related to the key decisions, (4) select the "game" elements that best suit the decision areas and the related data, (5) design and prepare the game materials, and (6) play the game several times as pilot exercises, to uncover gaps, clear up ambiguities, and train personnel who will assist in administering the game.

Herder describes further considerations for each of these steps. A few of his suggestions—those relating to step 5—are mentioned here. He categorizes six different kinds of materials needed to play the game. These are (1) environmental material, telling how the game is set up and how it is to be played, a brief description of the "dynamics of the company and the industry," the objectives of the game, and the rules and instructions to players; (2) referee material, describing all the players' materials but in more detail and with notes as to the principles to be applied and driven home, copies of all data and work sheets to be used by the players and the control staff, and the special instructions concerning the referee's duties; (3) input sheets, on which the referee presents the results of the previous cycle of play, and the data relating to the cycle or play period being started; (4) decision-making work sheets, on which the players work up the data needed for their decisions; (5) output sheets, on which the players record their decisions and submit them to the referee at the end of that playing period; and (6) summary charts, on which running progress of

---

[1] Short descriptions of scores of games are given in the books by Kibbee et al., *op. cit.*, by Greenlaw et al., *op. cit.*, and by Graham and Gray, *op. cit.*
[2] Described in Dr. Herder's article, "Do-It-Yourself Business Games," *Training Directors*, Sept. 1960, pp. 3–8.

each team, its decisions, and results may be recorded. The summary chart, augmented with the input and output sheets for each cycle of play, makes up the action record of the game; these are the basic documents used in the final critique sessions.

Fairhead of the United Kingdom lists the "apparatus" used by players and the umpires or control staff for a simple Executive Management Exercise as (1) the Decision Form, the market information it requires, the units it schedules, the prices at which the product is selling, where its salesmen should call, and so on; (2) the Operating Statement, used by the umpire to report the result of the quarter's trading by means of a simplified profit and loss account and a consolidated account, the inventory position, territories in which sales were made, etc., any market intelligence the company has bought, together with any general information such as whether there is a boom or slump in trade, competitors' prices, etc.; (3) the Exercise Board, a visual aid to decision making—a kind of Gantt chart, half devoted to the manufacturing and company headquarters operations, half to the market, salesmen, territory, etc.—by which the general status of activities can be visualized; and (4) the Balance Sheet, a year-end summary of the financial state of the company. A balance sheet on each company goes to all competing teams or companies.[1]

## GENERAL-MANAGEMENT GAMES

Under this heading are included games which simulate the functions of the top management or executive level of business enterprises and constitute the total enterprise, without being concerned with the details which lower echelons of management must consider in carrying out the program of top management. Such terms as executive, top management, or decision making appear in the names of many of these games. At this level of game, it is immaterial whether the industry and/or the product is specified. In some games they are specified; in others they are subsumed under a generalized term, such as "widgets," and in other games the products are aggregated into such inclusive terms as "dollar volume," "car loads," or "units."

General-management games of these types are represented by the AMA Top Management Decision-making Simulation, the Andlinger/McKinsey and Company Business Management Game, the IBM Management Decision-making Laboratory, the General Electric

---

[1] Fairhead et al., *op. cit.*

Business Strategy Simulation, the Michigan State University Business Policy Game, the University of Washington Top Management Decision Game, Clarkson College of Technology Executive Action Simulation, the Management Decision Simulations developed independently at Indiana University and at the University of Oregon, the Pillsbury Company's Management Decision Exercise, the Indiana University Executive Decision Game, and the General Management Simulation developed at the Tokyo Center for Economic Research. Some general-management games carry the name of the organization under which they were developed: the UCLA Executive Decision Game(s); the Harvard Business School Game, HARBUS; the Carnegie Institute of Technology (CIT) Management Game. A few brief descriptions of general-management games follow.

The UCLA Executive Decision Games

Several business games were developed in the Graduate School of Business Administration of the University of California, Los Angeles. One of the best known and most widely used games was the UCLA Executive Decision Game No. 2, which was introduced in 1958. It superseded game No. 1 of 1957 vintage, whereas game No. 3 which followed in 1959 was a multi-product game. Game No. 2 simulates a multi-firm, one-product industry, with two to nine companies competing, each represented by a team of three to six members of the "top management" of each firm. Management decisions are made quarterly on price of product, production volume, advertising and selling budget, research and development budget, investment in plant and equipment, and dividend to stockholders. The game may run for 4 to 7 years of operations, which can be played at the rate of 1 or 2 hours per year. The whole game can be played in one day. It is fully quantified and deterministic, and it can be played with or without a computer. These features undoubtedly contributed to its wide use.[1]

---

[1] The game was described in some detail by James R. Jackson at the National Symposium on Management Games, held at the University of Kansas in 1958 and published in *Proceedings of the National Symposium on Management Games*, pp. VI-9 to VI-14, Lawrence, Kans., The University of Kansas, Center for Research in Business, May 1959. Another description, by Jackson, is published in the *Proceedings of the Sixth International Meeting of the Institute of Management Sciences*, which appeared in a two-volume book under the editorship of C. West Churchman and M. Verhulst, *Management Sciences, Models and Techniques*, vol. 1, pp. 250-262, New York, Pergamon Press, 1960. Brief summaries appear in Kibbee et al., *op. cit.*, p. 330; in Greenlaw et al., *op. cit.*, pp. 284-285; and in Graham and Gray, *op. cit.*, p. 250. More complete information can be obtained from the university.

### The Carnegie Tech Management Game

This game is one of the most complex business games, involving from 100 to 300 decisions by each team each month of simulated operations. It is a specific-industry game, the three companies making up the packaged detergent industry. It is a total-enterprise, multi-product, multi-market, fully quantified, deterministic, computerized game, in which companies are represented by teams of seven to ten participants who develop their management roles within the team. Two to three hours of decision time are allowed for each of the 30 to 50 months of simulated operations included in a full run of the game. Special efforts have been made to bring realism into the game. For example, the prices of the seven raw products used in the manufacture of the detergents fluctuate during the game, and consumer reactions to product characteristics, such as sudsiness, vary in the different markets. The game was designed and used for second-year graduate students in industrial management, and one playing period or cycle (month or quarter) is scheduled each week throughout the school year. The game is used both for instructional and for research purposes.[1]

### INTOP

One of the more unusual business games is INTOP, the University of Chicago International Operations Simulation developed in the early 1960s. Its originators state that it "is the first major business simulation exercise oriented toward the specific problems of international trade and overseas operations."[2] Developed in the Graduate School of Business at the University, the game was designed to increase understanding of the problems of international operations in general, and those of a multi-national corporation in particular, as well as to yield substantial payoff in general-management training.[3]

---

[1] The game is described by Kalman J. Cohen and Eric Rhenman, "The Role of Management Games in Education and Research," *Management Science*, 7(2):139–140, Jan. 1961, and by Elliot Carlson, *Learning Through Games*, pp. 26, 32–44, Washington, Public Affairs Press, 1969. Brief summaries are also given in Kibbee et al., *op. cit.*, p. 318, in Greenlaw et al., *op. cit.*, p. 288, and in Graham and Gray, *op. cit.*, p. 286.

[2] *INTOP, International Operations Simulation. Player's Manual*, p. 1, by Hans B. Thorelli et al.

[3] *Ibid.*, p. 1. The description which follows was drawn from this source. The game has been reported in other publications by the same authors; see *Journal of Business* of the University of Chicago, July 1962. It is also described by Elliot Carlson, *op. cit.*, pp. 44–49.

The game is suited to play with from three to as many as forty-five teams, and with three to seven (student) executives composing each team. The "companies" may operate in one, two, or all three areas: Brazil, the European Economic Community (EEC), and the United States; these operations may involve any one or any combination of the functions of manufacturing, marketing of one's own product, serving as distributor, and licensing. Each company may manufacture and/or market one or both of two different products: medium-priced electrical appliances. In addition to the international aspects, the usual management functions are involved in the operations. The game requires the support of a computer. It was designed to pose classic management dilemmas and to force participants "into a stream of truly entrepreneurial (top management) decisions of business philosophy and objectives as opposed to the heavy strategy-tactics emphasis of most other games," claim INTOP's developers.[1]

### Small-business Games

General-management games have not been restricted to "big business." Some games have been developed for limited and specialized business activities and for small business enterprises of the type engaging in local or retail marketing and services. These games usually involve the total enterprise, but the scope of the business may be limited.

Dayco Corporation developed the Dayton Tire Simulation in which two tire manufacturers compete for the replacement and special-brands markets in passenger-car tires.[2] The game calls for quarterly decisions by top and middle management on matters affecting marketing: price, national or local advertising, market research, and field sales force, in two markets. Three teams participate. The game is simple enough to be scored by two umpires in a matter of minutes and is well suited to use in management courses in business schools.

The Cornell University Management Decision Game—Small Business is a general-management, total-enterprise, competitive, interacting, manually computed game, entirely quantified and deterministic. Two to five companies compete (with teams of two to six participants) in a multi-product multi-market, dealing with unspecified goods.[3]

---

Thorelli et al., *op. cit.*, p. 1
Greenlaw et al., *op. cit.*, p. 291; Kibbee et al., *op. cit.*, p. 319.
Greenlaw et al., *op. cit.*, pp. 290–291.

The Kroger Company developed the Supermarket Decision Simulation in which three retailing company chains compete in three areas, but against only one other chain in each of the three areas. This is a total-enterprise, manually computed, interactive game, entirely quantified but with major random aspects, and with all sales aggregated in dollar amounts. The game requires the competing company teams of four to seven participants to make decisions concerning purchasing, advertising, product research, promotion, and pricing.[1]

The Burroughs Corporation's Supermarket Battle Maneuvers game involves four competing companies operating in a multiple market. It is an interactive, entirely quantified, deterministic, computer-assisted game. A noncomputer game on the operation of a service station has been developed by the Esso Standard Oil Division of Humble Oil and Refining Company. An Automobile Dealer Simulation was developed at Wayne State University to train supervisory and middle-management personnel, as well as for use with business school students. It simulates two to five companies, and the market involves new and used cars, parts, services, and accessories.[2]

## FUNCTIONAL GAMES

Functional games are those which are designed to focus on one function of a more complex business enterprise. They usually relate to one of the subdivisions of a business, such as procurement, production, inventory control or warehousing, marketing, advertising, or personnel management. They may be intended for training the managers of these functions and/or employees in the activity or for teaching purposes in business schools. These games can be quite simple. Some are intended to focus on the management aspects of a functional department; others on the procedural aspects with which employees of the department must deal. The In Basket exercise, for example, is one of the simplest of functional games and can be designed to simulate almost any desk job, whether manager or clerk. The player simply acts upon the papers which represent those which would come to him in the process of performing his job. A number of functional games involve operations entirely within a business enterprise, such

---

[1] Kibbee et al., *op. cit.*, pp. 170, 322–323; Greenlaw et al., *op. cit.*, p. 308; Graham and Gray, *op. cit.*, pp. 449–450.

[2] Kibbee et al., *op. cit.*, pp. 170–171; Greenlaw et al., *op. cit.*, pp. 284, 339; Graham and Gray, op. cit., pp. 445–446.

as production scheduling or inventory control, and as such are not directly affected by the actions of competitors. Thus games of this kind are not interactive in nature. Games that fit in this category include In Basket; Inventrol; DISPATCH-O; Maintenance Management; Materials Management Simulation, Model 2; Physical Distribution Simulation, Model 1; Manufacturing Management Simulation, Model 1; Airline Sales Game; Marketing; Operation Federal Reserve; SOBIG.

The In Basket Exercise

The In Basket Exercise, originally designed as a test,[1] has been employed as an individual decision-making exercise, or a one-person business game. In Basket represents one facet of a manager's job. The participant is confronted by a collection of some twelve papers in his desk basket for incoming mail. Each paper requires some appropriate action. The papers resemble those which the simulated manager would encounter in a real-life situation. They consist of memorandums, letters, reports, etc. The participant is allowed a limited amount of time to work through his in-basket materials in sequence, deciding what action he will take on each item and writing a suitable action response, with reasons. The materials can be designed to represent almost any manager's or clerk's job and to call for the use of such technical knowledge and application of principles as may be considered important in that job. The In Basket device has been adapted to a wide variety of applications in the United States and in the United Kingdom.[2]

UCLA's Inventory Game

UCLA's Inventory Game is designed for individual play in courses in operations research, but any number may play it independently at

---

[1] Designed by Norman O. Fredericksen and described as "The In-Basket Test," by Norman Fredericksen, D. R. Saunders, and Barbara Wand in *Psychological Monographs: General and Applied*, vol. 71, no. 9, New York, American Psychological Corporation.

[2] See as examples: "In-Basket Business Game" by Andrew A. Daly in the *Journal of the American Society of Training Directors*, 14(8):8–15, Aug. 1960; a similar report from the United Kingdom published in the same periodical 16(4):27–30, April 1962, under the title, "'In Tray' Training Exercises" by H. E. Frank and S. J. Pringle. In Basket is also briefly described in Greenlaw et al., *op. cit.*, pp. 11–12, and in Kibbee et al., *op. cit.*, p. 40.

the same time. It is entirely quantitative but with major random aspects. It involves a single product, television sets, in a single market and is manually scored. Players are confronted with balancing the cost of placing orders against the cost of carrying reserve stocks.[1]

Inventrol

The General Electric Company's Inventrol is intended for production and purchasing specialists to provide insights into problems of ordering, stocking, and lot size. The single product, alloy castings, is sometimes in limited supply but needed for continuous production in a larger product. It includes minor random aspects, is entirely quantitative, and is interactive, with five companies competing for procurement.[2]

Maintenance Management Game

Allied Chemical Corporation's Maintenance Management Game is non-interactive, entirely quantitative with major random aspects, and manually scored. Participants are organized into teams of four to six persons in each of four companies. Each team is given a number of "work orders" for a week and must make out weekly maintenance schedules.[3]

DISPATCH-O-Game

The General Electric Company's DISPATCH-O game calls for production scheduling for job-shop dispatchers, using Gantt charts for the purpose of acquainting participants with new methods of releasing and sequencing work to the factory floor. It is non-interactive but deterministic and entirely quantitative and is manually scored. Teams of three or four persons make up each group, and any number of groups may participate simultaneously.[4]

MIT Marketing Game

The MIT Marketing Game is deterministic; has some qualitative features relating to advertising, sales promotion, and product design;

---

[1] Greenlaw et al., *op. cit.*, pp. 286–287.
[2] *Ibid.*, pp. 292–293.
[3] Greenlaw et al., *op. cit.*, pp. 273–274.
[4] *Ibid.*, p. 292.

and requires a computer. It is an interactive, multi-product, multi-market game for four companies represented by teams of six to eight persons. The game focuses attention on problems of product development, distribution, pricing, and sales promotion of two models of a specific product, floor waxers. Qualitative judgments are evaluated by senior umpires.[1]

The Airline Sales Game

The Airline Sales Game of Trans-Canada Air Lines sets up the objective of managing the system at a profit. Two or more companies compete, with teams of three to five participants. These teams are confronted with problems of scheduling of flights and determination of and optimum mix of first-class and tourist-class seats on each route to meet the passenger loads created by the sales and marketing activities. The game is interactive, entirely quantitative but with minor random aspects, and requires a computer.[2]

Operation Federal Reserve Game

The Boeing Airplane Company's Operation Federal Reserve game involves banks and the Federal Reserve System. Decisions relate to the amounts of callable and non-callable loans to be made, the amount to be called in, purchase or sale of government bonds and securities, amounts to be transferred to or from the Federal Reserve Bank reserve account, the amount of loans to be rediscounted, etc. The goal is to maximize the amount of non-callable loans and to maintain a good balance of other assets, all in conformity with Federal Reserve requirements. The game is manually computed, entirely quantitative and deterministic, but non-interactive. Players are organized into four companies of five to seven participants.[3]

SOBIG

SOBIG is a game developed at Princeton University, involves security negotiations in four corporations, and is used as a research instrument to study behavior of participants in conflict situations. It is entirely quantitative, deterministic, and interactive and usually is organized with more than five companies of three participants each.[4]

---

[1] *Ibid.*, pp. 309–310; Graham and Gray, *op. cit.*, pp. 327–329.
[2] *Ibid.*, pp. 332–333.
[3] *Ibid.*, pp. 282–283.
[4] *Ibid.*, pp. 322–323.

## SOME CONCLUDING THOUGHTS

Maximizing Learning

A basic characteristic of all learning methods which provide for student participation is the attempt by one device or another to get the student to identify with the situation in such a way as to assume some responsibility for the decisions to be made, or to react to decisions which have been made. He may be called upon to suggest alternatives or his own preferred decision, and perhaps to critique others' decisions or to justify or defend his own proposed decision(s). Such experience is valued for the functional understandings of the principles involved. In some of the less structured participation methods, such as role playing, emphasis may be pointed toward sensitizing the student to the problems of the official he enacts and to the viewpoints of the other protagonists involved in the situation, in the hope of developing some empathy with the real-life participants and insights into the problems represented. Games include most of the aforementioned values and, in addition, provide a continuous ongoing situation with built-in feedback approaching the dynamics of real life.

Although games stimulate strong identification of the participant with the situation or problems represented and have a marked motivational effect, some special efforts are required to assure that learning is maximized. The facts bearing importantly on the problem need to be singled out, and the principles involved need to be clearly identified and associated with their applications. These desired outcomes can be built into the game situation but are intensified and reinforced by pre-planned activities on the part of the participant (or student) and the game director, chief controller, umpire, or referee (or instructor). Among these pre-planned activities may be "work or data sheets," i.e., sets of data blanks and questions which the participant must fill out, or other forms of expression for his decisions. The latter forms may be reports, memos, letters, orders, or other means of taking actions which stand on the record. Such actions as these provide the inputs for the next set of circumstances to which other participants must react in appropriate ways, thus providing the feedback and the dynamics of the ongoing situation. After the game has reached a stage where its intended experience has been provided, the game may be stopped even if it has not reached its own completion point. At that time a review and critique session is a clincher and offers a propitious opportunity to drive home the principles involved, the pitfalls of omission and commission, and the lessons to be learned from the experience. A modification of the game director's critique is used in in-service management conference or seminar sessions involving the actual officials of an organization. In management training programs, institutes, and sessions of this type, the participating principals—the chief decision makers and their supporting staff associates—may be called upon to review their own actions and decisions, critique them, and give supporting reasons. Self-criticism and "lessons" are brought out in the discussion which normally follows. Often higher officials freely admit mistakes which were made, and differing opinions are expressed by their peers. In this way the session leader is spared the onus of direct criticism of a responsible official, but at the same time he is provided an opportunity to emphasize the underlying principles.

The motivational and competitive effects of business games can entice participants to strive to "win" to the point where the learning purposes of the game may be obscured. The game director or leader of the sessions in which games are used can, and should, make every effort to keep the spotlight on the principles involved and on the "learning" objective.

The decision-making process, whether used in management simulations or in other situations, is widely recognized as being composed of several successive steps. These have been variously expressed, but there is general acknowledgment that the following actions are involved: data gathering, sifting, and organization; diagnosis; consideration of alternative courses of action and their potential effects; and selection or formulation of the preferred decision. The business game provides a natural setting for these aspects to be employed and exercised. In this sense the game or simulation provides organized and systematized practice in these skills of decision making. The degree to which the components of the decision-making process are identified and emphasized in the game or simulation relates to the objectives and situations built into the exercise, to the points brought out in the game critique, and to the analyses and explanations required from the game participants in support of their game decisions. Recognizing this potential opportunity in properly structured games, many writers attribute to game participation a certain increased awareness and improvement in decision-making skills.

The astute game leader (or instructor) will strive to make the game participants increasingly aware of the forces shaping management decisions, recognize the principles of management science, exercise analytical skills, and systematize their decision-making processes.

**How Specific?** In discussing this question, Fairhead points out that it is difficult to reflect more than a few of the characteristics of a management situation in a manageable simulation. A simplified situation containing elements of a top-management problem can lay the foundations for effective learning, whereas details that evoke irrelevant associations may detract from the intended learning process and invite criticism focused on the game description and rules. The closer the resemblance becomes, Fairhead adds, the more complex and unmanageable the model becomes, and the less easy it becomes for the participants to learn from the experience. This, he believes, "is the essence of the criticisms leveled against the big American simulations [in the United Kingdom, presumably], for instance, one conducted by the American Management Association." He comments further, "as with all other instructional devices a simulation can only teach effectively if the ideas it presents are few enough to register in the timespan occupied by it."[1] The simulations developed by Fairhead were intended for use with graduate students in management science. His criticism is more apropos of the lesser business experience of students than of mature operating officials and executives for which the AMA simulations were designed. Nevertheless, Fairhead raises a valid point which deserves consideration in game selection or development.

One way of reducing the complexity of business games is to aggregate the data, as is done in many war games, thereby reducing the number of variables needing attention. The Kroger Company's Supermarket Decision Game aggregates the thousands of separate items in a supermarket in terms of total dollar volume. All this only emphasizes that games and simulations should be attuned to the intended objectives and to the participants to be involved.

Other Considerations

A well-structured business game, built upon a framework of fixed rules with mathematical formulas, permits flexibility only within limited boundaries. Such boundaries may encompass the normal range of business operations and reflect the recognized patterns on which business enterprises operate, and by which they flourish or fail. The occasional technological breakthrough, the surge resulting from a

---

[1] Fairhead et al., *op. cit.*, pp. 28–30. Fairhead's six business management exercises are all relatively simple and are likely well attuned to the experience of his students.

sudden fad, a devastating disaster not covered by insurance, or a takeover to effect major change illustrate some of the extreme events which can overshadow the normal business health of a company. In general, games do not include such extreme influences. Normal risks in business life are subsumed in the general rules of the game and included in operating costs. A business game is usually designed to reward the business which makes the most astute use of its resources and opportunities and its conservative ventures or prudent risks. To the extent that the game embodies those principles of business and management practice which have proved successful, game results will reflect wise application of those principles. In this way the game provides opportunity and even requires the participants to make management decisions which are then measured against the effects those decisions generally induce.

Chance events and chance effects are highly important in war games and are provided for in the game models, using some means of introducing this variability beyond the control of the players. Similar probabilities and stochastic injections can be provided in business games, if desired. The nature and purpose of the game will suggest what probability distributions are appropriate. In some cases real variability can be introduced in the game to reflect this variable if the game objectives are better served thereby. For example, the current interest rates on bank loans, indices of the current state of business, such as car loadings and retail sales, can be used instead of arbitrarily assigned values in the game situation or rules. It does not follow, however, that the purpose of the game will be enhanced by such tie-ins with the real market. In games used for educational or training purposes—and business games are predominantly used for such purposes—the game may be more effective if it is kept simple, with its focus on a few principles it is employed to teach. In that sense, the simple game is more like the laboratory experiment in which interacting variables are controlled.

In Summary

Management simulations and business games have emerged from their honeymoon period, have multiplied and settled into patterns of varied uses, and have been assimilated into the larger framework of business education and in-service training. Some scholars are confident of their expanded potentials, as for research, and these scholars are slowly but diligently forging ahead on several fronts. Active research of business games and management simulations is under way in more than a score of graduate schools of business administration, in engineering schools, and in the management-science field in the business and industrial world, professional societies, and government agencies.

Most participants and many observers believe games and simulations produce worthwhile gains. Some day, scholars hope to know what these gains may be, after more progress has been made in attempts to measure or evaluate the changes which games and simulations bring about in participants. These questions were raised in the blooming stage of business games, and many subscribe to the faith expressed by one practitioner of the art who said in 1958, "If we can't do it today, we will eventually."[1]

---

[1] Martin Shubik of the General Electric Company, at the National Symposium on Management Games at the University of Kansas.

# Strategic and Political Games, an Instrument of National Policy

The supremacy of civil authority over military forces is one of the firm traditions of the modern military profession despite the frequency of military juntas and dictatorships to refute this principle. The military profession always functions under political aegis. Karl von Clausewitz is widely quoted for his enunciation of the principle that war is "a continuation of policy by other means."[1] Although history records some durable dictatorships, those dictators who came to power by military takeover have tended to change their image from a military commander to a civilian chief of state.[2]

In the development of war gaming one might have expected the political and military aspects to develop concurrently and in parallel. This did not happen. War gaming was prosecuted for professional military purposes. Political aspects were recognized mainly as background information. In war games national policy or interests are indicated in the statement of mission and directives given the commander. Within the guidance of such directions which specify constraints (sanctuaries, free cities, and prohibition against use of certain weapons, such as gas or nuclear munitions), military operations are left to the military commanders.

---

[1] Karl von Clausewitz, *On War*, p. 16, Infantry Journal Press, 1950.
[2] Among recent examples of military leaders holding to political control are Gen. Francisco Franco, Spain, Aug. 9, 1939; Marshall Tito, Yugoslavia, Jan. 31, 1946; Generalissimo Chiang Kai-shek, Nationalist China, April 1948; the late Col. Gamal Abdel Nasser, Egypt-United Arab Republic, June 18, 1953 to Sept. 26, 1970.

There is another reason for emphasis on the detailed gaming of military operations in a setting of generalized political guidelines. Military and political actions develop at different rates. Military moves often draw reactions within minutes and achieve results in hours; political moves may take days or weeks to induce responses, and perhaps months or even years to achieve results.[1] This condition presents gamers with a persisting dilemma. Gamers can construct and operate the military aspects of war on an objective, systematic, and quantified basis, as in a rigid game, but the state of the art has not advanced to the point where political aspects can be dealt with except on a subjective and judgmental basis, as in a free game.[2] There is a growing need to game the political-strategic aspects with the military aspects, treating both as parts of the same international conflict situation. The problem is to find a means of considering the pertinent, subjective, and slowly maturing factors without distorting the objective data and quick-reacting effects. For many years military aspects of war games received concentrated attention, and significant progress was made. Political aspects received little attention, and developments lagged. Quite recently a surge of effort has been made to game political factors. These efforts have taken different forms and have employed a variety of methods.

## STRATEGIC MATTERS AND POLITICAL AFFAIRS

Strategic matters normally are wide in scope and concerned with various political, economic, psychological, and military resources. Strategy, involving national policy and national objectives, is determined at the highest governmental and political levels of a nation.

---

[1] As a military operation the Bay of Pigs "invasion" of Cuba (Apr. 17, 1961) was over in a matter of hours; the political antecedents and effects spread over years. The Cuba missile issue was a fast-moving political crisis, initiated and settled within 11 days (Oct. 22 to Nov. 2, 1962); the Suez crisis erupted July 26, 1956, and passed the threat of military intervention within weeks; the issue of the U-2 overflights (May 1, 1960) lasted months; and the residual sovereignty issue concerning the Ryukyu Islands continued as a political "hot potato" in U.S. and Japanese relations for many years. Slower-moving political developments do not fit in the time span of fast-moving military operations.

[2] David C. Schwartz, "Problems in Political Gaming," *Orbis*, Fall 1965, concluded: "The plain fact is that we presently know too little about the all-important limitations of the game as a research and strategy tool in political science." (p. 693.) See also David Easton's *A Systems Analysis of Political Life*, the third in a series of books on "empirically oriented political theory," 1965.

*Political affairs* may be considered an inclusive term embracing all types of forces and resources available to the state.

Numerous and varied strategic and political games have been developed: strategic war games, political games, and political-military games. Often these terms are used with overlapping rather than distinctly different meanings. A distinction is drawn here between strategic policy matters (the province of political leaders) and strategic warfare (the province of military commanders).

Strategic Warfare

The term *strategic warfare*, as used herewith, is warfare directed at capitulation of an enemy or rendering the enemy's military forces impotent, unwilling, or unable to conduct effective war. Emphasis is on the military operations used to achieve this purpose although political and economic factors usually contribute to the same end.

A basic assumption in strategic warfare is that the application of military force is politically directed; that the economy has provided military materiel; and that the economy can support the prosecution of the war. Military operations, in this sense of strategic warfare, are conducted "as an extension of politics."

Military people draw a distinction between tactical and strategic problems. Strategy has been defined as the art of war; tactics, the art of battle; yet a precise dividing line is not always clear.[1] Situations confronting a commander in combat, to which he must react, employing resources available at that given time and place,[2] are termed tactical situations. Situations that include resources normally beyond the tactical commander's area of responsibility,[3] and authority to use those resources, are termed strategic situations.

---

[1] A distinction used at the Naval War College some years ago follows: "If you can see the enemy you have a tactical situation; if not, the situation is a strategic one." Submarine use complicated this distinction. See Glossary: tactics, strategy, military strategy, and national strategy.

[2] Examples would be a division, corps, or army command. Units of this size usually operate as tactical units.

[3] A theater commander, such as CINCPAC (Commander-in-Chief, Pacific Command), is an example. Note the great area and the large number of national jurisdictions and the forces under CINCPAC, e.g., the U.S. Army, Pacific Command (USARPAC); the Fifth U.S. Air Force with Headquarters in Japan; the U.S. and UN Command and the Eighth Army in Korea; the IX Corps and Ryukyu Island Command in Okinawa; the Seventh Fleet in Southeast Asian waters; the Military Assistance Command in Vietnam, and others. CINCPAC, of necessity, is directly involved in strategic matters—plans, capabilities, operations—whereas subordinate commanders are directly involved in tactical operations.

## Strategic War Games

Devastating strategic warfare became a realistic threat with the nuclear plenty and the means for intercontinental delivery of nuclear weapons, achieved following World War II. The protection or defense of the homeland against nuclear attack by manned bombers or by missiles, and a retaliation capability, i.e., a "second-strike" capability, were a prime mission and concern of the U.S. Air Force. This concern was shared in part by the Army and Navy. In support of their respective sponsors—the Air Force and the Army—both RAND and ORO engaged in needed studies of simulated strategic warfare.

Next developments focused primarily on the strategic potentials and effects of air power, nuclear strike and retaliation capabilities, and strategic sea power—the gaming of strategic warfare without nonmilitary factors. Most of these simulations and games were performed "in-house" by military or contract personnel for the military services. Many of the games and simulations have not been released for reporting in the open literature, but a few have appeared.

The extent of "in-house" gaming is indicated by a statement of Gen. Earle G. Wheeler, USA, Chairman, Joint Chiefs of Staff. In a "Meet the Press" television program, early in 1967, General Wheeler was asked: ". . . going on with Secretary McNamara's posture statement, he said that if both sides deploy the ABM [antiballistic missiles] and both sides increase their offensive capability over roughly a decade, at the end of that decade casualties in the Soviet Union and casualties in this country would be in the order of 90 to 120 million. Do you think that is an overestimation?" General Wheeler responded: "I would say that the figures themselves are no doubt correct. But you have got to remember, Mr. Childs, that figures such as this are arrived at by war-gaming a situation. We have *run literally thousands of war games* of this type over the years. The results are very sensitive to the assumptions and the model, so-called, upon which the war game is based."[1]

Some large computer simulations were applied to air-defense problems in early studies. RAND developed and used such a simulation[2] in an experiment involving some forty persons and a computer. Human participation in the game consisted of aircraft observers who

---
[1] Italics supplied. Quotation refers to thousands of plays of war games and emphasizes the magnitude of strategic war gaming. Transcript of NBC, "Meet the Press," America's Press Conference of the Air, Feb. 26, 1967, p. 7, published by Merkle Press, Inc., Washington, D.C.
[2] Between 1950 and 1955.

spotted and reported aircraft sightings shown on radar screens activated by a mathematical model for air traffic; by "players" who ordered interceptor aircraft; and by assessors who plotted the paths of the interceptors. (9-1)[1]  See Figs. 9-1 and 9-2.

A strategic air warfare simulation called ZIGSPIEL,[2] developed and employed at ORO, was used in a comprehensive study on defense

---

[1] These numbers refer to Chapter Notes listed at the end of the chapter.
[2] ZIGSPIEL was developed and played in 1956 and 1957.

---

FIG. 9-1. RAND's System Research Laboratory Air Defense Simulation schematic. This schematic shows the relations among several functions performed by the air-defense direction center: surveillance (ellipses), identification (squares), and interceptor control (triangles). The senior director (responsible for supervision and decision-making), the adjacent direction center, and higher headquarters are represented by pentagons. The unshaded portion includes the embedding organizations (manned by experimenters) and the environment from which the simulated system information inputs come. (From RAND paper by R. L. Chapman et al. Also published in Management Science, 5:254. Used with permission of the RAND Corporation.)

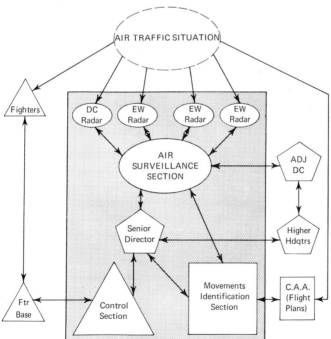

228  Venture Simulation in War, Business, and Politics

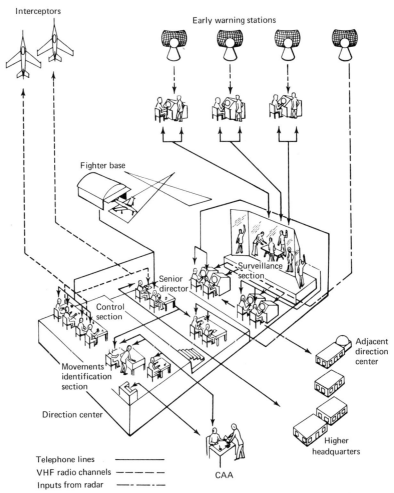

FIG. 9-2. Simplified model of Direction Center, RAND's System Research Laboratory Air Defense Simulation setup. This simplified model emphasizes aspects of the direction center's operation not readily seen in the previous schematic of functions: the physical arrangements within the center and the complex communication net (consisting of telephone lines, VHF radio channels, and radar inputs) that linked crew members to each other and to the external embedding organizations. Much of the communication within the center itself is either visual or by face-to-face conversations. *(From RAND paper by R. L. Chapman et al. Also published in Management Science, 5:255. Used with permission of the RAND Corporation.)*

of the United States against attack by aircraft and missiles. ZIG-SPIEL, an elaborate simulation, was subjected to hundreds of computer runs with varying combinations of attack and defense forces and weapons, to assess defense requirements and second-strike capabilities. The results of such gaming operations, and related analytical studies, have been considered in some of the most significant decisions made in United States (and Free World) defense measures.

Only military means and damage assessments were evaluated in these strategic air war games. The political, economic, and psychological implications and effects were left for consideration at the policy-making level. The other side of this coin is political games.

Political Games

Political games may be in the context of national self-interest in a peaceful but competitive family of nations or nations in a state of embroilment ranging from cold war through limited or even general war. Emphasis here is on the first condition, but with the capability of bringing any or all national power factors to bear to the extent required, although in grossly aggregated degree.

Dr. Herbert Goldhamer and Dr. Hans Speier, of RAND, are widely acclaimed as the pioneers of today's political gaming. F. M. Salligar, also of RAND, is credited with the "invention" or origin of the concept in which military and political factors are given simultaneous consideration in political simulations and games. (9-2)

"Exclusively political" games, attributed in concept to war gaming, have been considered by most military operations research organizations as beyond the province of the primary sponsors, the military services. The Department of State and the White House are the natural clients for this type of game once it reaches operational utility and can deal meaningfully with real-life situations. However, social and political scientists and colleagues engaged in other academic disciplines at several universities recognize political gaming as a promising field for research and teaching purposes. Games using hypothetical states are useful for concept development and the formulation of political theory. Accordingly, political gaming has had two arenas for development:

1. In research organizations, responsive to certain governmental agencies
2. In the universities

## POLITICAL GAMING IN THE UNIVERSITIES

Dr. Goldhamer and Dr. Speier aroused interest in academic circles. Dr. Speier reported his gaming experiences at the Social Science Research Council Institute at Denver (1956) and at the Center for Advanced Study in the Behavioral Sciences at Palo Alto, California (1957). Concurrently, Dr. Goldhamer briefed personnel at the Department of State, the Brookings Institution, the Army War College, and the Massachusetts Institute of Technology. Wide interest resulted from these conferences, and later presentations were sponsored by the Social Sciences Research Council at West Point and at the American Political Science Association, Washington, D.C. (9-3)

Applications of political gaming in teaching situations were reported by James A. Robinson and others in the *American Political Science Review* in 1966. (9-4) Political gaming in the form of simulations[1] used with college and high school students as subject participants was given critical review. These gaming experiences had a strong and enduring motivational effect on students. At this stage, research was not undertaken to measure resultant substantive learning.

Among the institutions conducting political simulations in 1966 and 1967 were Northwestern, MIT, Ohio State, Maryland, Michigan, Wisconsin, Johns Hopkins, Stanford, University of North Carolina, University of California at Irvine, Wayne State University at Detroit, Smith, Kansas State Teachers at Emporia, San Francisco State College, and high schools from Seattle to Baltimore.[2] Science Research Associates, Chicago, furnished a kit of materials called INSkit, for these efforts. Political simulations also were used abroad as teaching and exploratory techniques at Leeds University and Lancaster University in England, the University of Toronto and the University of British Columbia in Canada, and at Tokyo University in Japan.[3] (9-5)

### Inter-Nation Simulation (INS)

The Inter-Nation Simulation was developed and used extensively at Northwestern University under the leadership of Dr. Harold Guetz-

---

[1] Usually in the form of Inter-Nation Simulations. The use of the word "simulation" in political games usually refers to a representation rather than a mathematical model.
[2] This list is incomplete but represents the spread of gaming as a teaching technique.
[3] Work of Prof. Kinhide Mushakoji and Prof. Hiroharu Saki at the Institute of Behavioral Sciences, Tokyo.

kow and Dr. Richard C. Snyder, assisted by Chadwick F. Alger, Richard A. Brody, and others. This effort was started in 1957. Graduate students and colleagues of Dr. Guetzkow have carried the technique to other institutions. The INS uses a set of from five to nine "abstract countries." These artificial nations bear such names as Algo, Erga, Ingo, Omne, Utro, etc. Economic and political resources in the form of population, military forces, gross national product, production rates, and the like are specified for each nation. The players assume the role of decision makers (heads of state, foreign secretary, budget chief, envoys, and other officials). At the starting condition each nation is allocated a quantity of resources. Numerical values are used to express the units of each type of resource. Resources then can be committed on the basis of point values in support of desired national objectives. The play of the game proceeds with the manipulation of resources. Coalitions among nations may be negotiated by the decision maker, and assessments of total strength (points) are made on the growth or loss basis of a continuing inventory. Assessments are computed in accordance with established equations, rules, and quantified data (units of resources). (9-6) The assessments provide "feedback" to the decision makers to be used as a guide in employing current resources in terms of the international situation represented.

Oliver Benson, Professor of Government, University of Oklahoma, was inspired by the Guetzkow Inter-Nation Simulation and saw possibilities of developing computer models as research aids in the international field. With help from the University of Oklahoma Computer Laboratory and Richard Brody, he developed a set of assumptions, mathematical formulas, and flow charts for A Simple Diplomatic Game. Professor Benson first reported his simulation to the American Political Science Association (1959). A revised form of the paper was published in 1961.[1] Professor Benson suggested use of his simulation in connection with, or as augmentation to, Dr. Guetzkow's "most systematic all-human simulation game" and not as a replacement for it.[2]

In some centers simulations and games are related to real life. Dr. Andrew Scott at the University of North Carolina has used a hypothetical country, Simuland, and real nations—Brazil and Chile—to study the processes of change in developing countries. (9-7)

---

[1] *International Politics and Foreign Policy, A Reader in Research and Theory*, pp. 504-511, The Free Press of Glencoe, Inc., 1961.
[2] *Ibid.*, p. 509.

## The MIT POLEX Games

POLEX I (Political Exercise) was conducted at the MIT Center for International Studies, and at Endicott House, Dedham, Massachusetts, under the auspices of the United Nations Project. The game involved participation of senior faculty and research personnel from MIT, Harvard, Yale, and Columbia, between September 10 and September 12, 1958. Like earlier RAND efforts, the purpose was to explore questions of foreign policy, national interest, and national security. An international crisis was assumed precipitated by the demise of the head of the Polish government. The experiment was designed to indicate whether role playing by a group of leading scholars—highly qualified in knowledge of situations comparable to those represented—would yield predictive indications about a similar real-life crisis involving East-West relations.

Two follow-up exercises, these on the Berlin crises, were held in 1959. Ninety MIT undergraduates and forty political science majors from Harvard, Yale, Dartmouth, and MIT participated. Evaluation of these games did not confirm predicted values but the games were useful for educational purposes and to stimulate insights among experienced personnel. The political games were recognized as being subjective and less amenable to scoring than military games. In fact, in these games military aspects were limited to strategic estimates and minor military activity. Participants pointed out the desirability and feasibility of combining political and tactical military games in parallel games (9-8) directed to a common effort. POLEX II was played on the POLEX I pattern, in 1960, and these efforts are continuing at MIT.

## POLITICAL GAMING IN RESEARCH ORGANIZATIONS

The INS technique applied to real-life situations is represented by the work of Dr. Jerome Laulicht and colleagues. The work was performed in Sweden, the United Kingdom, and Canada, in association with university staff members and students, under the sponsorship of private peace research organizations. Jerome Laulicht, with Paul Smoker at the Peace Research Centre, Lancaster, England, and Norman Alcock at the Canadian Peace Research Institute, Clarkson, Ontario, used Vietnam as the base for simulated situations. Efforts were aimed at studying the forces at work to gain an understanding of how they may be employed to contribute to resolution of such conflict.

The simulations employed (9-9) real data from open literature sources.[1]

Dr. Laulicht characterized the emphasis of these simulations by saying, "The political and diplomatic factors are far more important than [details of] military factors; therefore the exact details of battles are disregarded." Simulation requires simplification. Dr. Laulicht continued to give an example of how economic factors are treated with the statement, "Capital, labour and raw materials, all the things which go to make a country's productive capacity are summed up in a single unit called basic resources. The greater a country's productive capacity, the more units it will possess in the simulation." (9-10)

Under the leadership of John R. Raser and Wayman J. Crow, political simulations of the INS type are being developed at the Western Behavioral Sciences Institute, La Jolla, California. Gaming and simulation are being applied to local and municipal activities such as land use and urban planning, as conducted under the leadership of Richard D. Duke of Michigan State University for the Institute for Community Development and Services, East Lansing, Michigan.

The Goldhamer-Speier, Bloomfield-Whaley, Guetzkow-Alger-Noel, Snyder, Brody, Coplon, Raser and Crow, Scott-Lucas, Laulicht-MacRae-Smoker-Alcock, Robinson-Burgess, Benson, Schelling, Coleman-Boocock, Abt, and Ivanoff studies are pioneering research approaches in the field of political simulation and gaming. Some researchers are hopeful, and a few are confident, that these methods will lead to systematic and effective analyses of international situations. Research groups and organizations continue to explore means of conducting political situations and games.[2]

The challenge presented by the simulation and gaming of political situations, international relations, and national and international crises has been so strong, the potential impact so great, and the dynamics of interactions so intriguing that motivation has been high. Some of the most imaginative and creative social scientists have been attracted to this field of research. In the past 10 years a new field of specialization and a rapidly growing literature have evolved from politically oriented simulations and games. The developments and the literature pertaining to them are too extensive and varied for description and summarization here. One persisting objective of these

---

[1] Jerome Laulicht, unpublished notes, undated, circa 1965-1966.
[2] RAND, CENIS (The Center for International Studies at MIT), ORO (now RAC), the Hudson Institute, and the Western Behavioral Sciences Institute are examples.

researchers is to treat political factors in a complete milieu of strategic forces—economic, sociological/psychological, and *military*—as postulated by Salligar, and Ellis and Greene.[1]

## INDUSTRY-SPONSORED RESEARCH

Certain large industrial corporations support active operations research and systems analysis programs related to the corporation's interests in national affairs.[2] (9-11) In these programs systems analysts attempt to apply scientific and objective methodology to the political problems represented in gaming situations. A study of threat analysis, a subject of basic consideration in higher-level political and military decisions, is supported by the Douglas Aircraft Company. The kind and magnitude of threat represent important input data for strategic planning, and for political or strategic games played to assess means of dealing with the threats. (9-12)

## THE DOUGLAS THREAT ANALYSIS MODEL

The Douglas Threat Analysis Model,[3] supported by analytical studies of real-world events including a global range of conflict situations since World War II, has been developed in detail. Future uses of such a model could be to suggest impending or emergent threats and means of preventing, avoiding, or meeting those threats. Once a potential threat is identified and evaluated, contributing conditions or

---

[1] In "The Contextual Study," Ellis and Greene credit Fritz Salligar, also of the RAND staff, as having originated the contextual concept; J. W. Ellis, Jr. and T. E. Greene, "The Contextual Study," *Operations Research*, p. 639, Sept.–Oct. 1960. The reader interested in pursuing this field further is referred to books and journals particularly helpful upon entry to such study. Among these are Guetzkow et al., *Simulation in International Relations*, 1963; Scott, Lucas, and Lucas, *Simulation and National Development*, 1966; the *Journal of Conflict Resolution*, issue of Dec. 1963 (article by Brody in particular); *Behavioral Science*, July 1959 issue, article by Guetzkow; and the *American Behavioral Scientist*, October and November issues, 1966, "Simulation Games and Learning Behavior" by Coleman, Boocock, and Schild; and other international relations and world affairs journals such as *World Politics* and *Orbis*. See also Schelling, "Experimental Games and Bargaining Theory," *World Politics*, Oct. 1961.
[2] R. D. Specht, "RAND—A Personal View of Its History," *Operations Research*, Nov.–Dec. 1960, p. 825.
[3] The material reported here is excerpted, with permission, from Dimitri N. Ivanoff's *Threat Analysis Briefing*, vol. I, Santa Monica, Calif., Douglas Aircraft Co. (now McDonald Douglas Corporation), June 1966.

supporting measures could be planned to advance national interests and free-world security.

A systems analysis procedure has been followed in the development of the Douglas model. An effort has been made to replace intuitive judgments with systematically evaluated considerations derived from maximum use of objective data and values. The data and values, recorded on prescribed criteria or analysis forms, are scored by teams of area and subject experts.

The magnitude of the threat-analysis technique can be illustrated with an example from the conceptual framework. Assume there are 135 "nations" in the world.[1] Further assume each nation has individual national policies. Then consider all possible relations of each nation with the others. Some 18,090 possible independent policies result. A large number of these policies can be disregarded on the basis of being irrelevant, inconsequential, or nonexistent in the real world of international relations. For example, Mongolia, Paraguay, Upper Volta, and Yemen would be little concerned with national-policy incompatibility with each other.

The maximum number of "interfaces" or possible conflicts of interest between pairs of the 135 countries would number 9,045. That is, among 135 nations there are 9,045 channels for communication or conflict. These interfaces may be considered analogous to telephone lines. What might be conveyed over these lines between any two nations may result in scores or hundreds of potential problems: political, economic, military, diplomatic, etc. The field of international problems is practically infinite. Some problems may prove to be advantageous, as trade agreements to exchange overabundant exports for needed imports. Other problems may be concerned with tension- or conflict-provoking issues as, for example, water rights to the Jordan River (an Israeli-Arab issue).

Figure 9-3 is a graphic portrayal of the Douglas Threat Analysis Model. A flow chart of this kind is used to program a series of steps or actions for a computer. Each decision or branch point is reduced to two alternatives.

The diagram shows the following alternatives in an international issue:

1. The issue may be resolved (whence it becomes a historical case for study if pertinent).

---

[1] Based on a definition of nations represented by United Nations members.

236   Venture Simulation in War, Business, and Politics

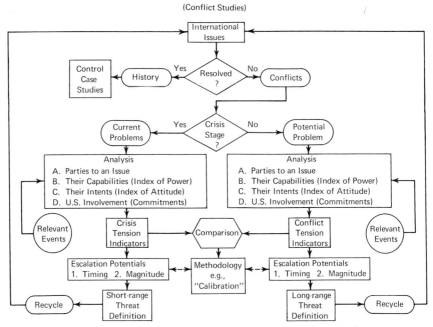

FIG. 9-3. Graphic portrayal of the Douglas Threat Analysis Model. (McDonnell Douglas Corp.)

2. The issue may be an area of disagreement that can result in a crisis situation.

If the situation reaches a crisis stage, it becomes an acute current problem requiring priority attention; if not, the issue continues unresolved with the potential of erupting into a full-blown crisis.

To provide a pragmatic check on the threat model, an intensive study was made of major international conflicts experienced in the world community since World War II. Each conflict was cataloged in one of six categories. A summary of these crises or tensions is shown as Fig. 9-4. Some crises continue to erupt in varying intensity and form.

Almost all international issues are dynamic. Most of them heat up or cool down from time to time; few are settled for all time. Threat analyses as represented here must be studied, analyzed, and assessed on a continuing basis. For this reason nations maintain diplomatic

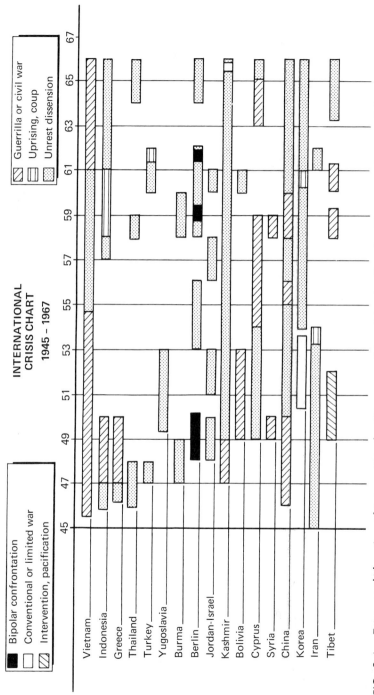

FIG. 9-4. Types and duration of international crises, as considered in the Douglas Threat Analysis Model. (McDonnell Douglas Corp.)

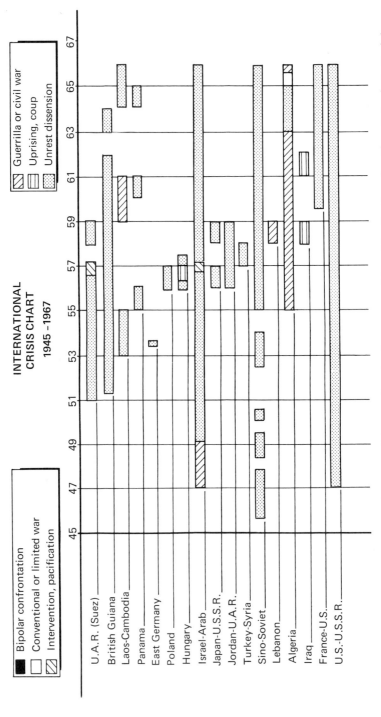

FIG. 9-4 (Continued). Types and duration of international crises, as considered in the Douglas Threat Analysis Model. (McDonnell Douglas Corp.)

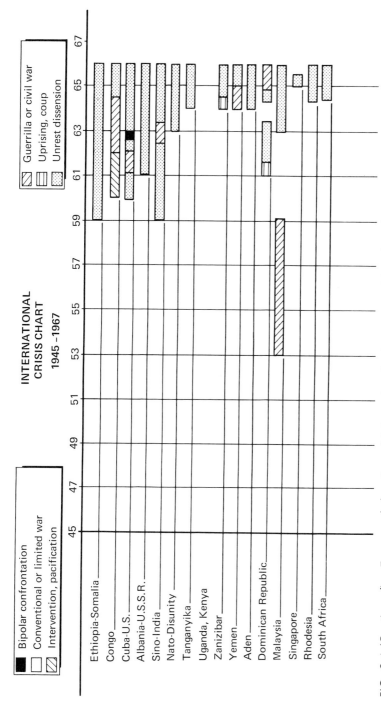

FIG. 9-4 (Continued). Types and duration of international crises, as considered in the Douglas Threat Analysis Model. (McDonnell Douglas Corp.)

239

240   Venture Simulation in War, Business, and Politics

contacts as lines of political communication. State Department, or Foreign Office, Desk Officers[1] are needed to keep abreast of the respective national interests and issues.

The issues and problems confronting the Desk Officer are not in computer input form. Information must be collected, selected, evaluated, and interpreted. This duty is in the province of the Desk Officer. A thorough student and expert analyst for a special field, the Desk Officer accumulates input data. Desk Officers rely on individual expertise to furnish appropriate interpretations and to suggest courses of action.

To reduce the human errors of bias, exaggerated evaluation of certain factors, and general inconsistencies, the use of Observation Work Sheets is prescribed by the Douglas method. Each category of accumulated data is significant; each is assigned a numerical value. The numerical values then are recorded on the work sheets. Thus the model provides a quantification scheme.

National-power indicators take into consideration potential power; effective home power, i.e., the power the nation can exert at its own geographic base rather than at a distance (as the United States in Southeast Asia); nuclear power; and other elements. The other elements also are tied to specific means or routines for evaluation.

Certain human-judgment determinations remain to be made even after all the working "tools" a resourceful analyst can devise are employed. These determinations, too, are subjected to a procedure that reduces the likelihood of wide extremes. The technique[2] is to submit the description of circumstances to a panel of experts—a jury of peers—each of whom makes an evaluation on a scale with a standard set of guidelines. An example: A scale of values relates to an "index of national power" in which "Dominant World Powers or Actors" (the United States and the U.S.S.R.) each are scored 100; the scale drops by steps to "Primitive Military Capability," or "Local Actors" (e.g., Laos, Costa Rica, Luxembourg) which are scored at 10

---

[1] Desk Officers specializing in the affairs of a particular foreign country or area are not confined to the State Department. The military services and other governmental agencies have Desk Officers, as do certain international organizations such as the International Bank or International Monetary Fund, and private business or industrial organizations such as the Aramco (Arabian-American) Oil Corporation.

[2] This technique is patterned after a procedure, used at RAND, dubbed the Delphi Method.

Strategic and Political Games, an Instrument of National Policy    241

each.[1] A portion of a more complete tabulation and assigned ranks of the home-base national power of selected nations is shown in Fig. 9-5. Individual crises can be scaled in intensity over a period of time. See Fig. 9-6. The U.S.-U.S.S.R. cold-war tensions shown accrued from some 3,000 incidents that occurred during the period studied. The seriousness of a threat is a product of the intensity of the crisis and the national power of the parties involved. For each threat, tension levels also can be represented in numerical units. The process illustrates the complexity and extent of study and analysis necessary

---

[1] The steps in this scale include "Dominant World Powers," 100 (e.g., the United States and the U.S.S.R.); "Great Military Powers" or "Major International Actors," 60 (e.g., France, United Kingdom, West Germany, and Communist China); "Secondary Military Powers" or "Minor International Actors," 40 (e.g., India, Japan, Sweden); "Tertiary Military Powers" or "Major Regional Actors," 30 (e.g., Nationalist China, Israel, Belgium); and "Minor Military Capability" or "Minor Regional Actors" 20 (e.g., Cambodia, Albania, Ireland).

FIG. 9-5. Home-base national power rankings of nations. (*McDonnell Douglas Corp.*)

| Stages of political and economic progress: <br> Military capabilities | (E) Traditional primitive societies | (D) Traditional civilizations | (C) Transitional societies | (B) Industrial revolution societies | (A) High mass-consumption societies | Rank-order categories based on PEMS capabilities |
|---|---|---|---|---|---|---|
| A. Dominant military powers | | | | U.S.S.R. 2 | U.S. 1 | Dominant world actors |
| B. Great military powers | | Chicom 5 | | France 3 <br> U.K. 4 <br> West Germany 6 <br> Canada 7 | | Major international actors |
| C. Secondary military powers | | India 10 <br> Pakistan 12 | Indonesia 16 <br> U.A.R. 17 | Japan 8 <br> Italy 9 <br> East Germany 13 | Sweden 11 <br> Australia 14 <br> Switzerland 15 | Minor International actors |
| D. Tertiary military powers | | Congo (Leopoldville) 38 | Chinat 19 <br> Turkey 20 <br> South Korea 25 <br> North Korea 26 <br> Portugal 32 <br> Philippines 35 <br> Cuba 36 | Spain 18 <br> Poland 21 <br> Czechoslovakia 23 <br> Yugoslavia 24 <br> Austria 27 <br> Greece 28 <br> Israel 29 <br> Brazil 30 <br> Mexico 34 | Netherlands 22 <br> Belgium 31 <br> Denmark 33 <br> Norway 37 | Major regional actors |
| E. Minor military capability | Ethiopia 57 <br> Sudan 62 <br> Nepal 65 <br> Libya 66 <br> Burma 67 | North Vietnam 42 <br> South Vietnam 43 <br> Thailand 49 <br> Cambodia 58 <br> Outer Mongolia 64 | Iran 40 <br> Algeria 48 <br> Morocco 51 <br> Albania 54 <br> Iraq 55 <br> Tunisia 59 <br> Jordan 60 <br> Saudi Arabia 63 | Rumania 39 <br> Argentina 41 <br> South Africa 44 <br> Bulgaria 45 <br> Hungary 46 <br> Finland 47 <br> Venezuela 52 <br> Colombia 53 <br> Ireland 56 | New Zealand 50 | Minor regional actors |
| F. Primitive military capability | Uganda 81 <br> Laos 90 <br> Tanganyika 96 <br> Togo 97 | Nigeria 69 <br> Zambia 74 <br> Kenya 76 <br> Madagascar 78 <br> Liberia 79 <br> Bolivia 92 <br> Haiti 93 <br> Congo (Brazzaville) 95 | Peru 68 <br> Ghana 73 <br> Guatemala 80 <br> Dominican Rep. 84 <br> Ecuador 85 <br> El Salvador 86 <br> Honduras 87 <br> Nicaragua 89 <br> Paraguay 94 | Chile 70 <br> Lebanon 71 <br> Malaysia 72 <br> Uruguay 75 <br> Jamaica 82 <br> Costa Rica 83 <br> Cyprus 88 <br> Panama 91 <br> Iceland 98 | Luxemburg 77 | Local actors |

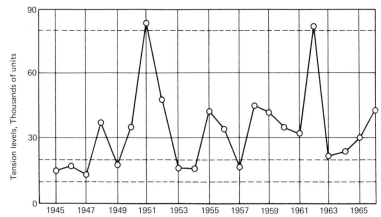
FIG. 9-6. United States–U.S.S.R. cold-war tensions (1945–1966). (McDonnell Douglas Corp.)

to reduce some of the problems encountered to an objective and quantified base to obtain a partial substitute for intuitive or judgmental evaluation. Only certain features of the Douglas Threat Analysis are reported here.

Threat analysis is one aspect of political gaming and represents one set of input information. Threat analysis does not cover what can be done about meeting or dealing with crises that are within the larger province of crisis gaming, political gaming, or strategic gaming.

## POLITICAL-MILITARY AND CRISIS GAMES

In early war games attention was confined to the meeting of enemy forces on localized battlefields. Civilian populations were presumed to have evacuated the battle area. Independent efforts were under way to develop political games. Meanwhile gamers found it unrealistic to restrict war games to military factors. The next step was to build political factors into military games. A number of so-called "political-military" games were played to explore means of including both aspects in the same game. These games were of two types:

1. Political games (including the potentials of military forces)
2. Military games that attempted to include political matters in the dynamics of play

The first type has almost always been a free-play, minimum-rule, predominantly political game. Military capabilities rather than operations are considered. Political games are represented by the RAND, MIT, and INS simulations. A recent effort, STRAT-X I, (9-13) was played as a research activity of RAC's Strategic Studies Department (1965). STRAT-X is concerned with developing alternative courses of action in the resolution of conflicts and is patterned on the MIT POLEX game. The rules are limited. There are no definitive mathematical models or quantified bases for the military, economic, and psycho-socio contributions to support political decisions or national objectives. This game was set at a high level, and the strategic objectives in the nations concerned were dealt with in the context of worldwide political and military considerations. (9-14)

## FAME, POMEX GAMES

The second type of military-political games introduced political factors in military games. In a series of seven RAC games, set in the Middle East in 1959 and early 1960 under the game title FAME, some play of civil affairs, military government, and psychological operations was included. (9-15) FAME was followed in 1962 with POMEX,[1] a THEATERSPIEL research game that experimented with a larger number of political inputs. See Figs. 9-7 to 9-10. Both FAME and POMEX were true war games, rigidly assessed and controlled. The political factors were infused on a free-play basis. Gamers who have been active in political gaming are shown in Figs. 9-11 to 9-13.

Both types of political-military game served a useful purpose, but neither resulted in a satisfactory end product. Political moves were not fused with military operations. Instead, a requirement was established for transition to the fuller spectrum of political gaming including economic, sociological, and ideological or psychological factors, along with political and military aspects. This effort toward full-blown strategic games continues. Meanwhile the cold war grew hot in Southeast Asia, and insurgency was stepped up to organized military operations. With these pressing incentives, emphasis shifted

---

POMEX (Political Military Exercise) has been referred to under several associated terms: SEA I, for the theater of play (Southeast Asia); TRIAL RUN, for its preliminary stage of development; THEATERSPIEL's Fourth Research Game, to stress its methodological objectives; and THEATERSPIEL's Cold War Model, for the new model being tried.

FIG. 9-7. ORO's FAME game, developed in 1957 as an early tactical war game. An operations analysis team is shown at work, developing the FAME (Future Army Missions and Equipment) game. *Participants, left to right:* Col. L. S. Simcox, Duncan Love, Brig. Gen. John G. Hill, Paul Dunn (1957). From this game a theater-level game, THEATERSPIEL, was developed in 1959–1960 with political inputs.

to attempts to extend political-military gaming to guerrilla and insurgency situations.

## GAMING GUERRILLA AND INSURGENCY OPERATIONS

A special type of political-military gaming is required by guerrilla and insurgency operations. The ascending intensity and seriousness of "wars of national liberation"[1] present war gamers with a special challenge.

In Indo-China after World War II, the French tangled with the Chinese brand of guerrilla warfare under Ho Chi Minh's leadership of the Viet Minh, ending with the decisive battle of Dien Bien Phu in northern Vietnam. The late Bernard B. Fall (1926–1967), the American-based French scholar who published extensively on Vietnam, kept expounding the theme that conventional military operations of the Western type had a poor chance of success against Com-

---

[1] Term borrowed from the Communist lexicon but used as viewed from the non-Communist world to refer to Communist-inspired and -supported insurgencies, not to true, internally inspired civil wars.

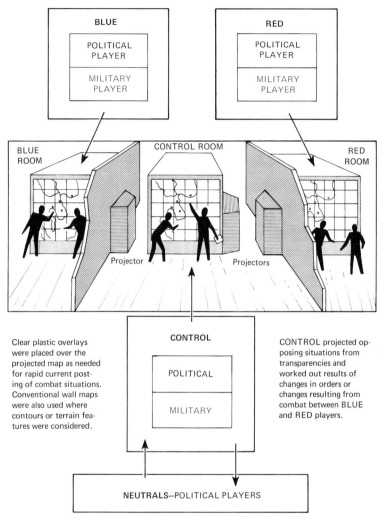

FIG. 9-8. Graphic representation of POMEX-I, a RAC-THEATERSPIEL research game to experiment with political inputs to a military game.

munist revolutionary war as advocated by Mao Tse-tung and practiced by Ho Chi Minh.[1]

Fall saw the turning point in U.S. field operations in Vietnam, epitomized by the extensive use of helicopters in application of the airmobile and air-cavalry concepts. After viewing the results of these

---

[1] See Fall's books: *Street Without Joy; Hell in a Very Small Place; Vietnam Witness, 1953–1956;* and *The Two Vietnams,* particularly chap. 16, "Insurgency: Myths and Facts."

246  Venture Simulation in War, Business, and Politics

FIG. 9-9. The THEATERSPIEL POMEX game. Representation of simulated UN Security Council meeting in POMEX game (Political Military Experiment), conducted as a THEATERSPIEL gaming activity in early 1962. *Players representing simulated Security Council members, left to right:* Anthony Fiacco, William Braxton, Eugene Visco, Dr. Howard Reese, Col. C. B. Murphy, Harry Crow. (February 1962)

concepts, supported by coordinated joint operations, in practice, Fall switched his position from that of "the No. 1 pessimist about a U.S. victory." He became encouraged that U.S. air- and firepower could carry the field.[1] Several independent attempts have been directed at developing models and simulations which would permit gaming of the evasive-type conflict branded as guerrilla warfare.

Bernard Fall called attention to an attempt by the French Colonel Gabriel Bonnett to define "guerrilla or revolutionary warfare by a quasi-mathematical formula:

$$RW = G + P \qquad (9\text{-}1)$$

where revolutionary warfare [$RW$] results from the application of guerrilla-warfare methods [$G$] and psychological-political operations [$P$] for the purpose of establishing a competing ideological system or political structure."[2]

---

[1] From *U.S. News & World Report*, Sept. 26, 1965. Fall's change in viewpoint also was revealed in personal conversations with the author. Fall was killed by a booby trap while with U.S. Marines (as a correspondent) near the Vietnamese "demilitarized zone," Feb. 21, 1967.

[2] Quoted from Gabriel Bonnett's *Les Guerres Insurrectionnelles et Revolutionnaires*, p. 60, Paris, 1958, by Bernard Fall in *The Two Vietnams: A Political and Military Analysis*, pp. 349, 484, 1963.

FIG. 9-10. The THEATERSPIEL POMEX game. Simulated political discussion among members of the UN Security Council, representing Ecuador, Ghana, U.S.S.R. *Left to right:* E. Visco, W. Braxton, H. Crow. (February 1962)

Alfred Blumstein of the Institute of Defense Analysis also recognized the utility of mathematical models and their application to problems of guerrilla warfare. Blumstein, along the lines pioneered by Lanchester with conventional war, suggested some approaches to models of counterinsurgency.[1] (9-16) He referred to two earlier efforts: (1) by Bonnett in 1958 and (2) by Ngo Dinh Nhu in Vietnam at about the same time.[2] Blumstein described Nhu's "personalism" theory as

$$TT + TG = TN \qquad (9\text{-}2)$$

In this formulation $TT$ represents three independent levels of self-sufficiency:

1. Ideological self-sufficiency
2. Logistical self-sufficiency
3. Technological self-sufficiency

$TG$ represents three degrees of personal vigilance, each of which is necessary to and interrelated to the other two:

1. Health
2. Conduct and behavior
3. Creative initiative

---

[1] Alfred Blumstein, "Strategic Models of Counterinsurgency," a paper presented at the 13th Military Operations Research Symposium, Washington, D.C., April 1964.
[2] Ngo Dinh Nhu was a brother of, and counselor to, President Ngo Dinh Diem. Nhu also was chief of the hated secret police; husband of the famous Madame Nhu, first lady of Vietnam. Both brothers were assassinated in the military coup of Nov. 1–2, 1963.

FIG. 9-11. Some of RAND's principal war gamers (*left to right, top to bottom*): N. C. Dalkey, M. Dresher, O. Helmer, J. R. Lind, E. W. Paxson, G. C. Reinhardt, M. G. Weiner, L. H. Wenger. (*Photos: RAND and Victor Barnaba.*)

FIG. 9-12. Some leading political gamers in the Inter-Nation Simulation (INS) and POLEX types. *Left to right, top to bottom:* INS gamers: H. Guetzkow, R. A. Brody, J. A. Robinson, P. Smoker, N. P. Gleditsch, T. Narden, W. J. Crow, J. R. Raser, G. Shure; POLEX gamers: L. Bloomfield, N. J. Padelford, T. C. Schelling.

# Strategic and Political Games, an Instrument of National Policy 249

FIG. 9-13. Political-military war gamers in other types of games, *left to right:* C. Y. Adams, JCS; C. Abt, TEMPER, Raytheon. THEATERSPIEL Cold War gamers, *left to right:* J. A. Elmore, H. H. Figuers, D. Pappas. TACSPIEL Counterguerrilla Model gamers, *left to right:* L. S. Simcox, D. Parker, L. J. Dondero, P. F. Narten, N. W. Parsons.

*TN* represents the "personalism" aspect, i.e., three attributes of regard for the individual human person:

1. Respect of person
2. Community destruction (recognition)
3. Collective rise (station, position)

One can recognize in Nhu's analysis certain elements of Confucian philosophy and a relation to some of Mao Tse-tung's principles of revolutionary (guerrilla) war. All recognize the importance of "winning the minds and hearts" of the people.

Alfred Blumstein goes beyond. He proposes a complicated set of mathematical equations, while recognizing the inadequacies and the difficulty of obtaining data in reliable measures to provide for precise analytical purposes. He applauds the effort to attempt precise models and measures of counterinsurgency and apparently regards these beginnings at least as promising as Lanchester's equations were in the search for models of conventional warfare.

Two different approaches at gaming guerrilla and insurgency operations have been tried. One has been to develop a Counterinsurgency Game per se. The second approach attempts to develop models of counterinsurgency as adjuncts to established games.

The Counterinsurgency Game

The first approach is illustrated by the efforts of Col. James Y. Adams, USA (Ret.) while at the Stanford Research Institute (SRI). Colonel Adams, building upon early experience as a principal designer of the JCS, free-play political-military, crisis or desk games, worked on a free-play type of guerrilla game[1] he described as follows:[2]

> A synchronized man-computer game, with two-sided live player planning and reaction to opponent created situations; and computerized assessment, game history and summary analysis. Procedures for live play were developed along functional lines (Political-Economic, Intelligence-gathering, Planning, Fighting) deemed susceptible to incorporation into computer routines.

---

[1] In 1964 Colonel Adams estimated that it would take about 5 years to achieve a functional and operating counterinsurgency game at the level anticipated.
[2] Unpublished notes provided to the author by Col. James Adams in 1964.

Several manual "tactical" games were played to check the methodology and to catalogue the types of activity the computer routines would simulate. From the combat situations developed in these free-play games, three "typical" confrontations involving flat-trajectory direct fire, high-angle area fire and aim miss-distance in ground-to-air-fire were subjected to detailed, multiple-option analysis. The basis of the analysis methodology was formulation of the combat type action-reaction into a series of sequential steps, each step describing a separate element of the man-weapon combat environment relation. The action-hypothesis could then be expressed in mathematical form, and several mini-games hand played to obtain mathematical insights and sample outcomes. The sample outcomes could be matched against available real data, the differences analyzed, and hypotheses revised as required.

The plans of Colonel Adams included participation by senior research personnel from SRI, faculty members and students from the political science and related departments of Stanford University, and personnel from other nearby research activities. Hypothetical scenarios, related to real terrain and other input data, were used.

Computer support for the guerrilla game was undertaken as a two-phased SRI-RAND project, with Dr. George W. Evans II of SRI and William Monroe Jones at RAND[1] working with Colonel Adams.

The program was planned in 1964, and an operational game to investigate the principles of Communist-inspired insurgencies was chosen for development. The fact was recognized that guerrilla warfare resembles certain features of a naval or aerial battle consisting of a limited number of sharp, individual actions localized in space and concentrated in time. The expenditure of a limited amount of munitions has a proportionately large immediate effect. The pattern is in sharp contrast to other forms of land warfare where contacts and interaction between opposed forces continue over extended periods of time, expend large quantities of small and medium-sized munitions, and have an effect which is often obscure and uncertain.

The program called for an operational gaming system to study insurgency/counterinsurgency in three steps:

1. Environmental gaming

---

[1] The SRI approach was to rely on Colonel Adams to develop models needed for the human-directed game for which Dr. Evans would build computer programs. Jones was to work independently at RAND and design computer simulations directly from the conceptual framework of the game.

2. Tactical gaming
3. Assessment gaming

Environmental gaming refers to strategic gaming. Assessment gaming was visualized as a means of providing indicators or measures on an accumulative, long-term basis, to assess the degree to which insurgency and counterinsurgency efforts were winning. Effort was concentrated in the area of tactical gaming. A manual for play of small guerrilla tactical operations was developed, and a number of insurgency situations were played. The insurgency situations depicted the types of activity for which computer subroutines were to be developed. Some progress has been made on programming this system, but many more manual games need to be played. Additional activities require the development of subroutines. The data, to be acquired from actual field operations such as those in Vietnam, must be identified. (9-17)

Difficulties were encountered in pioneering such a nebulous and complex situation, and the system developed for the manual game took "an inordinately long time to assess." Efforts to program the models for computer runs continued but as far as is known were not completed.[1]

The second approach to guerrilla gaming is represented by the THEATERSPIEL Cold War Model and the TACSPIEL Guerrilla Model.

### The THEATERSPIEL Cold War Model

The THEATERSPIEL Cold War Model attempts to simulate conditions of insurgency in Vietnam and effect an orderly relation between military and nonmilitary factors. The model, tested in a research game called TRIAL RUN, comprises THEATERSPIEL's Intelligence, Military, and Logistics Models for conventional war (revised for differing aspects of insurgency), combined with a new nonmilitary model and terminal models. The nonmilitary model assesses the economic, political, psychological, and sociological aspects of cold war; the terminal models evaluate the output of the other models and determine political changes.[2]

---

[1] Information from Colonel Adams indicates that the original project was discontinued in Nov. 1965.
[2] Billy L. Himes, Sr., Dino G. Pappas, and Horace H. Figuers, "An Experimental Cold War Model, THEATERSPIEL's Fourth Research Game," RAC-TP-120, Dec. 1964, 165 pp.

The Intelligence Model is oriented to military operations although certain output applies to nonmilitary areas. The Intelligence Model is not presumed to portray the elaborate information apparatus so necessary to complex guerrilla operations.

The Military Model portrays the operations of formal and organized internal and local security forces arrayed against organized full-time, and part-time, guerrilla forces. The model determines the results of conflict based on firepower, as modified by the element of surprise, terrain, etc.

The Nonmilitary Model assigns arbitrary values to depict the impact and effects of military and nonmilitary personnel in the political, economic, and psychological-sociological fields. Each individual is assigned a numerical value. These values, identical for both Red and Blue forces, are modified in control assessments by a formula which assumes a uniform distribution of variables and unknowns. Further, for nonmilitary personnel, a base formula is subjected to additional modifying factors.

The Logistics Model exerts control over both military and nonmilitary models by limiting actions to the support available. This realistic approach has a significant impact on political problems. (9-18)

The Terminal Model, which could have been called the Political Model, takes the output of the other models and evaluates the gain or loss of control of population for both sides. The model input consists of the elements of political decisions and the progress toward attainment of political objectives. Although the Cold War Model was played in the Southeast Asia theater, it may be adapted to other areas of the world by the substitution of related factors and the introduction of elements unique to the specific environment.

The model recognizes five power factors that contribute to the success or failure of insurgency warfare: geographic, military, economic, psychological-sociological, and political. The quantification of these factors is a major research problem, but the Cold War Model moved boldly into these areas and applied arbitrary units of measure based on informed opinions and limited data. The effort represents only a beginning in devising a gaming model for cold war.

Basically, the THEATERSPIEL Cold War Model was built on the objective of winning the support of the population of an area for political authority (Blue or Red) while assuring security and contributing to the economic and social development of that area. To do this, both military and civilian efforts are required. Demanded for this effort is the investment of trained people who will bring services, supplies, and know-how to an area, while military forces ensure security through the reduction or elimination of insurgent activities.

The Cold War Model was applied in geographic units corresponding to the political subdivisions of Vietnam. See Fig. 9-14. Each subdivision is represented with its own combination and status of situational factors: population, degree of area control by Red or Blue, developmental needs, etc.

The initial Blue and Red military situations used in TRIAL RUN are shown in Figs. 9-15 and 9-16. The Blue supply-net routes, and route capacities, are shown in Fig. 9-17. The Red supply-net routes, and distances, are shown in Fig. 9-18. The strengths and deployment of Blue and Red forces were indicated on maps similar to those shown. Force composition and the size of concentrations at each point or as dispersed (in each area) were given. For example, Blue (Republic of Vietnam) military personnel in Region V on October 1, 1963 (game time) totaled 71,600: 46,000 Army, 16,000 civil guards, and 10,000 Self-defense Corps. Red (Viet Cong) military forces in Region V first consisted of 4,500 guerrillas. Red forces subsequently were used (in the game) to harass and ambush Blue's military and nonmilitary efforts. In addition, Red forces were used as a means to gain influence (control) over the local populace in neutral areas and to support Red's military actions. Military engagements and logistic problems were adapted to the normal game models of THEATERSPIEL.

Strategic hamlets play a part in the game, partly as military and partly as nonmilitary factors. The status of the hamlets (as of game time) is shown in Table 9-1.

The play of nonmilitary factors included a wide variety of services furnished to the local population. Some indication of the variety and scope of services is shown in Table 9-2. Values (numerical points) were assigned each of three personnel categories: psychological and sociological, political, and economic. The efforts of the personnel were assessed according to their time period in the area, the absence or presence of military personnel, and the pressures of a minority group (this later only in the highland areas). By a series of formulas the nonmilitary and terminal models measure the values of the nonmilitary and military personnel for Red and Blue. Political change in the area is assessed by the change of value from the original political status to the results of actions by Red and Blue players.

The knowledge gained from this research experience represents a definite step toward the objective of development of a political appa-

256   Venture Simulation in War, Business, and Politics

FIG. 9-14.   South Vietnam: provinces by geographic area.

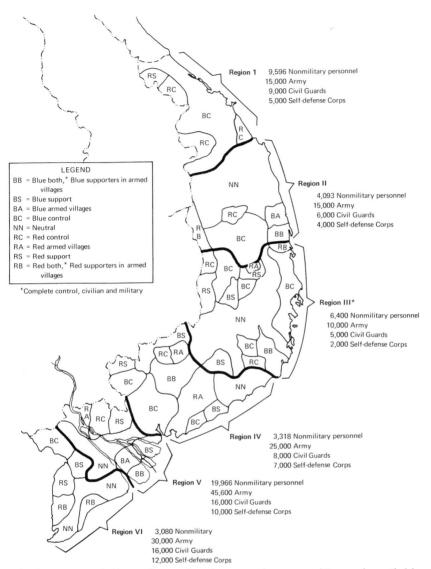

FIG. 9-15. Initial Blue military situation: popular sympathies and available forces, by region.

258    Venture Simulation in War, Business, and Politics

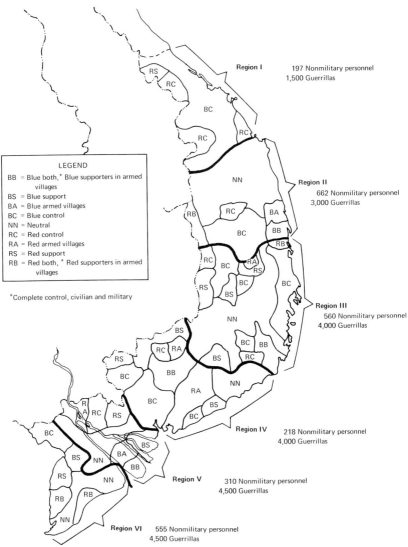

FIG. 9-16. Initial Red military situation: popular sympathies and available forces, by region.

## Strategic and Political Games, an Instrument of National Policy 259

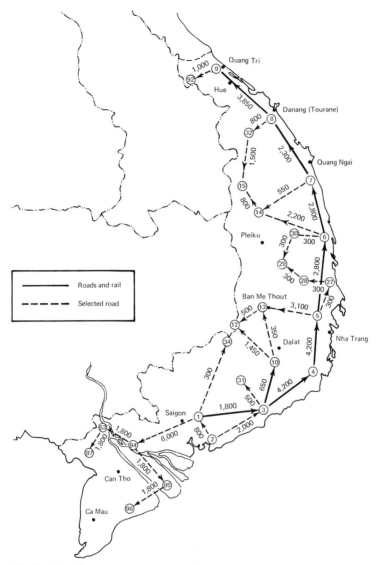

FIG. 9-17. Blue supply net. Maximum tonnages per day between supply points for coastal road and rail system and selected interior roads.

260   Venture Simulation in War, Business, and Politics

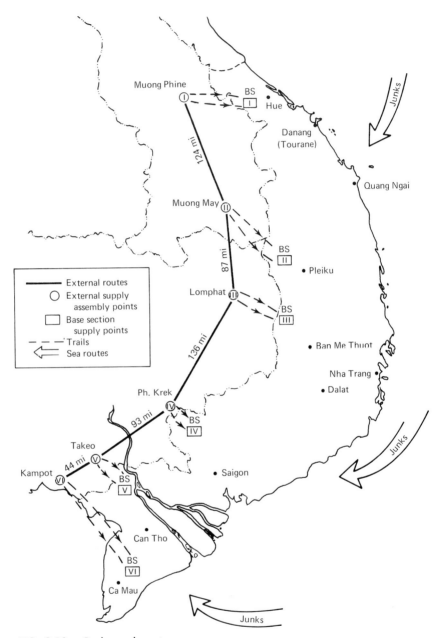

FIG. 9-18.   Red supply net.

Table 9-1. Data on Strategic Hamlets
Strategic-hamlet Program by Geographic Area (As of September 1962)

| Geographic area | Strategic hamlets | | | Population | |
|---|---|---|---|---|---|
| | Planned | Completed | Under construction | Total | In completed strategic hamlets |
| **Southern:** | | | | | |
| Saigon | 433 | 105 | 115 | 1,275,000 | 261,470 |
| Eastern provinces | 1,595 | 291 | 501 | 1,948,930 | 423,060 |
| Western provinces | 4,728 | 1,236 | 702 | 5,841,658 | 1,874,790 |
| **Total** | 6,756 | 1,632 | 1,318 | 9,065,588 | 2,559,320 |
| **Central:** | | | | | |
| Delta | 3,630 | 1,490 | 682 | 4,378,559 | 1,654,470 |
| High plateaus | 930 | 103 | 217 | 988,041 | 108,244 |
| **Total** | 4,560 | 1,593 | 899 | 5,366,600 | 1,762,714 |
| **Total program** | 11,316 | 3,225 | 2,217 | 14,432,188 | 4,322,034 |

ratus for use in gaming. The apparatus is susceptible to computerization and to use in gaming studies pertaining to intensive levels of the conflict spectrum. All models are susceptible to refinement and modification.

## The TACSPIEL Guerrilla Model

When TACSPIEL was employed to game portions of the airmobile concept, a series of games were planned and played. The final game was intended to explore use of air cavalry units in counterinsurgency operations such as existed in Vietnam. To conform to guerrilla-insurgency situations, the detailed, rigid, TACSPIEL rules, and the models to which these rules apply, were reviewed and modified.[1] (9-19) Late in 1964 planning got under way, and by mid-1965 a new model and rules were used in gaming operations of an Air Cavalry Squadron (ACS). The central highlands of South Vietnam were selected as the locale. The ACS was assumed to be supporting units of an ARVN (Army of the Republic of Vietnam) Division against guerrilla-type forces of hard-core Viet Cong and recruits from the local populace. The scenario called for a gradual escalation of intensity from terrorist raids to battalion-size coordinated attacks against ARVN forces. (9-20)

---

[1] The TACSPIEL manual at that time had approximately 400 pages of game rules. The guerrilla/counterinsurgency addendum to TACSPIEL required 66 extra pages of rules.

262  Venture Simulation in War, Business, and Politics

**Table 9-2. Regional Allocation of Nonmilitary Personnel**

*Regional Allocation of Blue Nonmilitary Personnel*

| Personnel | Region | | | | | | |
|---|---|---|---|---|---|---|---|
| | I | II | III | IV | V | VI | Total |
| **Psychological and sociological:** | | | | | | | |
| *Medical:* | | | | | | | |
| Physicians | 45 | 20 | 30 | 15 | 75 | 15 | 200 |
| Dentists | 4 | 2 | 5 | 2 | 10 | 2 | 25 |
| Pharmacists | 6 | 1 | 5 | 1 | 16 | 1 | 30 |
| Nurses | 300 | 150 | 200 | 100 | 800 | 50 | 1,600 |
| *Health and sanitation:* | | | | | | | |
| Sanitary engineers | 1 | 0 | 0 | 0 | 5 | 2 | 8 |
| Health technicians | 35 | 10 | 25 | 15 | 110 | 15 | 210 |
| Sanitary agents | 40 | 15 | 45 | 15 | 115 | 15 | 245 |
| Health workers | 40 | 10 | 50 | 5 | 65 | 5 | 175 |
| *Educational:* | | | | | | | |
| Teachers, secondary | 900 | 500 | 800 | 250 | 2,200 | 250 | 4,900 |
| Teachers, primary | 6,000 | 2,000 | 2,700 | 1,500 | 12,000 | 1,500 | 25,700 |
| **Political:** | | | | | | | |
| Propagandists | 500 | 350 | 850 | 500 | 1,000 | 250 | 3,450 |
| Political agents | 400 | 300 | 400 | 250 | 600 | 250 | 2,200 |
| **Economic:** | | | | | | | |
| *Construction personnel:* | | | | | | | |
| Road builders | 500 | 250 | 400 | 250 | 1,500 | 200 | 3,100 |
| Clinic and hospital builders | 300 | 150 | 200 | 150 | 400 | 50 | 1,250 |
| Bridge building | 200 | 100 | 250 | 100 | 300 | 200 | 1,150 |
| Canal and drainage-ditch builders | 100 | 100 | 200 | 100 | 400 | 200 | 1,100 |
| School builders | 200 | 100 | 200 | 50 | 300 | 50 | 900 |
| Agricultural extension agents | 15 | 20 | 20 | 10 | 35 | 15 | 115 |
| Livestock, poultry, fishing extension agents | 10 | 15 | 20 | 5 | 35 | 10 | 95 |
| **Total** | 9,596 | 4,093 | 6,400 | 3,318 | 19,966 | 3,080 | 46,453 |

*Regional Allocation of Red Nonmilitary Personnel*

| Personnel | Region | | | | | | |
|---|---|---|---|---|---|---|---|
| | I | II | III | IV | V | VI | Total |
| **Psychological and sociological:** | | | | | | | |
| *Medical:* | | | | | | | |
| Physicians | 2 | 2 | 5 | 1 | 0 | 20 | 30 |
| Nurses | 10 | 20 | 30 | 10 | 10 | 15 | 95 |
| *Educational:* | | | | | | | |
| Teachers | 100 | 400 | 285 | 100 | 125 | 300 | 1,310 |
| **Political:** | | | | | | | |
| Political agents | 60 | 130 | 130 | 60 | 115 | 120 | 615 |
| Propaganda teams | 10 | 85 | 80 | 35 | 50 | 70 | 330 |
| **Economic:** | | | | | | | |
| Clinic builders | 5 | 10 | 15 | 5 | 5 | 15 | 55 |
| School builders | 10 | 15 | 15 | 7 | 5 | 15 | 67 |
| **Total** | 197 | 662 | 560 | 218 | 310 | 555 | 2,502 |

The standard TACSPIEL game involved unit movements of mechanized and armored forces, at company level, in 30-minute cycles of game time. For guerrilla operations conducted largely by participants moving on foot through rugged jungle or mountainous terrain, a slower-paced game was required, with the unit size estimated at lower levels. Resolution was at platoon size (with patrols of squad size). Orders were issued on a 6-hour cycle. Assessments were made each 15 minutes of play and reports rendered with appropriate communications delays. Terrain grids were reduced from 1 to ¼ square kilometer. A new set of rules, equivalent to those of a civilian intelligence model, included instructional guidelines for agent operations and terrorist actions. Enemy information collected from friendly villagers was played for both sides, in areas under control of one or the other side. Neutral villages were used as a limited source of information collection. As in conventional warfare the maxim, "Find, Fix, Fight and Finish" was found valid, but with changed emphasis on Find and Fix and methods of fighting. The rules had to reflect the guerrilla propensities for frequent movement, sharp attacks, harassment and diverting actions, infiltration, and "melting away" (disappearing among the indigenous population). Rules for the counter-guerrilla force were added to provide for special problems such as detection, support for friendly villages and hamlets, patrolling, protection of communication lines (military and civilian), concentration and security of own forces, responsiveness and rapid movement to points of contact, and search and destroy operations.

At the jungle roots level, in active insurgency operations the more esoteric political, sociological, and economic aspects have little opportunity to function, and these factors were not played. But the TACSPIEL group expressed confidence that such factors could be built into the models concerned with larger areas and longer time periods.

The measurement of effectiveness, as in most war games, was an acute and complex problem, but even more elusive in the guerrilla-type game. An obvious solution is a system for upgrading or downgrading the effectiveness of each side in relation to the factors of guerrilla warfare. By this means terrorist acts such as the elimination (assassination) of a village chief, a teacher, or medic; the burning of an undefended hamlet; the destruction of crops or livestock; and the abduction of village males or females are actions that could be built into expanded models. The guerrilla game director wrote in his critique: "It is particularly noteworthy that during the conduct of the game many, many reports sifted through newspaper and other media pointing up the similarity of combat operations occurring in the central highlands of Vietnam and those occurring on the game board of TACSPIEL.[1] (9-21)

The model and game were developed for research purposes. The research objective was to test problems involved in air cavalry operations against insurgent forces. Additional research is needed to advance the art of gaming guerrilla and insurgency operations.

## STRATEGIC GAMES AND SIMULATIONS

What sets the truly all-encompassing strategic game apart is that the strategic game is primarily neither military nor political, each with some consideration of the other. Instead it deals with the entire international problem in consonance with all national power factors. There are two basic types of strategic game, free and rigid. Both types of game deal with the real world and real issues.

Free-play Strategic Games

A number of independent efforts to develop free-play strategic games have been made under various names, for example, cold-war games, limited-war games, desk games, crisis games, and strategic games. All rely on the expertise of the participants and limited reference data instead of on extensive rules, models, masses of detailed functional data, quantified values, and involved calculations. Although a recent book[2] (9-22) uses the title *Crisis Game* to embrace many of the game types named above, the more restrictive meaning of the term crisis will be used here. The Crisis Game will refer to an acute stage in international relations, such as the Cuba missile crisis, or threats as considered in the Douglas Threat Analysis Study. Extensive research is being conducted on analysis and simulation of crisis situations. (9-23)

In purpose and content, strategic games may be of utmost seriousness even though the game methodology or mechanics may be simple. The least formal physical setup for a game would be the conference table where a staff meeting or seminar-type discussion[3] of pos-

---

[1] Colonel L. S. Simcox, USA (Ret.), Game Director, RAC, May 18, 1966.
[2] Brigadier General Sidney F. Giffin, USAF (Ret.), *The Crisis Game*.
[3] The seminar method of free gaming Paxson describes as "Red, Blue and Control working together around a war map (the Napoleonic method) for the entire duration of the game." Operations are opposed, feasibility determined, and results assessed, all in the open. (Paxson, *War Gaming*, p. 25, RAND RM-3489-PR, Feb. 1963.) "Free" gaming has no relation to the uncontrolled "brainstorming technique," aimed at bold new ideas "no matter how far out."

sible actions and reactions can be held. Conferences such as these are hardly within the meaning of gaming, even though they may be strategy sessions.

Typically, free-play strategic games are closed, two-sided or multi-sided, human-decision, controlled or umpired games having a minimum of rules. Experts or actual officials function in the roles played. In general, the knowledge and expertise possessed by the participants, rather than use of a mass of rules and data, are relied upon for ultimate game decisions. Strategic games of this type have been played in the Pentagon, at Camp David, and elsewhere.[1] The Office of the Joint Chiefs of Staff (JCS) initiated games of this type in 1961.[2] Only an occasional mention of this activity has appeared in the public press. (9-24) Top-level officials from the White House, State and Defense Departments, the military services, and other agencies of government participate.[3] (9-25) The National Security Council and the Joint Chiefs of Staff are directly involved in matters treated in strategic games. The President and the White House staff took an active interest in political-military gaming during the Kennedy administration and were kept informed of the outcomes of the JCS games.[4] The games have been reported as JCS Politico-Military (PM) Desk Games, Political Military Games, and Cold War Games.

The mechanics of the JCS PM games are simple. Two or more teams, each of five or six officials, plan actions and reactions and submit these data through a Control Group of similar size. The teams meet for several hours daily, or on alternate days, to review the scenario and determine objectives, strategy, and plans. Conferences are held with senior officials who come to the game rooms for an hour or so each day.[2] With the concurrence of these senior officials, team moves and strategies are documented and submitted to Control; here they are analyzed in relation to moves of other teams and world influences. Control records the updated world situation and designates the intervening period of elapsed time. The game clock and calendar then are advanced a few hours to several months, and information is submitted to the teams for another cycle of play the following day.[5]

---

[1] Reported by Col. William T. Minor, USAF, former Director, Cold War Division, Joint War Games Agency, in an unpublished, unclassified Pentagon briefing, June 7, 1966.
[2] Giffin, *op cit.*, p. 69.
[3] The games are sufficiently matured to merit the direct participation of senior government officials. In these games such officials do not engage in role playing; they perform as the responsible officials they are in reality. The only suppositions are in the hypothetical situations or contingencies.
[4] Norris, *The Washington Post*, April 28, 1963.
[5] Lieutenant Colonel Thomas J. McDonald, "The JCS Politico-Military Desk Games," *Second War Gaming Symposium Proceedings*, pp. 70–71, 1964.

The POLEX games at MIT followed a similar procedure. Hypothetical or projected situations were used for the scenarios. Participating government officials came from higher levels of responsibility and represented their superiors without recourse to daily reviews and conferences. The primary objective of the POLEX games was to develop a methodology for research and study, not to suggest courses of action or possible consequences for the situations and scenarios played. (9-26)

The POLEX technique was used in a new series of research-oriented games at CENIS, MIT. These studies were intended to explore constraints and opportunities presented to top policy makers by certain new naval weapons systems. The games, designated DETEX (deterrence exercises)(9-27), were conducted between 1963 and 1965.

Rigid Strategic Games

Considerable research has been conducted to develop rigid strategic games. Starting with early RAND and ORO games, this activity continued with the INS, MIT, and other efforts. As yet, the goal has not been reached, although progress has been made.

The most ambitious rigid, strategic game project, potentially the most complex, was started at the Raytheon Company, Bedford, Massachusetts, by a resourceful group of model builders under the leadership of Clark C. Abt. The project, named TEMPER (Technological, Economic, Military and Political Evaluation Routine) (9-28), was first financed by the Raytheon Company. Later the project was supported through the Joint War Games Agency (JWGA), Office of the Joint Chiefs of Staff, and recently was acquired by the JWGA for further development.[1] (9-29)

### THE TEMPER SIMULATION

TEMPER is a coordinated system of computer-programmed, quantified simulations aimed at simulating international conflicts of cold or limited war. Eventually, the simulation could be useful as an analytical tool to study and perhaps suggest means of dealing with international crisis situations. Possible effects of national alternatives may

---

[1] Pentagon briefing by Colonel Minor, June 7, 1966.

Strategic and Political Games, an Instrument of National Policy    267

be suggested.[1] The models represent three world blocs and as many as thirty-nine nations or nation-coalitions (to thirteen in each bloc) spread over a globe segmented to twenty conflict regions.

Data concerning 117 nations have been accumulated. Covered are such items as population, resources, GNP, growth rate, military forces, and significant matters of strategic and political-economic-military survivability. These nations, or "actors," can interact and modify in seven categories: military, economic, political, scientific, psychological, cultural, and ideological. National motivations can be played in each of the last three categories. Decision making by "actors" (nations) is aided by a process of ideal sensing; reality sensing; ideal-to-real discrepancy measuring; resource allocating; international and interbloc bargaining; and alliance formation, operation, and dissolution. A time control provides an overriding master system to coordinate dynamic interactions in accordance with simulated response times. The time reaction intervals range from 1 week to 1 year; the total model operates on a 10-year time plan.

A central decision-making submodel exercises decisions that commit opposing-force blocs to combat. The action may range from insurgency or paramilitary actions to full-scale nuclear exchange. The cultural and ideological submodels may exert inhibiting factors on escalation. Combat results are assessed by the quantitative and qualitative characteristics of engaged forces relative to such features as terrain, attitude of local populations, intelligence, air support or superiority, weapon systems employed, etc. The usual means of providing for variable factors beyond the control of commanders—weather, for example—are built into the models on a probabilistic basis. Similarly, the economic and bargaining submodels provide for inclusion of their most important functional elements.[2]

Some Values of TEMPER

The TEMPER model attempts to consider *interactions* of economic, military, and other factors. Analytical studies of such interactions may help to advance the art of political gaming. There is also a very considerable educational by-product to game participants in deeper insights and fuller understanding of the salient forces at work

---

[1] TEMPER cannot predict future events or outcomes. With further development it may contribute insights or may even indicate possible courses of action.

[2] Clark C. Abt, "War Gaming," *International Science and Technology*, Aug. 1964, pp. 20–37.

in the model. Such values as these, coupled with "hoped-for" increase in validity, encourage continued research and development of extensive simulations of the TEMPER type. An expression of faith in these potential values was indicated by the acquisition of TEMPER for possible use in educational activities of the Industrial College of the Armed Forces in 1966.

Constraints of TEMPER

TEMPER as a full-fledged strategic game is in the developmental stage. The game objectives are pretentious, and the preparation of an appropriate methodology and the input data requirements set an almost impossible task. A few of these limitations, well known to the researchers and readily acknowledged by them, will be mentioned here.

All simulations require a simplification of realities. In international relations, it is difficult even to identify the essential factors. When the factors are identified, there is no means of testing their validity, other than reliance upon the judgments of experts. Thus precise, quantified inputs rest on the judgment of values in many cases and are not the direct result of objective measurement nor statistical analysis. For example, the range, accuracy, and reliability of missile systems can be ascertained with reasonable levels of confidence. The will to use missiles and the circumstances under which they will be used cannot be determined with nearly the same confidence level. Yet simulations require the programming of just such alternatives, decisions, and circumstances. Today they are programmed, but with quite uncertain degrees of realism.

The tyranny of the computer requirement that all actions be reduced to alternatives is a formidable handicap. It is true that programmers can break down a single Black or White, Yes or No alternative to a scale of different stages and provide for sensitive differences. For example, the computer can give a Yes or No reaction easily to any of the following questions: Is the quantity (1) between 1 and 100?; (2) between 60 and 70?; (3) between 61 and 62?; or (4) between 61.01 and 61.05? The difficulty lies in the obstacles presented to the model and program designer. He must foresee the range of questions to ask; he must keep the number of alternatives needed for full representation of essential features and values, as they function in the real world, within reasonable limits. The foregoing is only an indication of the nature and possible extent of difficulties encountered in bringing basic facts and nuances of international relations into meaningful, quantitative terms. A considerable literature has built up on just such problems.[1] (9-30)

After discussing a number of ramifications, Schwartz points out that "man-gaming may be said to be most useful in researching complex phenomena where concepts are not yet formalized (nor quantified), where more direct methods of observation are difficult or impossible to apply, and where the character of available models does not admit of simple, rapid, nor direct manipulation of variables."[2] It is these aspects of political–diplomatic–international-relations crisis situations with which mathematical models and computer simulations are least able to cope. Most decision makers in responsible positions welcome and even seek summarized factual data which a computer can be programmed to deliver, but it is doubtful if any high official would trust the computer for final evaluations.

This view of the incomplete development of TEMPER and the complexity and difficulty of quantifying the thousands of factors involved in *rigid* strategic games and simulations makes clear the reasons why the *free-play* strategic game is used. There is no better means as yet of exploring alternatives and indicating possibilities beyond those suggested by intelligence reports, analytical studies, the knowledge and judgments of experts, and the counsel of advisors. All efforts of these kinds, however, are only a contribution to the process of problem solving and decision making. The responsibility of the man who must make the final decision is not reduced. At best, the decision maker's choices are narrowed, and alternatives and their possible consequences are more clearly identified.

## Chapter Notes

9-1. Anders Sweetland and William W. Haythorn, *Behavioral Science*, 6:105–116, 1961.
Robert L. Chapman et al., "The Systems Research Laboratory's Air Defense Experiments," *Management Science*, 5:250–269, 1959.
David G. Hays, *Simulation: An Introduction for Anthropologists*, pp. 9–10, RAND, P-2668, 1962.

---

[1] Joseph T. Hart of RAC, in a doctoral dissertation submitted to the American University, School of International Service, reviews a number of the problems in a specific critique of the TEMPER game, pp. 192–200. The TEMPER designers point out like shortcomings.

[2] David C. Schwartz, "Problems in Political Gaming," *Orbis*, Feb. 1965, pp. 684–685.

9-2. J. W. Ellis, Jr., and T. E. Greene, "The Contextual Study," *Operations Research*, Oct. 1960, p. 639.

9-3. Information to the author from Dr. Harold Guetzkow, Northwestern University, who participated as a Fellow at the Center for Advanced Study in the Behavioral Sciences at Stanford in 1956–1957; from various published sources, including Goldhamer and Speier in *World Politics*, Oct. 1959; and from writings of Guetzkow, Bloomfield, Raser and Crow, and Schwartz. A condensed review is reported in Hans Speier, *Some Observations on Political Gaming*, pp. 18–19, RAND, P-1679, Jan. 20, 1960; and in Giffin, *The Crisis Game*, pp. 64–69.

9-4. James A. Robinson, Lee F. Anderson, Margaret G. Hermann, and Richard C. Snyder, "Teaching with Inter-Nation Simulation and Case Studies," *American Political Science Review*, 60:53–65, March 1966.

9-5. Information to the author from Dr. Harold Guetzkow, Northwestern University; Prof. Andrew W. Scott, University of North Carolina; and Dr. Jerome Laulicht of the Canadian Peace Research Institute.

9-6. From observation of an INS in operation and from the writings of INS practitioners.

9-7. Information supplied to the author by Professor Scott.

9-8. The POLEX games are discussed later in the chapter, and in more detail in the *American Political Science Review*, Dec. 1959, under the title, "Teaching Note; 'Three Experiments in Political Gaming'" by Lincoln P. Bloomfield and Norman J. Padelford, and "The Political-Military Exercise . . ." by Bloomfield and Barton Whaley in *Orbis*, 8:854–870, Winter 1965 (also condensed in *Military Review*, Nov. 1965, pp. 65–71). See Selected Bibliography under Bloomfield.

9-9. Unpublished information supplied to the author by Dr. Laulicht plus "The Vietnam Peace Game: A Simulation Study of Conflict Resolution," "Vietnam at Leeds," manuscript by Dr. Jerome Laulicht with John Martin, and "The Vietnam War Game" by Dr. Jerome Laulicht, first published in *New Society*, Jan. 27, 1966. These and other studies were in progress at the Peace Research Centre, Lancaster, England, and the Canadian Peace Research Institute, Clarkson, Ontario, Canada. A Reuters dispatch from Lancaster, England, reporting this study, appeared in U.S. newspapers (e.g., *The Washington Post*) in early January 1966.

9-10. Source materials included news reports; economic data from government and United Nations publications; general reference works such as the *Statesman's Year Book*, Whitaker's *Almanac*, Jane's *Fighting Ships*, and Jane's *All the World's Aircraft;* the works of P. J. Honey, Bernard Fall, David Halberstram, Truong Chinh, former Secretary-General of the Vietnamese Communist party, and Dr. B. S. N. Murti, Indian diplomat and member of the Secretariat of the

Strategic and Political Games, an Instrument of National Policy    271

Geneva Conference. Laulicht, "Vietnam War Game," unpublished notes.

9-11. When the Air Force established Project RAND in March 1946, the Douglas Aircraft Company was chosen as contractual agent for the project to conduct "a continuing program of scientific study and research on the broad subject of air warfare." Project RAND subsequently was transformed into the independent RAND Corporation (1948). Other industrial companies having OR or systems analysis programs to study trends, projections into the future, and problems of concern to the highest levels of government include Boeing, General Electric (TEMPO), Westinghouse (MELPAR), and Raytheon.

9-12. The Threat Analysis briefing material developed at the Douglas Aircraft Company is part of a larger effort under way in the Environmental Analysis Department. Its purpose is to develop techniques for forecasting future demand and requirements for aircraft and missiles. The techniques and the substance of these studies represent the type of input data needed for certain types of political games and simulations. The data presented here are illustrative of the kind needed and of the process by which they were obtained. The data should not be considered valid in itself.

9-13. Dino G. Pappas, "STRAT-X, A Gaming Concept for Regional Crises," unpublished manuscript, Research Analysis Corporation, Jan. 1966.

9-14. John G. Hill, "Notes on Political Decision Models," unpublished paper, Research Analysis Corporation, Military Gaming Division, Oct. 8, 1965.

9-15. Unpublished study by Paul F. Dunn, John G. Hill, Edward H. Huggler, J. Duncan Love, William Sutherland, Richard E. Zimmerman, and Benjamin A. Bache, Operations Research Office, Dec. 1960.

9-16. Alfred Blumstein, "Strategic Models of Counterinsurgency." The Blumstein paper, itself unclassified, was published in the classified volume of *Proceedings of the 13th MORS Conference*, pp. 164–168. The conference was held at the Industrial College of the Armed Forces, Fort McNair, Washington, D.C., under sponsorship of the Office of Naval Research. Blumstein credits the Bonnett reference to Bernard Fall's *The Two Vietnams*.

9-17. Information to the author from James Adams at Stanford Research Institute, Menlo Park, Calif., December 1964. Also in Giffin's *The Crisis Game*, p. 69.

9-18. Brigadier General John J. Hill, U.S. Army, Retired, Game Director and Chairman, THEATERSPIEL, commented further on the significance of the initiative. He said: "It is historically true that no guerrilla war has succeeded without the support of the population. However, in recent counterinsurgency operations in which the United States has been involved, always with great logistical advantage to our side, the enemy has been permitted to place our side in the position of reacting to his initiative. The results of this lack of [our] own initiative appear to have been erosion of the friendly political stability and loss of popular control." A realistic counter-guerrilla game model must recognize the value and importance of all the principles of war, and particularly intelligence, security, mobility, and willing support of the local population. Logistics, firepower, communications, quick reaction time—other great advantages—are degraded by the guerrillas' ability to disappear in the locale or among the population. Terror is a weapon which guerrillas use to control the population when more subtle means lose their effectiveness.

9-19. The basic TACSPIEL rules are contained in RAC-TP-111, dated Nov. 1963. The added guerrilla rules were issued as RAC-TP-223, Aug. 1966. Both documents were published by the Research Analysis Corporation, McLean, Va.

9-20. L. J. Dondero et al., "TACSPIEL War Game Procedures and Rules of Play for Guerrilla/Counterguerrilla Operations," draft manuscript, Research Analysis Corporation, Nov. 1963.

9-21. L. S. Simcox, "Lessons Learned in Guerrilla War Gaming," unpublished paper, Research Analysis Corporation, May 18, 1966, 7 pp.

9-22. General Sidney F. Giffin, *The Crisis Game. Simulating International Conflict*, Doubleday & Company, Inc., 1965. General Giffin reports two "games" or crisis situations. The first was a reconstruction of the actual Cuba missile crisis of 1962; the second, a hypothetical Kashmir crisis projected in the future. Anyone interested in the detailed actions and counteractions in a free-type political-military strategic game is referred to the Giffin book. In an original, if unorthodox, reversal of validation methods, Giffin concludes his "game reconstruction" account by a critique of the real Cuba-missile-crisis events. He writes, "One would be hard put to develop valid criticism of the manner in which the Washington participants handled this crisis. Considering the embarrassment of their position, the Moscow participants must also be given high marks for the skill with which they disengaged." (p. 116.)

9-23. Professor James A. Robinson of The Research Foundation, Ohio State University, reports a series of studies (part of Project Michelson) under the title *Crisis Decision Making* . . . and lists some sixty-seven crisis hypotheses and findings of other investigators in this field. John L. Enos of RAND reports work on an analytical model of political allegiance and its application to the Cuban revolution in which Castro overthrew the Batista regime. The revolution began on July 26, 1956 and culminated in Batista's resignation on Jan. 1, 1959.

9-24. Rear Admiral Clyde Van Arsdall, Jr., USN, "Joint War Games," *Second*

Strategic and Political Games, an Instrument of National Policy    273

*War Gaming Symposium Proceedings*, pp. 117–128, WORC, Washington, D.C., March 16, 17, 1964. Giffin, McDonald (see Chapter Note 9-25), Minor, and Norris in *The Washington Post* (see Selected Bibliography) also make brief reference to these games.

9-25.  Lieutenant Colonel (later Colonel) Thomas J. McDonald, "JCS Politico-Military Desk Games," *Second War Gaming Symposium Proceedings*, March 16, 17, 1964, pp. 63–74, particularly p. 65. Also mentioned in the Minor briefing, "Political-Military Gaming," p. 8; Giffin, *The Crisis Game*, p. 70; and Norris in *The Washington Post*.

9-26.  L. P. Bloomfield and Barton Whaley, "The Political-Military Exercise...," *Orbis*, 8:854–870, Winter 1965; also condensed in *Military Review*, Nov. 1965, pp. 65–71.

9-27.  "The Detex-Exdet Politico-Military Exercises on Naval Weapons Systems during Crises," Cambridge, Mass., MIT, Center for International Studies, Final Report, June 1965.

9-28.  Clark C. Abt, "War Gaming," *International Science and Technology*, Aug. 1964, pp. 29–37. The article includes a sampling of general background and descriptive information on war gaming and a very sketchy description of TEMPER. Another brief account was given at the Third Symposium of the East Coast War Games Council by Abt, Gorden, and Hodder. Little of the voluminous detail of the TEMPER simulation has appeared in the open literature. Manuals prepared at the Raytheon Company fill several thick volumes and give flow charts, analytical details, and other data used in the simulation.

9-29.  In early 1961 a Joint War Games Coordinating Group, headed by Brigadier General Calhoon, USA, was established in the office of the JCS. Subsequently the group became the JWGA. Purpose: "To provide the JCS with a capability for applying war gaming methods to some of the problems confronting the JCS," as stated by Rear Adm. Clyde J. Van Arsdall, USN, Chief, JWGA, in a paper presented in a War Games Symposium in March 1964. The JWGA was organized into three major divisions, each with a special province: General War, Cold War, Limited War. The Cold War Division supervised the TEMPER and the Politico-Military Games and was headed by Col. W. T. Minor, USAF, superseded by Col. T. J. McDonald in 1967. See Chapter Notes 9-24 and 9-25; also see Selected Bibliography items under Minor, and Washington Operations Research Council for McDonald and Van Arsdall.

9-30.  Joseph T. Hart, "Gaming as a Research Tool in International Relations," dissertation, American University, Washington, D.C., 1967, pp. 192–200 and 148–178; Morton Gorden, *International Relations Theory in the TEMPER Simulation*, Abt Associates, Inc., Sept. 1965. The Defense Communications Agency contracted with mathematicians (Balinski, Knorr, Morgenstern, Sand, and Shubik) at Princeton University to review and evaluate the TEMPER Model(s). In September 1966, this group concluded that TEMPER could not succeed in its objective and could not be redesigned to do so, and that no program as ambitious as TEMPER is within the capabilities of the present state of the art. See Balinski et al. in Selected Bibliography. See also Selected Bibliography entries under Abt, Bloomfield, Brody, Gorden, Goldhamer and Speier, Guetzkow, the Hermanns, MacRae and Smoker, Raser and Crow, James Robinson, Schwartz, and Scott.

# Gaming in Retrospect: an Appraisal

The most significant period in war-gaming development has been since World War II. In this period the techniques of war gaming and its cohort, simulation, have become sophisticated practices, increasingly employed in analytical studies and decision making at the highest national levels. Spurred by the power of computers, the surging fields of management and research "discovered" gaming and simulation. War-gaming techniques are being energetically extended to fields where little or no previous experience with these techniques in sophisticated form exists.

Enthusiasts, impelled by visions of gaming and simulation potentialities, may try too much, too quickly, too soon. Such enthusiasts may overlook the long, slow, tedious, and extensive data-collection efforts needed to make models usable. War gaming has been at this job for a century and is hard pressed to keep abreast of advancing technology. Gaming and simulation techniques, when adapted to fields other than war gaming, can yield quick values helpful for education and training, but for analytical purposes the rate of advance must be slower. Progress is being made, and will continue to be made, as a result of the drive now begun. It therefore seems fitting—if not indeed necessary—to take a look at gaming and simulation in its pristine field of application, war and national security, to see the obstacles encountered, the difficulties unsolved, and the promise of the future.

## LIMITATIONS OF WAR GAMES

Those who have read the preceding chapters may have gained the impression that, since gaming is used in arriving at significant decisions, the results of war games always can be regarded as valid. This chapter is explicit in cautioning the reader against any such conclusion. War-gaming capabilities have been reported. The effort here is to emphasize the limitations. As the art of gaming has advanced—and advance it has—the limitations have been reduced in degree but have not been eliminated.

The study of war may never become an exact science despite the striving of scientists and military specialists. There are many reasons. A most important reason is the inability of man to predict how he or any other individual will react in stressful and dangerous situations of warfare. This inability to predict human action and reaction is a major problem and a fundamental reason why the social sciences are generally less exact than physical sciences. Another reason is the vast number of variables present in almost any combat situation, and the even greater array of interrelations and combinations of factors involved. Furthermore, these variables do not recur in fixed amounts, degrees, or weights of relative importance. In some combat situations weather may not introduce noticeable interference; in other situations of combat snow or rain may have a paralyzing effect on force mobility. Conversely, clouds, fog, or mist may reduce visibility and detection of activities (own or enemy) and enhance mobility while reducing exposure to firepower.

Man's understanding of the process of warfare is incomplete and inadequate. In many respects certain elements known to have an important bearing on the outcome of battle can neither be predetermined nor measured. For example, no one can predict or measure whether a unit will stand and fight to the last man or will break. Neither can "break" points be predicted. Training, morale, leadership, fatigue, fear, courage, and stress, all are major factors in warfare, all wield strong influence, and yet the extent of this influence remains intangible.

The will to fight is unquantified and little understood. History records many examples of great differences among contending forces in the will to fight. Classic battles of the past, such as between ancient Greek city-states as Athens and Sparta; defense of the pass of Thermopylae; Gettysburg in the American Civil War; selected encounters in the Spanish Civil War; more recently, clashes between Communist and anti-Communist forces in Korea and Vietnam; and the

## Gaming in Retrospect: an Appraisal 277

remarkable advances of Israeli troops in the 1956 and 1967 Sinai campaigns all illustrate heavy fighting and strong will. A far less virulent fighting attitude was displayed in some of the pseudo-battles between bands of mercenary forces (led by the Condottiere) of the ducal principalities of Italy in the fifteenth century; the armed forces of the trichotomy of the three princes of Laos (Royal Lao, Pathet Lao, and neutralist forces) in the 1960s; and the opposing forces of the second uprising in the Congo from 1964 to 1965.

Even measurable physical forces, such as firepower, rate and accuracy of fire, lethal areas of individual munitions, amount of fire (ammunition expenditure), and the effects of these factors on surviving troops in battle are largely unknown. For example, one of the functions of an artillery barrage, as during an attack, is to suppress fire from the defending force. Yet valid quantified data are not available on just how much suppression is effected with a given amount of fire.

Alternatives open to commanders may be many. Decisions can range over a wide spectrum of possible actions. Moreover, any simulation, model, or war game is incomplete when measured against all the factors involved in a real combat situation.

## LOGIC OF BATTLE

Despite man's incomplete understanding of the process of warfare, most experienced military commanders recognize some precepts or rules applicable to combat operations. These rules fall into a logical sequence.

Each combat situation, as learned in military schools or through experience, falls into a sequential pattern that premeditates and accommodates to the step-by-step progress of a battle. This more or less normal sequence may be considered the "logic of the battle." For greatest effectiveness the logic of battle calls for the performance of certain activities at optimum times, e.g., air strikes and/or artillery shelling and barrage preceding an attack. In gaming and simulation such sequence also must be recognized to comply with the logic of the battle and accommodate the calculations of the model. An example of model adjustment would be to calculate the effects of interdiction in reducing the firepower, mobility, and possibly the size and nature of the force the enemy will (or will be able to) employ in battle and then subtract such losses from the enemy force prior to the battle effects of the opposing forces—effects assumed to result from contact and battle.

The models in a war game must conform to the logic of battle and yet provide the flexibility the Commander requires in meeting chance events that occur as opportunities or adversities. Manual or man-machine games can be responsive to quick decisions of the Commander; complete simulations are limited in this respect. However, free-play games are inadequate for serious analytical gaming, and rigid or semi-rigid games cannot provide the complete range of choice. In this respect games fall short of the hoped-for similarity to real war.

The Fog of War

The fog of war cannot be fully duplicated in a war game. War has been described as organized violence. Each opposing force attempts to confuse, disrupt, and annihilate the other force. Each force tries to maximize destruction and chaos in the environment of the opponent. Such destruction and chaos occur in combat on a chance or probabilistic basis. It follows that neither side is, nor can be, fully informed on what is happening to the opponent. In addition, the commander cannot fully know what is happening to the various units of his own force or in the various segments of the area in which he is operating. Reports come to the commander—if communications function properly—and are graphically displayed on operations maps and overlays. But in war nothing works exactly as planned; always there are some unanticipated successes and failures; always there is "someone who doesn't get the word," and frequently the commander is ignorant of what is happening in one or another unit. Hence, there is always an incomplete picture of the ongoing operations. This is part of "the fog of war."[1] General von Moltke the elder stressed this "fog of uncertainty" as one of the considerations for which war gamers must make provision.

A game model or simulation may be too perfect. Realistic provision may not have been made for information that does not get through, or arrives too late to be used, or for the mistakes and misunderstandings that occur in real life. The model builder must recognize the existence of such shortfalls and, if possible, provide a compensating adjustment. The best he can do is to approximate the condition, but in truth there is no way that this uncertainty in warfare may be accurately represented.

---

[1] Expression attributed to General von Moltke by Lieutenant General von Cochenhausen in booklets translated from the German and published under the title, "War Games for Battalion, Regiment and Division," Command and General Staff School, *Military Review*, March 1941, p. 51.

Doctrine and Obsolescence

Over the years military experience has been organized into a set of guidelines, an established way of doing things. In this connection the criticism often is heard that each war is fought with the weapons and tactics developed in the previous war. This charge has been made following the longer intervals between major wars, and in a way the interval between changes in guidelines represents the span of time it takes to develop new doctrine and weapons to the point where they are operational. Thus, the tank, aircraft, and aircraft carriers introduced in World War I did not come into full and effective use until World War II; and the replacement of the battleship, started shortly after World War I, did not become complete until after World War II.

With the onrush of technology, progress and new developments are a normal expectation. Many of these developments affect military doctrine. For this reason doctrine starts on the road of obsolescence as soon as it is formulated. Periodic or continuous revision is a standard practice in the military services. Input data for war games must keep pace with updated doctrine, even contribute to it. For this reason, among others, game input data must go through periodic or continuous updating to reflect new developments and new test data. In this sense, it is almost impossible for a war-game model, and its input data, to remain static through the years and continue to serve current uses.

## ASSUMPTIONS AND CONSTRAINTS

Assumptions

The assumptions to be made in a game or simulation are related to the objectives or purposes of the game and always should be stated and made specific to those ends. Any game starts with assumptions regarding forces meeting under specified conditions in a specific locale (geographic area). These factors are usually stated in the scenario and the general situation. Beyond these general assumptions or settings for the conflict are a number of specific assumptions, relating to forces, weapons and "ground rules," made so that the game will conform to intended purposes. Specific assumptions should be so designated and listed under that heading.

Ground rules that could be specified as assumptions in a war game are:

1. A nuclear strike will be detected by the Early Warning System (DEW Line, etc.) with an alert and warning time of 15 to 30 minutes.
2. A nuclear strike by missiles (ICBMs) will be accompanied by decoys.
3. An initial nuclear strike by ICBMs will be targeted on U.S. military installations, missile sites, and six token metropolitan areas with up to 100-megaton-yield H-bombs.
4. An attack by Warsaw Pact forces will strike against NATO forces with conventional weapons, but with capability of employing nuclear weapons.
5. Red has sixty divisions in Eastern Europe to employ in a drive to the sea, with twenty divisions as potential reinforcements. The Red force is composed of _____ armored divisions, _____ mechanized, _____ infantry, and with _____ air squadrons composed of _____.
6. NATO forces have $x$ days of combat resupply stocked and positioned in the combat zone.
7. France will authorize NATO forces to resupply through French ports and use existing communication lines in France.

Games need not be confined to those conditions considered probable. In fact, for planning and analytical purposes any conceivable set of assumptions may be appropriate if they relate to the objectives the game is to serve.

Constraints

Constraints and assumptions are related but not identical. The constraints on each Commander usually are stated in the special situations or instructions to the Commander, or, in some cases (if general), the constraints are presented under assumptions. Constraints forbidding the use of nuclear weapons stating whether or not overflights of neutral territory are permitted, whether cities over 25,000 are to be sanctuaries, whether refugees are or are not to be supplied from military stocks in the combat zone, and similar matters may result in identical or different constraints for each Commander. The circumstances depend upon the design of the game. For example, one Commander may not be constrained in his treatment of refugees and displaced persons, whereas the opposing Commander is. These differences may be consistent with the two cultures from which the opposing forces originate. Whatever the facts of the situation, constraints on each Commander need to be stated but need not—and often should not—be revealed to the opposing Commander.

Reference has been made to circumstances where assumptions are built into games in lieu of tangible and real data. A universal practice is to build war games and simulations of combat situations upon certain "givens" or assumptions, as neither war games nor situations of combat are ever complete in themselves. The use of assumptions can be viewed both as an advantage and as a liability. Assumptions are advantageous in ruling out of the game things the game cannot study. Much fruitless study, doomed to unsatisfactory results, is thereby avoided. Many war gamers are happy to have standard rules establish assumptions to avoid unmanageable factors if data are not available or if the state of the art has not advanced to cope with certain factors in a game.

Any war game, simulation, or model, save the most rudimentary, is a simplification of the world it represents. To the degree that all factors are not included there must be implicit or explicit assumptions regarding the omissions. Among omitted aspects will be factors such as morale, or will to fight. Accepted scientific procedure suggests that such conscious omissions be specified. Admittedly, there are other factors in real war about which we may know next to nothing at the current stage of knowledge, and these cannot be stated although they are implicit. Fanaticism may be something we may learn to induce, as with drugs, or counteract, as with a gas. Who knows what we may learn in time about reducing or overcoming fatigue, stress, fear, panic, etc., under combat conditions? War games include all these human reactions to combat only in indirect ways, if at all.

Parity Assumptions

One assumption of advantage in simplifying a model without introducing liabilities of any serious concern is called the *parity assumption*. The commonplace statement "other things being equal" is an assumption of parity, or equality for whatever is covered by "other things." This is a very useful type of exclusion from a comparative situation and is justified by reducing the number of variables or factors in a game, model, or simulation where the omissions do not invalidate nor unduly bias the system. Also it is a neat and legitimate way of eliminating factors for which means of measuring have not been developed (leadership, training, and morale of contending forces). The assumption of equality or parity in the sagacity of the opposing commanders and/or of comparable training and morale can be considered (indeed, should be) as the best or most reasonable assumption in many situations. In fact, without these elements in approximate parity, opposing forces most likely would not persist in a state of active combat.

Simplifying Assumptions

Another type of assumption serves to simplify the model or game without introducing any distortion of the results in terms of the game objectives. Suppose one purpose of the game is to determine the quantity of supplies—fuel, ammunition, etc.—that would be consumed (and lost) in a particular combat situation. The assumption may be made that all supply points have unlimited quantities of the supplies. The combat units need only to draw upon these supply points and transport the supplies (employing vehicles and aircraft of the unit) to the site of use. The problems of replenishing or maintaining the supply levels at supply points are not germane to the game objectives and, if included, would complicate and extend the mechanics and computations involved in game operation. The game designer must exclude irrelevant matter with a zeal equivalent to that exercised to include all essential factors.

Significant Assumptions

Significant assumptions are most important. Game results are entirely dependent upon them. Wrong, or invalid, assumptions of this category distort findings so much that play of the game would be wasted effort. (This statement refers to analytical gaming, not gaming to train players.) No matter how realistic and accurate the data, force, and weapons characteristics are for one side, unless the data pertaining to the opposing side are equally valid the findings will be grossly distorted. For example, if data for the Blue force are meticulously correct, but some significant aspect of Red force data, such as a type of antitank missile, is omitted or if the armament, number, or characteristics of Red's aircraft are grossly underestimated, these significant factors are likely to alter the findings completely. If any miscalculation of enemy capabilities is permitted in a game, an overestimate is preferred to an underestimate.

How can valid data on a potential enemy's military weapons and capabilities be obtained? The answer to this question is one of the most critical in war gaming. A function of military intelligence, a search for this answer is an ongoing activity in peacetime as well as during war. Each nation attempts to keep information about weapons development under security and authorize release only of such information as may serve political (internal or international) purposes. In a democracy information must be released to the law makers (at least to those who authorize expenditure of funds) to win support for appropriations. Under a totalitarian regime this information can be confined to fewer people.

All governments have a similar purpose in maintaining secrecy: to enhance *surprise* and *shock* on the enemy when a new development is employed. The development of the atomic bomb and its surprise use against Japan in World War II are an example. Conversely, release of certain information to a tempted, adventurous antagonist can serve as a deterrent to warlike action. The development of the Polaris submarine and ICBMs and nuclear weapons (the latter on a multi-national basis) illustrates this point.

How, then, do gamers deal with the unknowns of potential enemies in war games? By *assumptions*. Anything may be assumed. Games conducted with known assumptions yield results that can be analyzed and interpreted along such lines as: Assume potential enemy B possesses X, Y, and Z in capabilities; under a specific set of conditions the effect on opposing force A would be the result derived through arithmetical reduction of the input data.

The need to make assumptions regarding other nations' military capabilities (and intentions) increases the requirement for war games and simulations. Instead of a known set of specific data, a whole spectrum of possible capabilities of the antagonist and opponent must be gamed to plan an adequate defense and an effective national security program.

Assumptions need to be examined in terms of any tendency to distort data to a degree that would produce invalid results. In evaluating the results or findings of games, and analytical studies based upon data derived from games, both the *objectives* and the *assumptions* need to be scrutinized with utmost care. The latter, along with the input data, should be subjected to the most rigorous examination and validation possible. Findings are no better than the validity of the data and the assumptions from which they are derived, and they should not be generalized beyond these limitations.

Probability and Statistical Assumptions

Analysis of data produced in games and simulations presents mathematical and statistical considerations that are complicated in the extreme. Professional analysts, mathematicians, and statisticians have been giving serious attention to these problems, and although very sophisticated techniques have been developed, the problems are not regarded as solved.

The ultimate purpose of these considerations is to answer such questions as, Do games and simulations yield valid results? If so, how valid? Innocent looking, these questions are loaded with complexities. These concerns, quite characteristic of operations research problems, are important in gaming.

In model building and analysis of operational data many of the data are derived from observation, test results, or "best-estimate" values. Probability distributions are constructed and employed if the data exist. When needed data do not exist or cannot be developed, "best estimates" made by qualified persons are acceptable.

Wide deviations may occur within the data from which average or best-estimate values are derived. The ranges of variation of these values, within which the results will remain valid, i.e., within which the values can vary without changing the nature of the result, are determined by sensitivity analysis.[1] A justified but somewhat waggish appraisal of sensitivity analysis has been summarized as follows: "Sensitivity testing of assumptions does little more than to tell how important are the things the analyst doesn't know."[2] Sensitivity analysis does not correct the data or assumptions used; it does give the analyst some measure of the level of confidence that can or should be associated with derived findings or results.

## THE ELEMENT OF CHANCE

Many events occur by chance, others with certainty. The time the sun will rise and set can be predicted with confidence. Weather forecasts are made with a considerable element of uncertainty. The layman calls it chance; the scientist, probability.

In the lifetime of one person, death is certain; the time of death, quite uncertain. Life insurance companies use actuarial tables based not just on averages, but on whole sets of statistics, arranged in tables built from the actual experience of thousands of cases. A sample of 100,000 people from the United States population shows that, on an

---

[1] Ackoff, *Progress in Operations Research*, pp. 175–176.
[2] *Armed Forces Management*, Dec. 1966, p. 4.

average, a relatively constant number died at each age for each year during the past 25 years. The insurance risk for each age has been calculated from such statistics, and premium rates are set in relation to the risk. Many of the events in life are similarly subject to chance or uncertainty. Insurance of all types—fire, accident, health, burglary— recognizes such uncertainties.

War and many other situations in real life are replete with chance events. In war games consideration must be given to the probability that particular conditions will occur and to the various possible effects that will result from given actions. Two tanks may be engaged in a duel. Which tank will fire first? Which tank will get the first hit? A first hit will follow how many rounds of fire? Will the hit glance off and inflict little damage? Will the first hit cripple the tank track causing tank immobility? Will the tank be disabled in other ways? Will the hit "kill" the tank? What will happen to the opposing tank meanwhile? Even such a simple action has a whole set of probabilities, or chance outcomes. Tables have been constructed to record what experience—garnered from war, field experiment, and test data—has shown.

A two-tank duel is a very elemental situation that rarely occurs alone. More often, two groups of unequal numbers and different kinds of tanks meet on the battlefield in an action involving many other elements. Infantry armed with recoilless or other antitank weapons, artillery fire, napalm from aircraft, mines, terrain hazards, and smoke or dust all are factors that affect each tank's own fire and casualties. War games must attempt to duplicate this complex of elements and factors that may occur in almost infinite variety in real war.

Impossible? Not quite. As in life insurance, it is less important to compute the probability of each possible cause of death (and find some way to combine the individual probabilities into an aggregated or composite probability) than to assemble statistics on the important end results and establish a gross overall probability. The important point is to determine what end product is needed for each purpose of the study, and to attempt to measure that product as an aggregated whole.

A problem that has plagued war gamers for years is the question of the reliability of results. Because many assessments are based on chance factors (to which probability values are assigned in conformity with normal expectations) there always has been a haunting uncertainty about the results of any one game. Are the results those most likely to happen, or are they a "fluke," i.e., a highly improbable result? When shortcuts are taken, there may be an even greater opportunity for improbable results. Although there is a logical basis for discrediting the "fluke hypotheses" in most games, it would be reassuring if the game could be rerun several times, without variations, and the results of each run compared with an average of all runs made. Computer simulations provide such opportunities when the simulation does not involve human intervention.

## HUMAN DECISIONS

Games involve human decisions. The actions in war games are the result of decisions made by opposing commanders. The value and quality of these decisions are products of the expertise of the commanders. For this reason war gamers attempt to obtain experienced military commanders who are recognized as successful combat commanders at the level of command the game requires. If a division-level game is played, the Commander should have division-command experience (that of a Major General). If the player is serving as the division Commander's Deputy, Chief of Staff, or G3, he should have the experience of a Brigadier General or Colonel.

Operational games of high significance can draw upon active-duty personnel at the command levels involved. Analytical games that are, necessarily, of longer duration must draw upon sources other than active-duty officers of the most senior rank. No better source of such players has been found than retired officers of the desired ranks. RAC and RAND have been particularly successful in obtaining senior retired officers for game participation.

What has been said about the Commanders applies equally to the Game Director, the Controllers, the Staff Officers who augment the Commanders, and the Control Staff (in accordance with the game scope and nature of forces engaged). In RAC's THEATERSPIEL, for example, as many as twenty retired senior officers have participated at the same time. The majority of the players were in the general officer category.

Human-decision and human-assessment games lose some credibility for lack of measures of suitability, applicability, reliability, and validity. Further, human-decision games can rarely be re-played. Urgent, new problems usually displace intended or hoped-for replications. Computer-assisted games put the replication goal a long step closer by reducing the required playing time. Although reruns of human-decision games may be worthwhile for other reasons, from a scientific viewpoint the results of such replications are not comparable and do not yield a valid mathematical or statistical measure of reliability.

In every human-decision game the participants learn from experience. They cannot wipe this experience from their minds; therefore each rerun of a human-decision game is approached with knowledge not possessed by the participants at the time of the first run. Reruns of human-decision games, however, are useful to indicate a confirming consistency of results or lack of such consistency, and such results are reassuring to the decision maker who must evaluate the results with, or without, additional substantive evidence. In general, the reliability, validity, and applicability of the results of human-decision games are more a matter of judgment than of measurement. But this is true of many of the most important matters on which high-level decision makers must act.

Any human decision can go awry. Experience and expertise reduce the frequency of poor decisions, but some will occur despite all checks and balances that can be and are used. First, a game staff with *in-depth* expertise serves as a balance to catch poor decisions before they are injected as orders and played as actions in the game. Control and, if need be, the Game Director can and do screen Commanders' decisions before they take effect. Beyond these checks and balances is a Review Board of Senior Officers who follow each phase of the game and evaluate the play, the assessment of moves, the results and findings. Emphasis is placed on assuring militarily sound human decisions in the games. A Senior Review Board has functioned for several years to review game plans, game operations, and game results of the RAC TACSPIEL and THEATERSPIEL games. The Board consists of four four-star general officers, each of whom has had significant experience in recent and major wars. See Fig. 10-1.

In the realm of human-decision games, no one can assume that the best decision will be made, but with employment of the measures outlined considerable confidence can be placed in the general level of decisions. War itself can do no better.

Some fears may be expressed that the war game is played from obsolete experience or with less initiative, ingenuity, or resourcefulness than desired. The selection of personnel to participate in a particular game, augmented by timely retirement from gaming organization, can go far to assure that only those players who are able to render a suitable performance, in fact, do participate.

FIG. 10-1. Senior Review Board, Military Gaming Department. *Left to right:* Gen. Charles D. Palmer, USA (Ret.), Chairman; Gen. James E. Moore, USA (Ret.); Gen. Thomas T. Handy, USA (Ret.); and Gen. Carter B. Magruder, USA (Ret.)

## MANEUVERS VERSUS GAMES—WHAT GAMES CANNOT DO

Games have to be based on data. Maneuvers and field exercises can be used and indeed sometimes are necessary to generate data that are not available. Military operations under Arctic conditions are an example.[1] Some field-experience data exist but are not always applicable to a new set of circumstances. Some data may be drawn from military maneuvers and training exercises conducted by the Norwegian, Swedish, and Finnish forces. Arctic and Antarctic expedition reports and explorers are a further source of certain kinds of data.

The Russo-Finnish War of 1939 yielded useful information, particularly in the use of ski troops for purposes of harassment, sabotage, interdiction, and scorched-earth policy to deny the enemy shelter, food supplies, and other such indigenous support resources, scant though they may be. The net purpose and effect are to increase the

---

[1] Carl Mydans, "Cold Siege on the Tundra," special report, *Life*, Feb. 28, 1964, p. 21. This reference states that the isotherm (a line marking the edge) of the polar zone, where the temperature never rises higher than 50°F, comprises one-fifth of the earth's land mass.

enemy's logistics load and degrade his operational capability by prolonged exposure to the common enemy of unrelenting cold and an inhospitable natural environment. The U.S.S.R., having a vast expanse of Arctic terrain, has paid particular attention to the problems of military operations in such areas. Combined U.S.-Canadian exercises and maneuvers have been conducted, usually out of Fort Greeley, Alaska, since 1949. United States troops stationed in Alaska and Canadian Army troops assigned to duty in the northern provinces are used in these cold-weather investigations. The natives, who prefer to be called Americans rather than Eskimos, participate in the exercises as members of the Alaskan National Guard. Indigenous personnel provide important local know-how regarding Arctic living and, in turn, have profited. Scientific studies are conducted by the Arctic Institute, the U.S. Army, and the Office of Naval Research Station at Point Barrow, Alaska.

Examples of immediate benefits accruing from the cold-weather studies and exercises include the Quartermaster Corps' development of thermal boots, to replace Eskimo mukluks, and Arctic clothing that displaced the winter furs of the Eskimos. Fiberglass sleds, new-type snowshoes, and five-man tents are innovations replacing native items. In medicine the Medical Corps has demonstrated that the quick-heat method for treatment of frostbite is superior to the traditional method of rubbing the affected area with snow.

The Harvard University Fatigue Laboratory, for example, has conducted experiments to measure the exercise or energy output that can be expected from a man under varying temperature conditions. A game cannot produce such information. But this information is necessary as a gaming input and can be obtained through small-scale field exercises and maneuvers conducted under the realities of real-world conditions and under appropriate scientific control. Exercises yield input data for gaming which, in turn, can investigate simulations of much more extensive operations under like conditions. Polar Siege, conducted in the winter of 1963-1964 from Fort Greeley, Alaska, with about 10,000 U.S. troops and some Canadians participating, was one such field exercise.

Data on ground-troop detection by aircraft reconnaissance illustrate one sort of problem on which actual field data are required. In Polar Siege it was known that one team was working in a particular area, but it took six long helicopter flights over that area and careful scanning of what seemed to be every bush and hillock before the ground team was spotted. The ground teams also needed to learn the relative effectiveness of various evasive and camouflage techniques.

## THE BIG LIMITATIONS

The major limitations on analytical gaming and simulation in the current era (in the opinion of this author) appear to be four massive requirements: data, time, staff, and computer support. These limitations add up to heavy costs. Even if conditions such as national security override monetary considerations, the basic requirements remain.

Computer Support

Modern analytical gaming is impractical without support from rather sophisticated computers. This requirement tends to limit analytical games and simulations to those organizations which can arrange for computer support. In some cases computer services for game operation have been provided from remote centers. The Naval War College at Newport, Rhode Island, using its NEWS gaming installation, has supported fleet war games conducted from the Atlantic Coast to Hawaii.[1] In part as a feasibility study, a training game at the Army War College, Carlisle Barracks, Pennsylvania, was supported by the computer facility of the Research Analysis Corporation at Gaithersburg, Maryland.[2] At present this means of providing computer support appears to be more commonly employed in training and operational games than in analytical studies. With fully automated simulations computer support from remote centers have been used effectively by time-sharing techniques and by scheduling computer runs during off-peak periods or night shifts. The spread of commercial enterprises to provide computer services, much as telephone service is available, will facilitate wider use of computer-dependent games and simulations.

Impact of Limitations

For analytical purposes, operational answers often are needed before execution of the gaming process. The solution most frequently used in the past has been one of the following:

1. Use of a short-cut gaming procedure that yields only rough estimates

---

[1] McHugh, *Fundamentals of War Gaming*, 2d ed., pp. 5-47 to 5-50.
[2] A study conducted by a team of RAC analysts and U.S. Army officers of the Signal Corps and the Army War College in 1962.

2. The omission of gaming, with the substitution of expert judgment to supplement certain analytical techniques

The great need and the challenge of the future are to find means of improving the validity and of speeding up the analytical gaming process without sacrificing values and insights offered by, and unique to, gaming.

An ever-present limit of gaming and simulation, as with most analytic techniques in the field of science, lies in the state of the art. Research and innovation will continue to push back the frontiers in all fields. As improved technology is developed in gaming and in computer techniques, the gaming process will improve. Similarly, the state of the art in man's understanding of man, of warfare, and of other activities of man also may be expected to advance. When achieved, these hoped-for developments can be applied to reduce the amplitude of present limitations and improve the end validity of game results.

An Overall Evaluation

With the apparent limitations, how can any confidence be placed in the data generated in war games?

First, precise results as a set of outcomes cannot be determined by a war game. The number and nature of personnel casualties that individual units will sustain; the number of tanks, aircraft, and vehicles put out of operation; the objectives achieved within time periods planned; the amount of ammunition and supplies used and lost in a given period; and similar outcomes cannot be predicted with accuracy. These data obtained from the play of a game cannot be regarded as a true picture of what would occur in an actual battle.

Second, chance plays a big part. Results of particular actions may occur at almost any point and over a very broad scale. The outcome of military operations, like the turn of a roulette wheel, the dealing of hands in a card game, or the throw of dice, will turn up different results in different plays.

Third, much data put into a game are based upon military judgments, conjecture, and sheer guess. Data obtained or verified by test, experiment, or analytical study of records account for only a small part of necessary game input data. The number of gallons of fuel needed by a tank to cover a particular distance and the rations required by personnel are known. But how long will an infantry platoon retain its fighting effectiveness under heavy, sustained combat? When should a combat unit be replaced with a fresh unit? And under what circumstances should unit replacement be made? The answers to these questions now are matters of military judgment in which commanders differ. Military history and experience have contributed some guidelines, but the effects are variable and, at this time, cannot be established with the accuracy attending the determination of food and fuel requirements.

In the fluidity of the battlefield how much ammunition will be consumed? How much lost? How much abandoned? These questions are even less amenable to military judgment—these are creatures of the elements of chance—but not entirely beyond calculation in the step-by-step development of a game.

Fourth, war games often are used to evaluate the effects of projected weapons and weapons systems—systems under consideration but not yet produced. In these cases certain weapon characteristics, performance, and effects have to be postulated. Sometimes the characteristics may be derived from the QMDO[1] or QMR[2] for a weapon, such as a VSTOL[3] aircraft, before design, that indicate the lift capacity, forward air speed, overall weight and dimension limits, flight range, special gear, munitions and personnel and cargo space, and all such necessary details pertinent to the weapon. The war gamer assumes the specifications to have been achieved and the item to be in operational use and then plays the aircraft as part of the force employed in the game.

Fifth, it is asserted that war games can be used for (1) training; (2) assessing plans; (3) a basis for analysis of military problems, i.e., serving to establish common understanding between the military man and the analyst; (4) simulation of command and decision processes; (5) formulation of insights and intuition; (6) detection of flaws in assumptions; (7) an environment for innovation; and (8) as an aid in dispute settlement. There probably are other uses.

Do games give valid answers? As in all science, the question is dependent upon whether the model agrees with nature. There are three cases:[4]

1. If the model is deterministic, the validity of its functions may be fully determined.

---

[1] Qualitative Materiel Development Objective (QMDO).
[2] Qualitative Materiel Requirement (QMR).
[3] Vertical and Short Take-off Landing (VSTOL).
[4] Adapted from comments made by Prof. B. O. Koopman of Columbia University in an address at the Centre Internationale de Recherche Operationnelle, Paris, 1961.

2. If the model is not deterministic but the probability distributions of the various events are known, the answer may be valid in the actuarial sense or within statistical limits.
3. If the probability distributions are unknown, two choices are open:
    *a.* Admission that the validity of the answer has not been established.
    *b.* Intuition, judgment, and conviction as developed by the game can be accepted as limited evaluative assessments.

# Gaming in Prospect: A Look to the Future

Eli Whitney, the Yankee schoolmaster turned inventor, startled the world in 1798 when he took a contract to manufacture 10,000 stands of arms for the American militia. Weapons of that time were handcrafted by skilled artisans, each by an individual gunsmith. The Springfield Armory, one of the most experienced of military firearms factories, had produced only 245 muskets in the two preceding years. What folly was this fantastic obligation taken by Whitney!

The world was even more astonished when Whitney made good on his contract. By applying the principles of *machine production, standardization,* and *interchangeability* of parts, Eli Whitney became the father of mass production in the modern concept.

Analytical games, too, have been handcrafted affairs. Game models have been custom-built, designed and constructed by game teams, each team working independently of other teams. Although the skilled gamer can carry his trade with him from one employer to another, there has been no *standardization* and little model *interchangeability* between games. Differences in computers and machine language often prohibit the transfer of mathematical models, submodels, routines and subroutines, programs, and simulations from one programming facility to another. Lack of game computer-program interchangeability is a deterrent to the rapid computer programming of certain games.

## AUTOMATION

Automation and mass-production techniques have only begun to be applied to gaming. Computer simulation, conducted with or without human intervention, is the form most applications have taken (to 1969). The next step is to adapt to gaming newer techniques that parallel standardization and interchangeability of parts, yet leave the critical, decision-making aspects to the responsible game participants. Command and control of this type are accomplished, in part, in semiautomatic or computer-assisted games. The mass production and standardization of models and submodels and computer-produced models and programs have not been seriously attempted or achieved. The state of the art of gaming and computer technology verges on these frontiers. This virgin field for research and development beckons—even cries out for exploitation.

The scarcity of gaming and computer-model building skills suggests the diversion of some skilled practitioners to the type of efforts suggested. In future analytical studies of significant problems such activity may be expected to return immeasurable payoffs. Mass production of gaming applications may be brought to practical use, and dependence upon custom-crafted games for each analytical need can then be eliminated.[1]

Data Banks

The standardization of war-game input data is closely related to standardization of gaming models. Using war games as examples, let us recognize at the start that different purposes, different levels of resolution and aggregation, different geographic areas and terrain types, and different situations—including differences in forces and weapons mixture—all require different war-game input data. The requirements dictate the amount of needed input data that must be available. An analogous situation is the large inventory of parts, and great storage capacity to stock these parts, required by the large number of different makes and models of automobiles. Another example is that a large number of books in a library requires more shelf space and catalog entries than does a smaller number of books.

---
[1] It is not suggested that standardization of models and automatic computer techniques will substitute for or replace the skilled judgments and decision-making function of the experienced and resourceful game participant. Instead it may extend the use of such talent by relegating to the computer a larger and larger share of the routine chores now performed by game participants.

The modern electronic computer provides almost unlimited capacity for the storage and retrieval of data or information.[1] Massive sets of data may be recorded and stored at computer centers to be available, on call, as needed. Vast quantities of information may be drawn from business, industrial, governmental, and military operations; logistics records; laboratory, field, and proving-ground tests; plans, maps, weather, intelligence, history, and other sources. Such libraries of stored and available data do exist and are referred to as data banks.

Potential Data-bank Uses

Ultimately, through uniformity of categorization and classification, the data in one data bank may be called upon by a user in another area. When a system of data banks, with branch installations and retrieval systems, is developed, great amounts of time can be saved at any one gaming center that now must generate such data. Through use of remote-control electronic linkups, one may be able to transfer data more easily than books may be shuttled between libraries.

## IMPROVED COMPUTER MODELS

In complex, analytical games the pre-game preparation phase and the post-game analytical phase are extremely time-consuming. These time requirements can be reduced through use of improved computer models. Further, a substantial saving can be made in time used to prepare daily input data.

A third type of saving is model design that facilitates side analysis. This means the model is designed so that computation and storage of data for each submodel and for each interval or cycle of play result in a form that can be readily extracted and printed as a computer operation. The computer can be instructed to find and print particular detailed or aggregated data. If, as a side activity of analysis, for example, one desires to single out the extent of employment of helicopters for troop transport in a particular campaign, the data

---

[1] Computers have finite and limited direct storage capacities (a memory) in components called the *core, drum,* or *disk file.* All have means of "reading into" such components practically limitless data that have been stored on magnetic tapes and/or punch cards. There need be no serious restriction on the amount of data that can be stored in a library of tapes or decks of punch cards.

relating to all such uses can be extracted from the game records *by the computer.*

A fourth type of saving in player time can be accomplished by using the computer to display significant elements and summaries on a display screen. In a war game, for example, troop dispositions, movements, deployments, and the unit conditions (relative combat effectiveness) may be displayed and recorded. The display screen shortcuts the need for manual placement of unit designators, i.e., stickers or game pieces on a map or terrain board. Automatic display screens are in use at a number of gaming installations. One example is the Naval War College NEWS facility. Although the display technique is used extensively for training purposes, the technique can be used to greater extent in analytical games.

A fifth type of model improvement, also within the capabilities of computer technology, in war games is the automatic computer processing of terrain data direct from photogrammetric tapes. Photogrammetry has been used to map much of the earth's surface and is an excellent process for the mapping of remote terrain. Essentially it consists of taking a sort of stereoscopic picture of terrain from the air. The picture is such that it permits identification of terrain features and the reproduction of these features on maps. Only one step beyond is the translation of graphically recorded data to automatic computer representation and reproduction of terrain features. Military maps having not more than 1 foot of error in every 10,000 feet can be produced from aerial photographs in from 1 to 24 hours, through use of the Universal Automatic Compilation System (UNAMACE). The accuracy and speed afforded by this system (operational in 1967) were not possible hitherto in hand-drawn maps.

## SIMPLIFIED TERRAIN ANALYSES

Terrain analyses range from those made from simple representations of terrain to those made from complicated representations that include such items as trafficability of wheeled and tracked vehicles under different weather conditions, e.g., rain, snow, frozen ground, etc. Terrain analyses related to military problems have been undertaken by many nations. In the United States the U.S. Geological Survey and the U.S. Army Map Service conduct extensive research of this type. The Research Analysis Corporation has conducted research for the purpose of adapting recognized techniques of terrain analysis to gaming.

Specific research on the adaptation of terrain features to computer inputs has been conducted for the TACSPIEL game.[1] Terrain features have been codified in terms of effect on military operations. The functional or terrain "utility system" of terrain analysis simplifies the process but remains a major undertaking in the preparatory stage of war gaming. To whatever extent automation can substitute for human decisions and recording, major gains in speedup of game preparation and in reduction of drudgery will result. The automatic computer processing of terrain data direct from photogrammetric tapes would render valuable assistance.

## SIMPLIFICATION OF DAILY INPUT DATA

One approach to the simplification of daily input data is a concept developed for the THEATERSPIEL game. THEATERSPIEL provides for three different levels of computer input data:

1. Doctrine
2. The Commander's plan
3. Individual missions

First, the doctrine submodel provides for automatic employment of units according to established doctrine (standard employment under varying conditions). The automatic play of a doctrine submodel by the computer is a simulation, rather than a controlled human-decision game.

The second level of input, referred to as the Commander's plan, provides for the human-decision prerogative of the Commander. The computer submodel exercises routine procedures that normally would be carried out by subordinate Commanders. This input, while leaving the Commander in control of the situation and directing operations, does not consume his time or clutter computer operations with a great mass of specific orders to individual units. These orders are carried out in a standardized play. The Commander's plan usually extends over a series of intervals. Instead of specifying particular moves and objectives for each unit, in each interval or cycle of play, the plan might call for a force to carry through to a specific objective or complete a particular mission, such as the capture of a bridge across a river. A classic example is the exploitation of the lucky

---

[1] Under the direction of Richard G. Williams, RAC Military Gaming Department.

capture of the undestroyed bridge across the Rhine at Remagen during the final stages of World War II.[1] The submodel automatically keeps the units involved pushing ahead against such losses and resistance as may be encountered, until the mission is completed, or until the force is defeated or otherwise prevented from achieving its mission, at least within the prescribed time.

The third level of input is data concerning the individual specific mission the Commander may assign to selected units. These third-level inputs give the Commander full opportunity to override any submodel procedures of the first two data-input levels specified by doctrine or original plan. The Commander can issue specific mission orders to any or all units at each interval of play. This third level guarantees that the game can and will be conducted as a human-decision game under full control of the Commander and the Controllers. The time-consuming, unproductive standard procedures are completed through automatic computer submodel play.

Individual missions can be introduced at any interval or cycle of play according to the judgment of the Commander. Although the Controller has the authority to question, modify, or, rarely, even to override a Commander's individual-mission orders on the basis of game rules or superior knowledge of the situation, such actions occur infrequently in practice. The Commander is permitted to make his decisions, whether they be good or poor. However, if the Commander's decision violates the rules of the game or overlooks a game condition, such as ordering more aircraft in an operation than he has operable after the previous action, it is the Controller's responsibility to intercede.

When the foregoing three levels of computer input are applied to a particular set of game data, it might be asserted that the initiative, resourcefulness, and responsiveness enjoyed by the Commander in a human-decision game are reduced. Instead, these features permit even closer approximation of the conditions found in the realities of actual warfare. In any human-decision game there are many features that the Commander will choose to have "go by the book." These are

---

[1] Through some "snafu" the Germans failed to blow the bridge as expected. When Brig. Gen. William M. Hoge's Combat Command B captured the bridge, there were anxious hours of uncertainty whether this lucky break could be supported sufficiently by forces required from higher commands (General Leonard's 9th Armored Division, General Bradley's Army Group, and General Eisenhower's Supreme Headquarters) to enable a solid bridgehead to be established across the Rhine. Source: Maj. Gen. James G. Christiansen, then at 9th Armored Headquarters.

the operations relegated to the computer to play out in terms of the doctrine submodel.

Through use of this system in THEATERSPIEL,[1] the standard daily input instructions to the computer would be reduced from several thousand to those instructions contained on 100 to 200 punch cards. The Commander is allowed as much freedom, initiative, and resourcefulness as he had with the original thousand or more computer inputs per cycle of play. The system results in a speeded-up computer process. The time loss and drudgery imposed on human players are reduced without degrading the human direction of the game.

## PROGRAMMING IMPROVEMENTS

The three-level division of input data in computer-assisted human-decision games is an application of the well-established modular programming technique. Modular programming (defined as the subdivision of application programs to relatively independent functional modules, each to serve a single, recurring purpose)[2] facilitates problem analysis and documentation and is an aid in data-retrieval-system use.

Improved computer languages facilitate programming and communication between the human operator, or analyst, and the computer. Fortran is one such language. In many computer languages shortcuts are provided in the form of subroutines. Subroutines may be established for any repetitive operation for which the computer is suited and can be called to action by a simple command to the computer. An elementary example is the computation of the square root of a given number, or the extraction of a random number from a table of random numbers.

### Models Programmed for Data Extraction and Analysis

The designers of the STAG game LEGION and the JCS strategic game TEMPER, when developing these models, recognized the advantages of subroutine shortcuts. In their present form, however, these games are too complicated to employ such shortcuts. LEGION and TEMPER require massive computer operations to generate a limited amount of useful data for specific analysis. The failure to use

---

[1] Richard G. Williams and Jarvis S. Seely engaged in these studies at RAC.
[2] Roger A. MacGowan, "Modular Programming," *Data Processing*, Oct. 1964.

short-cut subroutines in these games might be analogous to searching through all listed telephone numbers in a large metropolitan-area telephone directory to count the number of subscribers named Smith. As ridiculous as this analogy seems, it has relevance to the extraction of data for a limited specific purpose from the great mass of data generated in an analytical game. The dissimilarity of this analogy between a war game and the telephone-subscriber search is that a relatively complete war game must be run to establish component-data validity.

Game technology needs to be advanced so that models can be operated as meaningful components without the requirement to operate all other components of the game.[1] No doubt progress of this type will be made during the next several years.

As society develops additional numbers of complex, specialized elements and as civilization becomes more interdependent, improved techniques for studying the totality of situations, i.e., syntheses, become urgent. Gaming is one method or technique by which the interactions among competing and cooperating forces and their impacts may be studied. But it is only one type of synthesizing methodology. Other techniques and advanced mathematical and statistical processes are being developed to serve analytical purposes.

International politics, as well as intranational politics, is in urgent need of improved tools for evaluation of power factors and techniques. Means to assess courses of action to advance national interests and human well-being need development. Political gaming is one such effort to create observable situations where gaming techniques may be applied and the results used to study the field of international relations. So far there is an area for gaming in *national* problems, largely untouched, even in state, metropolitan, and community situations.

## NEW TECHNOLOGY

Just as aircraft added a new dimension to warfare, so may space, oceanography, and other new technology open new developments in

---

[1] This refers to the detailed subdivisions of the model, which may be processed in the computer either in aggregated or detailed form and printed out in the specific form needed for the analysis. This process is a matter of information retrieval. It does not suggest that a game would be played with only one arm of the engaged force, such as the artillery alone without infantry, armor, and air actions as would normally be involved. A whole combat operation is played, but only certain types of data may be selected for analysis.

a wide band of applications. Although future developments in these environments are obscure, orbiting satellites have had considerable impact on detection, data collection, communications, and military planning and technology. Satellites have increased the reliability and speed of detection devices. The free world has been able to follow the progress, to estimate the stage of development of technology, and to forecast capabilities heretofore shrouded in tight secrecy. Aerial surveillance proved effective in the detection and identification of missile sites, in the limited Cuban and North Vietnamese areas. Satellites have stepped up this capability on a worldwide basis and enabled the free world to follow the development of Red China's nuclear-weapons production facilities. Worldwide weather observation and instantaneous global communication are products of space technology. Long-range weather forecasting appears within reach. Increased speed and accuracy in mapping and geodetic observation are obvious. Navigation for air and naval forces will be greatly improved. Similarly, naval forces of the type that attacked Pearl Harbor cannot hope to escape detection, and individual raiding forces or men-of-war no longer will be able to elude detection in the vastnesses of the world's oceans.

## NEW FIELDS FOR GAMING

The technique of gaming has been applied in the fields of commerce, finance, and industry with business and management games. The technique has been used for educational purposes in university classes in business, political science, and international relations. More recently experiments in the use of games have been conducted in social studies classes in secondary schools. These experiences demonstrate a utility for the technique of gaming beyond military and strategic uses and foretell the probable extension of the technique into other new fields.[1]

One example is the Metro gaming-simulation project conducted as a joint study by Michigan State University (MSU) and the University of Michigan in an effort to work "toward a new science of Urban Plan-

---

[1] For many years various forms of simulated war or war games have been pursued as hobbies or recreation. Such applications, although extensive, are not included in this book. Reference here is made to applications with serious purposes in the worlds of scholarship and reality.

ning."¹ Related games and simulations are Metropolis and the Cornell Land Use Game. Professor Duke of MSU says, "On the horizon, a considerable potential exists for utilizing an exercise of this type in conjunction with sophisticated urban data systems," and "we envision a time when the real world decision-makers . . . will assemble in teams . . . to consider . . . and . . . select some mutually agreeable course of action. When this is achieved, they will have, in effect, an 'instant plan.'"²

Another application of gaming is in the field of vocational guidance and occupational choice. James Coleman of The Johns Hopkins University, who designed the Life Career Game, has tried this application in the Baltimore public schools. A whole series of games in different fields is used in the San Diego public schools in connection with Project Simile of the Western Behavioral Sciences Institute, La Jolla, California. A Viet Nam War Game was played by Los Angeles high school seniors under the guidance of the University of Southern California. In combination with a kind of automatic, programmed instruction, grade school children in Westchester County, New York, play the Sumerian Game individually against a computer.³

Strategic matters, arms control, and disarmament have been gamed by means of the quick-gaming technique. Other applications of gaming may be related to the adjudication of international disputes; the monitoring or policing of truces and armistices; protective occupations such as in the Congo and the Dominican Republic; treaty negotiations and renegotiations such as the Kashmir or the Panama Canal; economic unions and common markets such as the Inner Six and the Outer Seven, and Britain in the European Common Market; aid and development programs for underdeveloped countries; and regional

---

[1] This study is directed by Dr. Duane L. Gibson, Director, Institute for Community Development at MSU, and coordinated by Dr. Lyle E. Craine of the School of Natural Resources, University of Michigan, with the cooperation of William C. Roman, Executive Director of the Tri-county Planning Commission, Lansing, Mich. The study is supported by a grant from the Department of Housing and Urban Development.
[2] See Prof. Richard D. Duke, *Gaming Urban Systems*, Michigan State University Institute for Community Development, Report 30, 1965–1966, and *Gaming-Simulation in Urban Research*, also by Professor Duke, same source.
[3] For more details, see "Games Students Play," *Time*, June 3, 1966, p. 51, and "The Proper Study of Mankind," *Newsweek*, Aug. 15, 1966, pp. 80–82.
[4] Members of the European Economic Community: The Inner Six are France, West Germany, Belgium, Luxembourg, Italy, and the Netherlands. Nations referred to as the European Free Trade Association, or the Outer Seven, are Austria, Denmark, Great Britain, Norway, Portugal, Sweden, and Switzerland.

development programs, such as the Indus River or the Mekong River programs. These are only some fields to which gaming techniques can, and probably will, be applied as soon as resourceful thinkers and scholars turn efforts thereto. Much preliminary, exploratory, and developmental study is under way in the form of crisis gaming and internation simulations. These studies can be expected to yield insights and experience that will carry to other fields. Other unnamed fields will occur to the imaginative reader. For example, riot and crime control, international trade agreements, strikes and labor disputes, sports, gambling strategies, transportation, and traffic planning may be fields in which gaming will prove useful.

## A NEW ERA IN WARFARE

The half century since World War I represents a new era in warfare. It is as though World War I was fought as a final climactic to old-style warfare. New weapons, equipment, concepts, and tactics made entry on stage. New developments such as tanks, aircraft, aircraft carriers, and long-range submarines were introduced in World War I but not fully exploited until World War II. All innovations, be they great or small, furnish grist for military analysts, historians, and old soldiers. But something more than argument and learned tomes is required to devise and assess the place of updated accoutrements in the art of warfare.

Real innovations may change almost everything about the art of warfare. Each new major innovation in warfare, such as the long bow, gunpowder, tanks, air power, or atomic weapons, calls for a new examination and evaluation of military tactics, strategy, force structure, equipment, and doctrine. Such evaluations require analytical studies made in the context of the interacting dynamics of warfare. Analytical war gaming is a technique developed to meet the expanded requirements of military studies.

Guerrilla warfare may be the wave of the immediate future, and it could force a complete reevaluation of all the current and older concepts of the military art. And what lies beyond guerrilla warfare?

In many other fields, thresholds are being reached where the precepts and practices of the past need to be reexamined in order to adapt to new developments. Simulation and gaming may provide a tool for use in such reassessments.

## NEW CHALLENGES

Coincident with the aftermath of World War I was the drive to restrict or outlaw war. The League of Nations,[1] in existence from January 10, 1920 to January 10, 1946, and the Limitation of Armaments Conference[2] were optimistic and visionary efforts. Although some steps were taken toward the goal of lasting peace, persisting national ambitions doomed these efforts to cataclysmic failure in World War II. Nevertheless, among the peoples of the world an increasing desire grew to find a means to prevent war, and from such hopes have grown attempts at peace gaming. Fortunately, this groundswell has continued, and aspirations toward peace have strengthened the desire, but the international climate remains unfavorable and the means of eliminating war continue to be elusive.

## PEACE GAMING

Peace gaming appears to be the antithesis of war gaming. In actuality the techniques are similar; only the objectives differ. In one case the goal is to defeat the enemy by military operations in actual warfare; in the other, the goal is to avoid warfare by political maneuvers. In one case military capabilities are employed; in the other, military capabilities are assessed as one of several national power factors in an equation that is to be kept in a state of balance. Peace gaming looks toward the health of nations in the international environment.

E. W. Paxson expressed a situation analogous to that above: "Like Chinese medicine, the military profession is most successful if the client does not become a patient." He adds that war gaming can assist in designing a balanced deterrent posture, and that mistakes can be made "on paper," i.e., in the game, instead of in real life. Paxson also says that, since the military posture of our country is no secret, the enemy can conduct his own war games "based on the inherent capa-

---

[1] The League of Nations signatories agreed to outlaw poison gas (an agreement observed) and submarine attacks on unarmed ships (an agreement not observed). When it was disbanded, physical assets of the League were given to the then new United Nations.
[2] The Limitation of Armaments Conference (1921–1922) succeeded in reducing and restraining the U.S. and British naval power more significantly than it did the Japanese.

bilities of that posture—not its intentions . . ." and, it is hoped (by all), detect no soft spots inviting attack.[1] This was not the case in Admiral Yamamoto's war gaming of Pearl Harbor.

The foregoing cases focus on the maintenance of peace by deterrence of war. Although it is not questioned that deterrence is a realistic approach to the prevention of armed conflict, there are many who believe peace should be sought by constructive efforts as an objective in itself, rather than as a by-product of the balance of ever-increasing military capabilities held in readiness to bring about a holocaust.

A number of investigators are applying gaming and other analytical techniques to the direct study of peace. Peace gaming, broadly considered, is political or strategic gaming although somewhat different in nature. Lincoln Bloomfield, active in gaming at MIT, is probing a direct approach to peace gaming. Professor Bloomfield reported his thinking on this subject and the role of the United Nations in an article entitled "Peacekeeping and Peacemaking."[2] Bloomfield emphasized the importance of regional organizations and arrangements to mitigate differences before they reach such stages of acuteness as are referred to the UN. Quincy Wright, and others,[3] have devoted much study to analytical approaches to peace in efforts to understand thoroughly the causes of war. Such information is needed, and more studies of this type are required to yield the information from which computer input data for peace-gaming models may be developed.

Another type of effort is represented by the work of Howard and Harriet Kurtz, who attempt to bring together diverse viewpoints: contributions from leaders in technology, religion, politics, communications, computer science, engineering, and the social sciences. The Kurtzes' approach emphasizes the use of instrumentation and satellite technology to provide surveillance and early detection. A quick-response communication and reaction system, reinforced by strategic power administered under world law and international ethics, is envisioned. Once the necessary factors and data are built into a functional system, gaming models could be constructed and employed to explore interactions. Various options then might be exercised in the hope of dealing with issues promptly enough to prevent the growth of crises, they believe.

---

[1] E. W. Paxson, *War Gaming*, p. 29, RAND Memorandum RM-3489 PR, 1963.
[2] Lincoln Bloomfield, *Foreign Affairs*, July 1966, pp. 671–682.
[3] See Selected Bibliography: Quincy Wright, Lewis F. Richardson, John C. Harsanyi.
[4] Organizers of War Control Planners, Inc., Chappaqua, N.Y., *War Safety Control Report*, 1964.

Peace Research Institutes have been established in several countries, sometimes but not always in connection with a university. Such centers are functioning in England, Scotland, Canada, Sweden, and the United States, and employ gaming and simulation techniques in their studies.[1] The INS type of simulation is the most common. Laulicht, reporting on work at Leeds University and the Peace Research Centre at Lancaster, England, comments that, although the simulation method is but one approach to acquiring better knowledge, "it is the only method developed within the last few decades which holds real promise as an addition to traditional analytical techniques."[2] These approaches presage a type of gaming built less on the advancing technology of war and more on the search for a technology of peace.

## THE SEARCH FOR AN END TO WAR

John C. Harsanyi, writing in *World Politics* in 1962, reviews the works of Lewis F. Richardson,[3] as edited by Quincy Wright, et al. Richardson's search included the construction of mathematical models for the genesis of war and statistical and psychological analyses. He hoped that a fuller understanding of the situations and conditions that lead to war would reveal the key to the settlement of quarrels among nations by peaceful means.

The model he sought to build, if fully developed, might be applied in gaming situations, either with the participation of or under the observation of potential antagonists, each having the objective of assessing probable costs (against the desired objectives) and the probabilities, or the improbabilities, of achieving the objectives by means of war. Carrying this visionary application of gaming to its ultimate rendezvous with destiny and service to mankind, war games might become the underwriters of peace, demonstrating to potential antagonists the folly of war as a means of achieving political goals.

Such application of war gaming might bring into reality, on a global scale, the process demonstrated by the legendary Chinese wise man

---

[1] See works of Guetzkow et al., Laulicht, et al., Scott and the Lucases, Benson, Bloomfield, and Robinson and associates for examples.
[2] Laulicht, "The Vietnam War Game," *New Society*, Jan. 17, 1966.
[3] English physicist and Quaker who applied himself to the study of the genesis of war, the psychological origins, statistical analyses of the causes of war, and a search for means of averting war.

many centuries ago. By a sort of map maneuver scratched in the sand, this ancient sage showed two contending princes the folly of warlike intents toward each other and thus staved off impending war.

It may be too visionary, but one might dare to conceive of a body of ambassadors, augmented by other high-level military and economic representatives of the nations of the world, who could employ gaming techniques to study national problems and interests in a realistic international setting. By such means the possible, or probable, consequences of alternative actions might be assessed. Add to this vision the hope that mankind, and the nations it divides into, may have the wisdom and compassion to favor constructive solutions over the disaster of war and the folly of destruction as a solution to national frustrations and appeasement of national ambitions. The United Nations, or a similar organization, could provide a center for such activity. This organization might facilitate the long step beyond a forum for dialogue among nations to a foundry in which peace and a new golden age for man may be forged.

Man stands on the threshold of such capability. The paradox of power conflict, with its historical technique of war, may provide an instrument of warfare—war gaming—as a contributing means to fulfill the words of prophecy,

> . . . *they shall beat their swords into plowshares, and their spears into pruning hooks; nation shall not lift up sword against nation, neither shall they learn war any more.*
>
> Isaiah 2:4

# A Gaming Glossary

Definitions of simulation and gaming terms appear in the dictionaries of the military services, but an official war-game glossary has not been published. The definitions given in this list have been gathered from four primary sources. Preference has been given those terms accepted by recognized groups of gamers, either by a consensus, as in the case of the Ad Hoc Working Group, or by common usage of terms among gamers. It must be kept in mind that all details of the definitions are not fully accepted even by the members of the Working Group itself, and less so by war gamers in general in the four countries represented. Where a definition that conveys meaning to non-gamers has not been provided by the primary sources consulted, an attempt has been made to simplify the statement. The selection and evaluation of definitions, the simplification of definitions, and the formulation of new definitions that are used in this book are entirely the responsibility of the author. The definitions are keyed to the four sources thus:

(M) indicates definition given in official military service dictionaries and documents.

(N) indicates the *Glossary of War Gaming Terms*, 1961, as the source. Formulated at the U.S. Naval War College, Newport, Rhode Island, these terms evolved from the longest continuous group of naval war games. The list is reported by McHugh in *Fundamentals of War Gaming*, published by the U.S. Navy Department.

(Q) indicates the 1964 tentative draft of the Ad Hoc Working Group on Gaming, Quadripartite Conference on Military Operational Research (representing principally Army gamers in the operations research organizations of the United States, the United Kingdom, Canada, and Australia).

(A) indicates author-formulated; drawn from various sources, including personal experience.

One important term is *war game* defined in the April 23, 1965 issue of the *Dictionary of US Army Terms*, p. 438, published by the Department of the Army as Army Regulations AR 320-5. A second such term is *war gaming* defined in the January 1961 issue, p. 600, of the same publication, but omitted from the 1965 issue. Other official terms are included.

To date, the standard dictionaries of the English language are not very helpful. They give conflicting, incomplete, or specialized definitions that do not fit the various gaming situations. The dictionary definitions reflect the earlier confusion that existed among gamers, a confusion not fully resolved at this writing.

**aggregation.** "The process of considering men, weapons, equipments, events, etc., in a **war game** in groups (usually corresponding to military units or formations) rather than singly. The term often is applied to the process of developing rules for a particular game from rules designed for a game of higher resolution." (Q)

**analysis period or phase.** The third and final stage of a game. The data generated in the play period is collated, grouped, tabulated, and treated statistically to provide meaningful summarized findings, comparisons, and interpretations. Insights and other bonus effects or "fallout" results of the gaming experience are also consciously sought. The whole process is directed toward maximizing the achievement of results bearing on the objectives of the game and the lessons to be learned from it. The analysis period includes the reporting process in which the analyzed game data and findings are recorded (and published) for use by others concerned with the purpose and outcomes of the game. (A)

**analysis, post-game.** "Use of data generated during a game or series of games to derive conclusions about the problems to which play was directed." See *analysis, side*, which may be a part of this post-game analysis. (Q)

**analysis, side.** "Use of data generated during a game or series of games to explore in detail certain aspects not fully developed during the main flow of the game." (Q)

**analytical game.** "A game conducted for the purpose of deriving information to assist civilian officials and military officers in reaching decisions or to evaluate [tactical or strategic] plans, doctrine, concepts, weapons systems, force requirements and capabilities." (N)

**assessment.** "An evaluation (usually mathematical) of the results of a cycle or period of combat activity leading to a current description of a unit's status and location." (Q)

**assessor.** "A member of a war game control group who performs the detailed work of determining the outcome of game interactions, on the basis of a book of rules, and collects and collates all data and record sheets relevant to the game in play." (Q)

**board game or board maneuver.** "A war game played on a horizontal terrain or map surface representing the area of operations and the forces involved. Another name for a tactical war game." (N)

**break point.** "That critical level of combat capability, measured by some explicit function of game variables, below which the unit is considered unable to persist in its assigned mission." (Q)

**business game.** "A sequential decision-making exercise structured around a model of a business operation in which participants assume the role of managing the simulated operation." (Greenlaw, Herron, and Rawdon, *Business Simulation in Industrial and University Education*, pp. v and 5, Prentice-Hall, 1962.)

**chance device.** A device used to determine which of two or more conditions of the simulation of an event may or may not happen. Used in connection with events which occur in a known or assumed probability, such as weather conditions. Common chance devices include regular dice, special multiple (isohedron) dice, tables of random numbers, roulette wheel, random-number generator. (A)

**command post exercise (CPX).** "An exercise involving the commander, his staff, and communications within and between headquarters." (M)

**conflict situation.** "One in which two or more individuals, organizations, nations, or allies are competing for the same goal, or having opposing objectives." (N)

**control group or personnel.** "The person or groups of persons who monitor (controllers) and assess (assessors) the results of players' orders, and provide guidance in accordance with the objectives of the game." (Q)

**controller.** A member of the control group who monitors and evaluates the decisions and actions of the players and the interactions generated by the decisions and actions. The controller may determine what information to inject into the game at a given time, such as changes in political constraints, which units will clash in movements, and intelligence reports. In a complex game there may be a chief controller who coordinates the work of each of the subordinate controllers or the control staff, and of the individual controllers responsible for certain functions such as an operation, intelligence, etc. (A)

**critical-event method.** "A technique used in games wherein the sequence of events is not continuous. The game advances from event to event rather than by fixed intervals of time. Also referred to as Skip Gaming." (N)

**cycle.** "The normal sequence of events in a game, i.e., a **player** plans his moves; the player passes moves to control; control assesses the outcome of resulting actions and interactions; and control then passes relevant information and intelligence back to the player. A single full sequence of such events is called a game cycle." (Q)

**director.** "The individual responsible for the whole process involved in the analysis and report of the game." (N)

**display.** "The visual presentation of a game situation on a viewing surface such as map, game board, or screen. The situation is depicted by symbols, charts, overlays, lists, etc., the posting of which may be done manually or by automatic means." (Q)

**drainage.** The natural channels—or lack of channels—for the movement of water, which determine the direction and rate of flow or runoff, if any. (A)

**educational game.** "A game conducted to provide military officers and commanders with decision-making experience involved in command with staff support; and to familiarize them with the operations and problems involved." (N)

**excursion.** "A re-run of a play with some variation made in starting or subsequent conditions." (Q)

**game.** "A simulation of a situation of competition or conflict in which the opposing players decide which course(s) of action to follow on the basis of their knowledge about their own situation and intentions and their (usually incomplete) information about their opponents." (Q)

**game against nature.** "A probabilistic simulation including a player considered to be in conflict with a malevolent Nature responsible for all adverse chance events and bad luck." (Q)

**game board.** "A surface used to represent the area of operations on which a war game is being conducted. The board usually is marked with a grid, and may represent terrain features by the use of contour lines or a relief model (terrain model)." (Q)

**game, closed.** "A game in which the player has only such knowledge of his own and his opponent's situation as is transmitted to him from the game control group." Compare *game, open.* (Q)

**game, computer.** "A simulation of a competitive situation carried out completely on a computer in which the only human intervention is made by the **players** themselves issuing orders." Compare *game, manual* and *game, computer-assisted.* (Q)

**game, computer-assisted.** "A manual game in which a computer is used to provide assistance in recording data, book-keeping, assessment, and so on." Compare *game, computer.* (Q)

**game, conference.** "An open, controlled-play game in which the player is presented with a sequence of situations for which he must determine the best course(s) of action." (Q)

**game, critical-event.** "A game that is designed to advance from significant event to significant event rather than by fixed intervals of time." Compare *critical-event method.* (Q)

**game directive.** A statement of the purpose, objective(s), forces and weapons, locale, constraints and conditions under which a game is to be played. (Usually originates with the decision maker to whom the game results are to be delivered.) Compare *starting conditions.* (A)

**game, free.** "A game in which the results of interactions between opponents are determined subjectively by the control staff on the basis of experience and judgment." Compare *game, rigid.* (Q)

**game, free-play.** "A game in which the player is free to make any tactical decisions he desires consistent with his resources and the game objectives." (Q)

**game, hand-played.** See *game, manual.* (Q)

**game interval.** "The length of a **game cycle** measured in **game time**. It may be either constant or, as in a critical-event game, variable." (Q)

**game, man-machine.** See *game, computer-assisted.*

**game, manual (war).** "A game in which pins, pieces, or symbols representing the forces involved are moved by hand by the players and control personnel, and in which all decisions are man-made." Same as *game, hand-played.* (Q)

**game, Monte Carlo.** One wherein the results of chance events are determined by use of a chance device. Repeated game trials provide a distribution of possible outcomes. Same as *game, probabilistic* or *stochastic game.* (A)

**game, one-sided.** "A game in which one of the opposing forces is played by the controllers." Compare with "N" definition: one-sided game. (Q)

**game, open.** "A game in which the opposing players have full knowledge of each other's positions and actions." Compare *game, closed.* (Q)

**game, operational.** Same as *game.* Compare *operational gaming.* (Q)

**game periods.** The major subdivisions of the whole process of conducting or operating a game (assumes the game facilities, staff, models, and programs exist in operable readiness). See also *preparation period, play period,* and *analysis period.* (A)

**game, probabilistic.** "One wherein the results of chance events are determined by use of a chance device. Repeated game trials provide a distribution of possible outcomes." Same as a *stochastic* or *Monte Carlo game.* (N)

**game, rigid.** "A game in which the results of interactions between opponents are determined objectively by controllers, simulation equipment, or computers in accordance with predetermined rules, data, and procedures." Compare *game, free.* (Q)

**game security.** Game security relates to information that must be denied to either or both player sides prior to or during play. Such information may or may not be classified within the usual meaning of the word in military and intelligence circles. (A)

**game, set-piece.** A game in which one (or more players) is constrained to follow a particular sequence of moves, or a specific strategy, selected in advance. (A)

**game time.** The time represented in a [war] game, i.e., the time of day and date on which the game action is supposed to take place. Compare *real time.* (A)

**game, war.** A game representing a military conflict. It may apply to any one military force, or combination of forces, i.e., land, sea, air. See also *war game.* (A)

**indirect fire.** "Gunfire delivered at a target that cannot be seen from the gun position or firing ship." Army Regulation AR 320-5.

**kriegsspiel (kriegspiel).** Literally, war play or war game. Actually the name of one of the earliest war games, devised, developed, and employed extensively in the Prussian Army and later in the German Imperial Army. Later adopted, and adapted, to army war games in many other countries. (A)

**level of game.** "The largest formation [division, regiment . . .] on the side of principal interest whose play is required for the objective of the game." (Q)

**management game.** A simulation, in accordance with predetermined rules and procedures, of a selected aspect of a corporate function or of a business or industrial situation, usually conducted under conditions of competition and uncertainty. (A)

**map game or map maneuver.** A manual war game employing a map (or chart, in naval games) to represent the area of operation(s). (A)

**micro-relief.** Special features and characteristics of the topography including the presence and distribution of protection-affording elements. (A)

**military strategy.** "The art and science of employing the armed forces of a nation to secure the objectives of national policy by the application of force or the threat of force." Army Regulation AR 320-5, April 1965, p. 250.

**model.** *1.* "A representation of a real situation in which only those properties believed to be relevant to the problem being studied are represented." (Q)
*2.* "A representation of an object or structure; an explanation or description of a system, a process, or series of related events." (N)

**model, combat.** "A model of battle, usually mathematical or symbolic." (Q)

**model, deterministic.** "A model in which the outcome is predictable and the element of chance is absent." Compare *model, probabilistic*. (Q)

**model, mathematical.** "A model in which properties of the things represented, and their interactions, are expressed symbolically by means of mathematical expressions." (Q)

**model, probabilistic.** "A model in which the outcome is subject to chance variations." Same as *model, stochastic*. Compare *model, deterministic*. (Q)

**model, stochastic.** Same as *model, probabilistic*. (Q)

**model, terrain.** "A model of terrain, either physical (e.g., using a trafficability or a three-dimensional relief map) or symbolic (e.g., using a computer program)." (Q)

**model, war-game.** "The procedures and rules required for the control and conduct of a war game." (N)

**Monte Carlo method.** *1.* "The use of random sampling procedures for treating probabilistic mathematical problems." (Q)
*2.* "The use of a chance device to determine the outcome(s) of chance events, or to approximate a probability distribution that is difficult or impossible to compute." (N)

**multi-sided or n-sided game.** "A game where more than two sides, competitors or player teams are involved in a conflict situation." (N)

**national objectives.** "Those fundamental aims, goals or purposes of a nation—as opposed to the means for seeking these ends—toward which a policy is directed and efforts and resources of the nation are applied." Army Regulation AR 320-5, April 1965, p. 263.

**national policy.** "A broad course of action or statements of guidance adopted by the government at the national level in pursuit of national objectives." Army Regulation AR 320-5, April 1965, p. 263.

**national strategy.** "The art and science of developing and using the political, economic and psychological powers of a nation, together with its armed forces, during peace and war to secure national objectives." Army Regulation AR 320-5, April 1965, p. 264.

**one-sided game.** "A game in which the opposition is furnished by the control group, or by Nature." Compare with "Q" definition: *game, one-sided.* (N)

**operational gaming.** "The application of gaming techniques to operational situations. The term is used to describe the simulation of either military or non-military operations, e.g., business and industrial operations." [Definitive terms are to be preferred, i.e., war gaming, management gaming, etc.] (N)

**piece.** "A movable symbol on a game board." (Q)

**play.** "A single run-through of a game. Also used to represent replications of a game under a single set of starting conditions." (Q)

**play period or phase.** The second stage of a game. That period of the whole game sequence in which the simulated operations are carried out, i.e., played, cycle by cycle until the actions are completed. During this stage the contending teams or commanders (e.g., Blue and Red) and their staffs plan, issue orders, and follow developments, cycle by cycle, while Control assesses and governs the play of the game. (A)

**player.** "A participant who represents or is part of a group representing one of the opposing sides in a game. In a war game, the players assume the roles of commanders and staff officers of military units or formations." (Q)

**playing-time ratio.** The ratio of real time (playing time) to game time (simulated time) for an event or series of events. (A)

**preparation period or phase.** The first stage of a game. The make-ready or formative time period in which the game staff works up the objectives, the research or analytic design, the scenario, general and special situations (in a war game these include such details as the order of battle, weapons, lines of communication, geographic aspects), and other relevant input data; acquires and prepares maps, markers, and other equipment and materials; and organizes and trains the game staff—all prior to actual initiation of game play. (A)

**probability.** "The probability of the occurrence of an event is the ratio of the number of equally likely ways in which the event can happen to the total number of equally likely ways in which the event can or cannot happen. Or, the chances that an event will occur, or a particular result will occur." (N)

**program.** "Noun: a series of instructions required to complete a given procedure. Verb: to prepare a program." (N)

**programming.** "Preparing a program." (N)

**purpose.** "The general and specific reason(s) for which a war game is planned and played. Sometimes described under objective(s)." (N)

**quick gaming.** Simplified, free gaming that moves relatively fast, is highly aggregated, and is played for most part by use of charts and tables of data with minimum calculations. Only principal factors needed for the intended purpose are included. (A)

**real time.** The normal local time as used throughout the world, universally tied to Standard Time of day and the Gregorian calendar of the Western World. Compare *game time*. (A)

**relief.** The topographic or surface features of a region, including hills, valleys, rivers, lakes, canals, bridges, dams, roads, railroads, airfields, pipelines, villages, cities, etc. (A)

**resolution.** "The basic units of force, distance and time used in a war game, e.g., company, kilometer, 30-minute interval." See also *aggregation*. (Q)

**resources.** The total means available to each contender represented in the game, including facilities, personnel, capital, logistics, manpower, firepower, mobility, communications, reconnaissance, and command. (A)

**room, control.** "The room in which the control personnel of a war game work, where assessment is carried out, and records kept." (Q)

**room, player.** "The room occupied by one of the opposing [players or] teams of players in a game. It is physically separated from the control room in a closed game, but is linked to it by one or more forms of communication." Also called side room, Red or Blue room. (Q)

**routine.** "A specific portion of a model, usually as programmed for a computer." (Q)

**rule.** An objective statement of the limits or results of any particular action, or interaction, between opponents in a game. This objective statement may be either deterministic or probabilistic in nature. (A)

**rule, deterministic.** "A rule that states precisely and uniquely the results of any particular action or interaction between opponents." Compare *rule, probabilistic*. (Q)

**rule, probabilistic.** A rule that states the results of any particular action or interaction between opponents in terms of a probability function. The precise result to be applied on a given occasion in a game is determined in the control room by random sampling from the specified distribution. Compare *rule, deterministic*. (A)

**scenario.** "A description of the general situation and a chronological listing of pre-planned situations, messages, etc., to be fed into the game by the control group to initiate a two-sided game, or to stimulate player decisions during the play of a one-sided game." (N)

**scope.** "The scope of a war game is expressed by the range of command levels represented, the military services involved, the contemplated types of operations, and the size of the area of operations." (N)

**simulation.** *1.* "An operating representation of events and processes." (N)
*2.* "A technique used to study and analyze the operation and behaviour, by means of models of systems conditioned by human decision and/or probabilistic natural influences." See also *model*. (Q)

**simulation, deterministic.** "A simulation in which the outcome is predictable and the element of chance absent." Compare *simulation, probabilistic*. (Q)

**simulation, probabilistic.** "A *simulation* in which the outcome is subject to chance variations." Same as *simulation, stochastic*. Compare *simulation, deterministic*. (Q)

**simulation, stochastic.** Same as *simulation, probabilistic*. (Q)

**starting conditions.** Instructions and information issued to players to provide background and detailed information, and to initiate play of a game. A statement of the mission to be achieved, forces available, boundaries, intelligence acquired, and related data is included. Compare *game directive*. See also *scenario*. (A)

**state of a unit.** "A set of values for the game variables that describe the condition or characteristics of a unit at a given game time." (Q)

**status file.** A summary of the condition of each unit involved in a game, taken at some fixed game time such as at the end of a cycle of play. (A)

**stochastic game.** "One wherein the results of chance events are determined by the use of a chance device. Same as a *probabilistic* or *Monte Carlo game*. Repeated trials of a stochastic game provide a distribution of possible outcomes." (N)

**strategy.** "The art and science of developing and using political, economic, psychological and military forces as necessary during peace and war, to afford the maximum support to policies, in order to increase the probabilities and favorable consequences of [success or] victory and to lessen the chances of defeat." Army Regulation AR 320-5, April 1965, p. 393. See also *national strategy*.

**submodel.** "A representation, through procedures and rules, of a specific segment of a game model, e.g., artillery support, TAC-AIR." (Q)

**subroutine.** "A program that can be stored in the main or auxiliary memory of a digital computer and used as part of other programs to perform a specific operation; e.g., square root." See also *routine*. (N)

**tactics.** "1. The employment of units in combat.
2. The ordered arrangement and maneuver of units in relation to each other and/or to the enemy in order to utilize their full potentialities." Army Regulation AR 320-5, April 1965, p. 408.

**time, game.** Clock or calendar time in terms of the simulated operation being gamed. Game time does not necessarily have any fixed relation to actual (real) time. (A)

**time, playing.** "The duration of an event or series of events in a game measured in real time." (Q)

**time-step method.** "A technique employed in computer games wherein the game advances by regular time steps. At the end of each time step the computer decides if interactions occurred and, if interactions occurred, the computer determines the outcomes. For any particular play of a game the time interval is constant, but may be changed from play to play." (N)

**two-sided game.** "A game in which there are two opposing players or teams of players." (N)

**umpiring, free.** "Term used in a free or open game, or in training games in which the results of interactions are determined by umpires (or members of the control staff) in accordance with their professional judgment and experience." (N)

**updating.** "The revision of game variables to newly determined, hence, current values." (N)

**war game.** "A simulation by whatever means of a military operation involving two or more opposing forces, conducted, using rules, data and procedures designed to depict an actual or assumed real life situation." (M)

**war gaming.** "1. An operations research technique whereby the various courses of action involved in a problem are subjected to analysis under prescribed rules of play representing actual conditions and employing planning factors which are as realistic as possible." (Q)

2. "For Research and Analytical Studies. A technique for simulating warfare as a means of studying developing situations and generating data to be used in certain types of military problems, e.g., investigation of current, future or hypothetical capabilities and requirements usually not observable under controlled conditions except in the context of the military operations represented in a war game." (Q)

3. For Planning. A technique for simulating warfare as a means of providing data to evaluate concepts or plans for military operations. (A)

4. For Training. A technique for simulating warfare as a means of providing experience and training in the conduct of military operations. (A)

# Selected Bibliography

The list of documentary sources which follows represents the results of the author's search of the unclassified literature related to war gaming and simulation. Also included are some works on the art and technology of war which contributed to the changing patterns of warfare. These developments, in turn, were reflected in the evolution of the variety of war games and simulations. Each item selected, the author believes, will contribute to understanding the foundations and practices of war gaming.

Some of the documents are not readily available. A number are out of print. A few are rare. Others were issued in limited editions with copies available only in special depositories. They are included to credit their authors, to indicate sources, and to serve scholars.

Citations include references to organizations which have been superseded. Among them are the following. ORO, the Operations Research Office of The Johns Hopkins University, became RAC, the Research Analysis Corporation, McLean, Virginia, 22101. SORO, the Special Operations Research Office, American University, Washington, D.C., became AIR/CRESS, American Institutes for Research/Center for Research in Social Systems, 10605 Concord Street, Kensington, Maryland, 20795.

Wherever an AD number is shown in a bibliography item, it indicates that the document is available from the Defense Documentation Center—OSR, Cameron Station, Building 5, 5010 Duke Street, Alexandria, Virginia, 22314, whether or not the document can be supplied by the original issuing agency.

Preparation of this list of selected references for publication was a joint effort of the author and Miss Alice Hirsch, Research Librarian of the Research Analysis Corporation library.[1]

Nonmilitary Games and Simulations

Many readers will be concerned with nonmilitary applications of simulation and gaming techniques. To accommodate readers whose interests fall into only one of the two categories, the reference lists have been kept separate. The nonmilitary applications focus on business and management situations. Being of more recent origin and of lesser extent, references on Business Games and Management Simulations are listed first. That list is followed by a section on Military and Political Games and Simulations and Related Source Materials.

References on Business Games and Management Simulations

Acer, John Whedon, *Business Games: A Simulation Technique*, Iowa City, Iowa, State University of Iowa, College of Business Administration, Bureau of Labor and Management, Information Series 3, Nov. 1960.

Adamowsky, Siegmar, *Das Planspiel*, Frankfort/Main, Agenor-Verlag, 1963.

Agersnap, Torben, and Erik Johnsen, "A Decision Game of Managerial Strategy as a Research Tool" (paper presented in Paris, Sept. 1959), C. West Churchman and M. Verhulst (eds.), *Management Sciences, Models and Techniques*, vol. 1, pp. 225–240, New York, Pergamon Press, 1960.

American Management Association, *Simulation and Gaming: A Symposium*, AMA Report 55, New York, American Management Association, 1961.

Andlinger, G. R., "Looking Around: What Can Business Games Do?," *Harvard Business Review*, 36(4):147–152, July–Aug. 1958.

Aubert, Bourges, Minthe, Anstett, and Tricaud, "Contributions et Expériences en Matiere de 'Management Games'" (paper pre-

---

[1] The master card file from which this list was prepared has come to be known as the Hausrath-Hirsch bibliography on war gaming.

The collections drawn upon most frequently included those of the Library of Congress, the National Archives, the Army Library, the Navy Library, Harvard University, the National War College, the Armed Forces Industrial College, the Army War College, the Naval War College, the U.S. Army Command and Staff College, the U.S. Military Academy, the U.S. Naval Academy, and the United States Naval Institute. Many other libraries, museums, archives, professional societies, and individuals, here and abroad, contributed special information.

Selected Bibliography 323

sented in Paris, Sept. 1959), in C. West Churchman and M. Verhulst (eds.), *Management Sciences, Models and Techniques*, vol. 1, pp. 209–224, New York, Pergamon Press, 1960.

Babb, E. M., and L. M. Eisgruber, *Management Games for Teaching and Research*, Chicago, Educational Methods, Inc., 1966.

Barton, Richard F., *A Primer on Simulation and Gaming*, Englewood Cliffs, N.J., Prentice-Hall, Inc., 1970.

Berglund, Jan E., and Robert W. Grubbstrom, *Foretagsspel—lek och verklighet*, Stockholm, Bokforlaget Aldus/Bonniers, 1968.

Carlson, Elliot, *Learning Through Games*, Washington, Public Affairs Press, 1969.

Churchman, C. West, and M. Verhulst (eds.), *Management Sciences, Models and Techniques* (Proceedings of the Sixth International Meeting of the Institute of Management Sciences, held at Paris, France, Sept. 7–11, 1959), 2 vols., New York, Pergamon Press, 1960.

Clark, Wallace, *The Gantt Chart: A Working Tool of Management*, 3d ed., London, Sir Isaac Pitman & Sons, Ltd., 1952.

Cohen, K. J., W. R. Dill, A. Kuehn, and P. Winters, *The Carnegie Tech Management Game: An Experiment in Business Education*, Homewood, Ill., Richard D. Irwin, Inc., 1964.

———— and E. Rhenman, "The Role of Management Games in Education and Research," *Management Science*, 7(2):131–166, Jan. 1961.

Crook, Richard, and Paul Wright, "Proof of Training by the Game Approach," *Training Directors*, 15(8):46–50, Aug. 1961.

Dale, A. G., and C. R. Klasson, *Business Gaming, Survey of American Collegiate Schools of Business*, Austin, University of Texas, Bureau of Business Research, 1964.

Daly, Andrew A., "In-Basket Business Game," *Training Directors*, 14(8):8–15, Aug. 1960.

Dill, William R., *Management Games for Training Decision Makers*, Studies in Personnel and Industrial Psychology, Homewood, Ill., The Dorsey Press, 1961.

————, "What Management Games Do Best," *Business Horizons*, 4(3):55–64, Fall 1961.

Dobles, Robert W., and Robert F. Zimmerman, "Management Training Using Business Games," *Training and Development Journal*, 20(6):28–30, 32, 34, June 1966.

Fairhead, J. N., D. S. Pugh, and W. J. Williams, *Exercises in Business Decisions, A Manual for Management Education*, London, English Universities Press, Ltd., 1965.

Feeney, George J., "The Future of Management Gaming," in C. West Churchman and M. Verhulst (eds.), *Management Sciences, Models and Techniques*, vol. 1, pp. 263–268, New York, Pergamon Press, 1960.

Flagle, Charles D., et al. (eds.), *Operations Research and Systems Engineering*, Baltimore, The Johns Hopkins Press, 1960.

Frank, H. E., and S. J. Pringle, "'In Tray' Training Exercises," *Training Directors*, 16(4):27–30, April 1962.

French, Wendell L., "A Collective Bargaining Game," *Training Directors*, 15(1):10–13, Jan. 1961.

———, "A Collective Bargaining Game at the University," *Training Directors*, 16(1):12–17, Jan. 1962.

Graham, Robert G., and Clifford F. Gray, *Business Games Handbook*, New York, American Management Association, 1969.

Greene, Jay R., and Roger L. Sisson, *Dynamic Management Decision Games Including Seven Non-computer Games*, New York, John Wiley & Sons, Inc., 1959.

Greenlaw, Paul S., L. W. Herron, and R. H. Rawdon, *Business Simulation in Industrial and University Education*, Englewood Cliffs, N.J., Prentice-Hall, Inc., 1962.

——— and Stanford S. Kight, "The Human Factor in Business Games," *Business Horizons*, 3(3):55–61, Fall 1960.

Hall, Arthur D., *A Methodology for Systems Engineering*, Princeton, N.J., D. Van Nostrand Company, Inc., 1962.

Hamburger, William, *Monopologs, An Inventory Management Game*, Santa Monica, Calif., The RAND Corporation, RAND Research Memorandum RM-1579, Jan. 3, 1956.

Haynes, W. W., and J. L. Massie, *Management Analysis, Concepts and Cases*, Englewood Cliffs, N.J., Prentice-Hall, Inc., 1961.

Henshaw, Richard C., Jr., and James R. Jackson, *The Executive Game*, Homewood, Ill., Richard D. Irwin, Inc., 1966.

Herder, John H., "Do-It-Yourself Business Games," *Training Directors*, 14(9):3–8, Sept. 1960.

Jackson, James R., "Business Gaming in Management Science Education," in C. West Churchman and M. Verhulst (eds.), *Management Sciences, Models and Techniques*, vol. 1, pp. 250–262, New York, Pergamon Press, 1960.

———, "Learning from Experience in Business Decision Games," *California Management Review*, 1(2):92–107, Winter 1959.

Johnston, Donald Richard, "An Evaluation of the Business Decision Games," *Training Directors*, 15(5):33–41, May 1961.

Karlin, Samuel, *Mathematical Methods and Theory in Games, Pro-*

*gramming, and Economics,* vol. II, Reading, Mass., Addison-Wesley Publishing Co., Inc., 1959.

Kedzie, Daniel P., "Decision at Zenith Life, A Unique Educational Experience," *Training Directors,* 16(4):3–6, April 1962.

Kibbee, Joel M., Clifford J. Craft, and Burt Nanus, *Management Games: A New Technique for Executive Development,* New York, Reinhold Publishing Corporation, 1961.

McGuinness, John S., "A Managerial Game for an Insurance Company," *Operations Research,* 8(2):196–209, March–April 1960.

McGuire, C. B., "Some Team Models of a Sales Organization," *Management Science,* 7(2):101–130, Jan. 1961.

McKenney, James L., *Simulation Gaming for Management Development,* Boston, Harvard University, Graduate School of Business Administration, Division of Research, 1967.

McMurray, Fred D., "Science and Group Dynamics," *Training Directors,* 15(5):5–13, May 1961.

McNair, Malcolm P., *The Case Method at the Harvard Business School,* New York, McGraw-Hill Book Company, 1954

Malcolm, D. G., "Bibliography on the Use of Simulation in Management Analysis," *Operations Research,* 8(2):169–177, March–April 1960.

──── (ed.), *Report of System Simulation Symposium* (cosponsored by AIIE, ORSA, and TIMS), New York, May 1957, New York, American Institute of Industrial Engineers, 1958.

Martin, E. W., Jr., "Teaching Executives via Simulation," *Business Horizons,* 2(2):100–109, Summer 1959.

Meier, Robert C., William T. Newell, and Harold L. Pazer, *Simulation in Business and Economics,* Englewood Cliffs, N.J., Prentice-Hall, Inc., 1969.

Miller, David, "The Government Management Game," unpublished paper, Gaithersburg, Md., U.S. National Bureau of Standards, 1966.

Moore, Larry F., "Business Games vs. Cases as Tools of Learning," *Training and Development Journal,* 21(10):13–23, Oct. 1967.

National Symposium on Management Games, M. E. Fessler, C. B. Saunders, and J. D. Steele, *Proceedings of the National Symposium on Management Games,* Dec. 12-13, 1958, Lawrence, Kans., University of Kansas, Center for Research in Business, May 1959.

Pierson, Frank C., *The Education of American Business Men,* New York, McGraw-Hill Book Company, 1959.

Radell, Nicholas, "Concepts of Management Gaming," *Systems and Procedures Journal*, 15(2):24–29, issue no. 64, March–April 1964.

Rapoport, Anatol, "Three Modes of Conflict," *Management Science*, 7(3):210–218, April 1961.

———, "The Use and Misuse of Game Theory," *Scientific American*, 207(6):108–114, 117, 118, Dec. 1962.

Renard, B., "Remarques sur les Expériences de Gestion," in C. West Churchman and M. Verhulst (eds.), *Management Sciences, Models and Techniques*, vol. 1, pp. 199–208, New York, Pergamon Press, 1960.

Ricciardi, Franc M., Clifford C. Craft, Donald G. Malcolm, Richard Bellman, Charles Clark, Joel M. Kibbee, and Richard H. Rawdon, in Elizabeth Marting (ed.), *Top Management Decision Simulation: The AMA Approach*, New York, American Management Association, 1957.

Rich, Robert P., "Simulation as an Aid in Model Building," *Journal of the Operations Research Society of America*, 3(1):15–19, Feb. 1955.

Robbins, Robert M., "Decision-making Simulation Through Business Games," *Training Directors*, 13(9):12–19, Sept. 1959.

Sanders, Donald H., *Computers in Business*, New York, McGraw-Hill Book Company, 1968.

Schellenberger, Robert Earl, *Development of a Computerized, Multipurpose Retail Management Game*, Chapel Hill, N.C., University of North Carolina, Graduate School of Business Administration, Research Paper 14, 1965.

Schrieber, Albert N., "A New Way to Teach Business Decision Making," *University of Washington Business Review*, April 1958, pp. 18–29.

Shubik, Martin, "Games Decisions and Industrial Organization," *Management Science*, 6(4):455–474, July 1960.

———, "Gaming: Costs and Facilities," *Management Science*, 14(11):629–660, July 1968.

Smith, Adair, Thomas B. Scobel, and Ronald J. Le Frois, "General Motors Institute Experiences with Business Gaming," *Training Directors*, 15(4):27–32, April 1961.

Steinmetz, Walter W., "Management Games—Computer Versus Noncomputer," *Training Directors*, 16(9):38–40, 42–45, Sept. 1962.

Stephens, Warren S., "A Business Game in Orientation," *Training Directors*, 16(10):55–57, Oct. 1962.

Symonds, G. H., "A Study of Management Behavior by Use of Competitive Business Games," *Management Science*, 11(1):135–153,

Sept. 1964.

Taylor, Frederick W., *The Principles of Scientific Management*, New York, Harper & Brothers, 1911.

———, *Scientific Management*, New York, Harper & Brothers, 1911, 1939, 1947.

Thorelli, Hans B., Robert L. Graves, and Lloyd T. Howells, "The International Operations Simulation at the University of Chicago," *The Journal of Business*, 35:287–297, July 1962.

———, ———, and ———, *INTOP: International Operations Simulation. Players' Manual*, The University of Chicago, Graduate School of Business, New York, The Macmillan Company, 1963.

Vajda, S., *The Theory of Games and Linear Programming*, New York, John Wiley & Sons, Inc., 1956.

Vance, Stanley C., *Management Decision Simulation, A Noncomputer Business Game*, New York, McGraw-Hill Book Company, 1960.

"'War Game' Will Train Managers to Make Good Business Decisions," *Factory Management and Maintenance*, 115(6):132–133, June 1957.

Winston, James Stewart, "The Controlled Exercise—An Industrial Application of a Dynamic Simulation Technique for Personnel and Organizational Development," dissertation submitted for the degree of Doctor of Education, Columbia University, Teachers College, New York, 1965.

Yost, Edna, *Frank and Lillian Gilbreth, Partners for Life*, New Brunswick, N.J., Rutgers University Press, 1949.

Zimmerman, John W., "Business Gaming" (an abstract), *Training Directors*, 14(7):25–27, July 1960.

———, "Non-mathematical Simulation—Dynamic Development Guide," *Training Directors*, 16(6):30–35, June 1962.

——— and Seymour Levy, "Decision Simulation for Top Management Training," *Training Directors*, 14(5):3–11, May 1960.

References on Military and Political Games and Simulations and Related Source Materials

Abt, Clark C., *Games for Learning*, Cambridge, Mass., Educational Services Incorporated, The Social Studies Curriculum Program, Occasional Paper 7, 1966.

———, "War Gaming," *International Science and Technology*, no. 32, pp. 29–37, Aug. 1964.

———, M. Gorden, and J. C. Hodder, "Strategic Model (TEMPER)

Description," *Proceedings for the Third Symposium of the East Coast War Games Council*, Feb. 27-28, 1964, Miami Beach, Fla., Bedford, Mass., Raytheon Company, Space & Information Systems Division Strategic Studies Department, April 1, 1963.

Abt Associates Inc., "Counter-insurgency Game Design Feasibility and Evaluation Study," a study for The Advanced Research Projects Agency, Washington, D.C., Nov. 1965.

———, "Six Demonstrations of the AGILE/COIN Games," Cambridge, Mass., ARPA Order No. 681, Oct. 1966.

Ackoff, Russell L. (ed.), *Progress in Operations Research*, vol. I, New York, John Wiley & Sons, Inc., 1961.

Adams, Hebron E., R. E. Forrester, J. F. Kraft, and B. B. Oosterhout, *CARMONETTE: A Computer-played Combat Simulation*, Bethesda, Md., The Johns Hopkins University, Operations Research Office, ORO-T-389 (AD 257012), Feb. 1961.[1,2]

Adams, Phillip L., and Strother H. Walker, "The Wholesale-level Logistics Game," Bethesda, Md., The Johns Hopkins University, Operations Research Office, ORO-SP-66, July 1958.

Adams, R. H., and J. L. Jenkins, "Simulation of Air Operations with the Air Battle Model," *Operations Research* 8(5):600-615, Sept.-Oct. 1960.

The Aeroplane Goes to Sea, in "Evolution of Aircraft Carriers," *Naval Aviation News*, 43rd Year, 22-28, Feb. 1962.

American Management Association, *Simulation and Gaming: A Symposium*, New York, American Management Association, Management Report 55, 1961.

Anders, Leslie, "Return of the North Sea Ghosts," *Military Review*, 46(10):19-26, Oct. 1966.

Anderson, Andrew, *The Game of Draughts Simplified* and illustrated with practical diagrams, 6th ed., revised and extended by Robert M'Culloch, Glasgow, Edinburgh, John Grant, 1878.

Anderson, Kingsley S., *The Game of War*, Burlington, Mass., Technical Operations, Inc., 1960. Information assembled for this booklet from works of John P. Young, *A Survey of Historical Developments in War Games*, The Johns Hopkins University, Operations Research Office, 1959.

Andrews, Marshall, *Disaster Through Air Power*, New York, Rinehart & Company, Inc., 1950.

---

[1] ORO became RAC, the Research Analysis Corporation, McLean, Virginia.
[2] An AD number identifies document available from the Defense Documentation Center, Alexandria, Virginia.

Antosiewicz, Henry A., "Analytic Study of War Games," *Naval Research Logistic Research Quarterly*, 2(3):181–208, Sept. 1955.
Archer, W. L., "The Technique of Modern War Gaming," *Canadian Army Journal*, 15(4):15–25, Fall 1961.
*Armies of To-Day*, A Description of the Armies of the Leading Nations at the Present Time, a collection by various authors, New York, Harper & Brothers, 1893 (copyright 1892).
Army Command and General Staff College, *War Gaming*, ST-105-5-1, Fort Leavenworth, Kans., Nov. 1965.
Army War College, "War Games, Brief Anthology," pt. II, course 5, *War Games*, 1957–1958 Curriculum, Carlisle Barracks, Pa., United States Army War College, Jan. 6, 1958.
Arthur, Robert, "Historical Sketch of the Coast Artillery School," *Journal of the United States Artillery*, (whole no. 134):15–48, July–Aug. 1915.
———, "Historical Sketch of the Coast Artillery School" (concluded), *Journal of the United States Artillery*, (whole no. 135):164–203, Sept.–Oct. 1915.
Asby, Raymond C., Jr., "Realistic Umpiring—Map Maneuver Mainspring," *Military Review*, 32(8):37–42, Nov. 1952.
Atkins, Gaius Glenn, *Procession of the Gods*, New York, Richard R. Smith, Publisher, Inc., 1930.
Atkinson, James David, *The Edge of War*, with Foreword by Adm. Arleigh A. Burke, Chicago, Henry Regnery Company, 1960.
Auger, Le Commandant breveté, *Trois Études Tactiques*, I. "Une manoeuvre a double action sur la carte;" II. "Une manoeuvre avec cadres sur le terrain;" III. "Une attaque décisive," Paris-Nancy, Berger-Levrault & Cie, 1901.
Aumann, Robert J., "The Game of Politics," *World Politics*, 14(4):675–686, July 1962.
Ausland, John C., and Hugh F. Richardson, "Crisis Management: Berlin, Cyprus, Laos," *Foreign Affairs*, 44(2):291–303, Jan. 1966.
Baerensprung, Erik von, *Einfuhrung in das Kriegsspiel*, Berlin, Siegfried Mittler und Sohn, 1913.
Balinski, Michel, Klaus Knorr, Oskar Morgenstern, Francis Sand, and Martin Shubik, "Final Report, Review of Temper Model," Princeton, N.J., Mathematica (AD 816457), Sept. 30, 1966.
Bard, Bob, *Making and Collecting Military Miniatures*, New York, Robert M. McBride Co., Inc., 1957.
Barnes, Stanley M., "Defense Planning Processes," *United States Naval Institute Proceedings*, (whole no. 736):26–39, June 1964.

Barnett, Jonathan, The Computer Revolution: How Does It Affect Architecture . . . , in "The New Age of Architecture, Part 1, Science and Technology as a Design Influence," *Architectural Record*, 140(1):168–170, July 1966.

Barringer, Richard E., and Barton Whaley, "The M.I.T. Political-Military Gaming Experience," *Orbis*, 9(2):437–458, Summer 1965.

Bartholomew, Fletcher L., Jerome Bracken, et al., "A Logistic-gaming and Simulation System: General Concept," McLean, Va., Research Analysis Corporation, RAC-TP-179, Jan. 1966.

Batchelor, James H., *Operations Research: An Annotated Bibliography*, St. Louis, Mo., St. Louis University Press, 4 vols., 1959, 1962, 1963, 1964.

Beach, Dwight E., "Combat Developments Command Seeks Answers for the Future," *Army-Navy-Air Force Journal and Register*, 101:20–22, June 13, 1964.

Beaufré, André, *Introduction to Strategy*, translated by Maj. Gen. R. H. Barry, Preface by Capt. B. H. Liddell Hart, New York Frederick A. Praeger, Inc., 1965.

Benson, Oliver, "A Simple Diplomatic Game," in James A. Rosenau (ed.), *International Politics and Foreign Policy*, New York, The Free Press of Glencoe, Inc., 1961.

Berkowitz, L. D., and Melvin Dresher, "A Game Theory Analysis of Tactical Air War," *Operations Research*, 7(5):599–620, Sept.–Oct. 1959.

Bernens, John C., "Field Exercises and Maneuvers as Data Sources," Technical Operations Inc., Combat Operations Research Group, Hq Continental Army Command, Fort Monroe, Va. Paper presented to the Fourth Annual Meeting of the Operations Research Society of America held in Washington, D.C., May 10–11, 1956. Abstract in *Operations Research*, 4(3):388, June 1956.

Blackett, P. M. S., *Studies of War, Nuclear and Conventional*, New York, Hill and Wang, Inc., 1962.

Blackford, John H., "Management Gaming and Its Applicability at the United States Naval Postgraduate School," master's thesis, United States Naval Postgraduate School, Monterey, Calif., 1964.

Blackwell, David, *Game Theory for War Gaming*, Chevy Chase, Md., The Johns Hopkins University, Operations Research Office, ORO-SP-9 (AD 236152), April 1957.

Bloomfield, Lincoln P., "Arms Control and World Government," *World Politics*, 14(4):633–645, July 1962.

―――, "Peacekeeping and Peacemaking," *Foreign Affairs*, 44(4):671–682, July 1966.

———, "Political Gaming," *United States Naval Institute Proceedings*, (whole no. 691):57–64, Sept. 1960.

———, *Report and Analysis of Political Exercise Held Sept. 10–12, 1958 Under the Auspices of the United Nations' Project*, Cambridge, Mass., Massachusetts Institute of Technology, Center for International Studies, Dec. 1958. (Mimeographed report.)

——— and Norman J. Padelford, "Teaching Note: Three Experiments in Political Gaming," *American Political Science Review*, 53(4):1105–1115, Dec. 1959.

——— and Barton Whaley, "The Political-Military Exercise (POLEX)," *Military Review*, 45(11):65–71, Nov. 1965.

Bluehdorn, Robert W., et al., *VALOR War Gaming Handbook*, vol. 1, Washington, D.C., Technical Operations Inc., OMEGA SM-62-4-2, Dec. 21, 1960.

Blumstein, Alfred, *R&D on Internal Conflict in the Underdeveloped World in the 1970's*, Arlington, Va., Institute for Defense Analyses, Research & Engineering Support Division, Internal Notes 275, Oct. 1965.

Bodart, Gaston (ed.), *Militar-historisches Kriegs-Lexikon (1618–1905)*, Wien und Leipzig, C. W. Stern, 1908.

Boehm, George A. W., "The First Battle of World War III," *Technology Review*, 69(8):14–17, June 1967.

———, "Toward Negotiation by Computer," *Current*, no. 88, pp. 55–56, Oct. 1967.

Boikov, Ivan, "At the General Staff Academy," *Soviet Military Review* (Moscow), 30(6):12–13, June 1967.

Boocock, Sarane S., and James S. Coleman, "Games with Simulated Environments in Learning," *Sociology of Education*, 39(3):215–236, Summer 1966.

Borko, Harold, *Computer Applications in the Behavioral Sciences*, Englewood Cliffs, N.J., Prentice-Hall, Inc., 1962.

Bowen, Harold G., *Ships Machinery and Mossbacks*, Princeton, N.J., Princeton University Press, 1954.

Bowers, Ray L., Jr., "The Beginnings of Armored Warfare," *Military Review*, 46(12):18–28, Dec. 1966.

Boyd, Charles T., *Criticisms upon Solutions of Map Problems*, published for Army School of the Line by The Collegiate Press, George Banta Company, Inc., Menasha, Wis., 1915.

Bradley, Omar N., *A Soldier's Story*, New York, Henry Holt and Company, Inc., 1951.

Brain, A. E., G. E. Forsen, N. J. Nilsson, and C. A. Rosen, Learning

Machines, in chap. 6, "Mathematics, Computers, and Control," in Robert Colborn (chief ed.), *Modern Science and Technology*, Princeton, N.J., D. Van Nostrand Company, Inc., 1965.

Brennan, D. G., "Review of Rapoport: Strategy and Conscience," *Bulletin of the Atomic Scientists*, 21(10):25–30, Dec. 1965.

Brent, Robert, "Mahan—Mariner or Misfit?," *United States Naval Institute Proceedings*, (whole no. 758):92–103, April 1966.

Bretnor, Reginald, "Vulnerability and the Military Equation," *Military Review*, 46(9):18–26, Sept. 1966.

Brinckloe, W. D., "Research Navy," *United States Naval Institute Proceedings*, (whole no. 670):97–104, Dec. 1958.

Brodie, Bernard, *The American Scientific Strategists*, Santa Monica, Calif., The RAND Corporation, P-2979, Oct. 1964.

———, "The Scientific Strategists," in Robert Gilpin and Christopher Wright (eds.), *Scientists and National Policy Making*, New York, Columbia University Press, 1964.

Brody, Richard A., "Some Systemic Effects of the Spread of Nuclear Weapons Technology: A Study Through Simulation of a Multinuclear Future," *Journal of Conflict Resolution*, 7(4):1–126, Dec. 1963.

Brooks, Harvey, "The Scientific Adviser," in Robert Gilpin and Christopher Wright (eds.), *Scientists and National Policy Making*, New York, Columbia University Press, 1964.

Brooks, Richard S., "How It Works—The Navy Electronic Warfare Simulator," *United States Naval Institute Proceedings*, (whole no. 679):147–148, Sept. 1959.

Brossman, Martin, "One-sided Logistic Games," *Proceedings for the Third Symposium of the East Coast War Games Council*, Feb. 27–28, 1964, Miami Beach, Fla., McLean, Va., Research Analysis Corporation.

——— (ed.), *Proceedings, Fourth Symposium on War Gaming*, East Coast War Games Council, Aug. 1965, McLean, Va., Research Analysis Corporation (AD 468994), 1965.

——— et al., "Computer-assisted Logistic-Planning-Program Descriptions," McLean, Va., Research Analysis Corporation, RAC-T-431 (AD 464014), Jan. 1965.

Brown, Bernice, and Olaf Helmer, *Improving the Reliability of Estimates Obtained from a Consensus of Experts*, Santa Monica, Calif., The RAND Corporation, P-2986 (AD 606970), Sept. 1964.

Brown, Richard H., *A Stochastic Analysis of Lanchester's Theory of Combat*, Chevy Chase, Md., The Johns Hopkins University, Operations Research Office, ORO-T-323 (AD 82944), Dec. 1955.

Brucer, Marshall, *A History of Airborne Command and Airborne Center*, text from the official *History of the Airborne Effort* by John Ellis, published by The Command Club for its members.

Bruijn, W. K. de, "Automation in Europe," *Datamation*, 12(9):25–27, Sept. 1966.

Bruner, Joseph A., "Simulation and Gaming," Lectures 15 and 16, Operations Research Course, presented by Research Analysis Corporation for the Industrial College of the Armed Forces, Washington, D.C., McLean, Va., Research Analysis Corporation, Oct. 15 and 19, 1965.

Burck, Gilbert, "The Boundless Age of the Computer," pt. I of a series, *Fortune*, 69:101–110 ff., March 1964.

―――, 'On line' in 'Real Time,' pt. II of "The Boundless Age of the Computer," *Fortune*, 69:141–147 ff., April 1964.

―――, The 'Assault' on Fortress I.B.M., pt. IV of "The Boundless Age of the Computer," *Fortune*, 69:112–116 ff., June 1964.

―――, Will the Computer Outwit Man?, pt. VI of "The Boundless Age of the Computer," *Fortune*, 69:120–121 ff., Oct. 1964.

Burns, Arthur Lee, "Prospects for a General Theory of International Relations," *World Politics*, 14(1):25–46, Oct. 1961.

Busch, Fritz Otto (ed.), *Taschenbuch fur die Kriegsmarine*, Hannover, published for German Kriegsmarine Oberkommando by Adolph Sponholtz Verlag, 1944.

Butterfield, H., *Napoleon*, New York, The Macmillan Company, 1939.

Cady, Alice H., *CHECKERS, A Treatise on the Game*, New York, American Sports Publishing Company, 1896.

―――, *Go-Bang*, New York, American Sports Publishing Company, 1896.

Cagle, Malcolm W., "Studying the Navy's Future," *United States Naval Institute Proceedings*, (whole no. 728):92–99, Oct. 1963.

Camp, Glen D., "Operations Research in Turkey," *Proceedings of the First International Conference on Operational Research*, Baltimore, Operations Research Society of America, Dec. 1957.

Caruthers, Osgood, "New Soviet Thinking Disclosed in Huge War Games," *The Washington Post*, Sept. 25, 1966, p. 12.

Casner, Lewis E., et al., "Study of Computers to Improve Command Post Exercises," Silver Spring, Md., Computer Applications, Inc., Nov. 19, 1965.

Casson, Lionel, *The Ancient Mariners*, New York, The Macmillan Company, 1959.

―――  and editors of Time-Life Books, *Ancient Egypt*, a volume in the series *Great Ages of Man, A History of the World's Cultures*, New York, Time-Life Books, 1965.

Chamberlaine, William, *Coast Artillery War Game*, Fort Monroe, Va., privately printed by William Chamberlaine, 1912.

———, *Coast Artillery War Game*, 3d ed., with revisions, Fort Monroe, Va., Coast Artillery School Press, 1914.

———, *Coast Artillery War Game*, 4th ed., with revisions, Government Printing Office, 1916.

Chapin, Ned, *An Introduction to Automatic Computers*, 2d ed., Princeton, N.J., D. Van Nostrand Company, Inc., 1963.

Chapman, R. L., *The Systems Research Laboratory and Its Program*, Santa Monica, Calif., The RAND Corporation, RM-890, 1952.

———, John L. Kennedy, Allen Newell, and William C. Biel, "The Systems Research Laboratory's Air Defense Experiments," *Management Science*, 5(3):250–269, April 1959.

Chapman, R. M., W. L. Hughey, E. R. Sharp, and D. B. Wallace, *A Concept for Service Support War Gaming*, Fort Monroe, Va., Technical Operations Inc., CORG-M-187, Sept. 20, 1965.

Cho-Yo, *Japanese Chess (Sho-ngi), The Science and Art of War or Struggle, Philosophically Treated; Chinese Chess (Chong-Kie) and I-GO*, New York, The Press Club of Chicago, 1905.

Churchill, Winston S., *The Second World War*, vols. 1–6, Boston, Houghton Mifflin Company, 1949–1953.

Churchman, C. West, "The X of X," *Management Science*, 9(3):351–357, April 1963.

———, R. L. Ackoff, and E. L. Arnoff, *Introduction to Operations Research*, New York, John Wiley & Sons, Inc., 1957.

Ciccolella, Richard G., "Maneuver Planning," *Army Information Digest*, 19(4):45–53, April 1964.

Clark, Alan, *BARBAROSSA, The Russian-German Conflict, 1941–1945*, New York, William Morrow & Company, Inc., 1965.

Clark, C. E., G. E. Clark, Jr., W. E. Cushen, F. W. Dresch, S. E. Forbush, J. O. Harrison, Jr., J. G. Hill, P. Iribe, E. M. Lee, P. F. Michelsen, P. R. Newcomb, W. B. Taylor, B. Urban, and C. A. Warner, *War Gaming, COSMAGON and ZIGSPIEL*, Chevy Chase, Md., The Johns Hopkins University, Operations Research Office, ORO-SP-12 (AD 235892), April 1957.

Clark, D. K., L. E. Keefer, and W. W. Walton, Jr., *FOE, A Model Representing Company Actions*, The Johns Hopkins University, Operations Research Office, ORO-TP-17 (AD 254156), Dec. 1960.

Clark, John J., "The Economics of Systems Analysis," *Military Review*, 44:25–31, April 1964.

Clausewitz, Karl von, *On War*, translated from the German by O. J. Matthijs Jolles, with a Foreword by Joseph I. Greene, Preface by Richard McKeon, Washington, D.C., Infantry Journal Press, 1950.

Selected Bibliography    335

———, *On War*, translated by J. J. Graham, new and revised edition with introduction and note by F. N. Maude, vols. I–III, New York, Barnes & Noble, Inc., 1956.

Clerk, John, Esq. of Eldin, *An Essay on Naval Tactics, Systematical and Historical*, Ann Arbor, Mich., University Microfilms, Inc., 1964. (Reprint.)

Cline, R. E., *A Survey of Mathematical and Simulation Models as Applied to Weapon System Evaluation*, University of Michigan for U.S. Air Force Aeronautical Systems Division, ASD Report 61-276 (University of Michigan Report 3681-16), Oct. 1961.

Clutterbuck, Richard L., *The Long Long War*, Counterinsurgency in Malaya and Vietnam, with foreword by Harold K. Johnson, New York, Frederick A. Praeger, Inc., 1966.

Cochenhausen, Lt. Gen. von, "Anleitung fur die Anlage und Leitung von Planübungen, und Kriegsspielen mit Beispielen und Aufgaben zur Fuhrerschulung," booklet, no date, translated by O. C. Michelmann, *Military Review*, 20:48–51, March 1941.

Cockrill, James T., "The Validity of War Game Analysis," *United States Naval Institute Proceedings*, (whole no. 755):44–53, Jan. 1966.

Cohen, Kalman J., and Eric Rhenman, "The Role of Management Games in Education and Research," *Management Science*, 7(2):131–166, Jan. 1961.

Cole, Hugh M., *The Ardennes: Battle of the Bulge, U.S. Army in World War II: European Theater of Operations*, vol. 3, pt. 7, Washington, D.C., Department of the Army, Office of the Chief of Military History, 1965.

Coleman, James S., Sarane S. Boocock, and E. O. Schild (eds.), "Simulation Games and Learning Behavior," Part I, *American Behavioral Scientist*, 10(2): entire issue, Oct. 1966.

———, ———, and ——— (eds.), "Simulation Games and Learning Behavior," Part II, *American Behavioral Scientist*, 10(3): entire issue, Nov. 1966.

Colomb, Philip H., "Le Duel ou Jeu de la Guerre Navale" (The Duel or the Naval War Game), translated into Italian by L. Rivet, *Revue Maritime et Coloniale*, Tome 48:363–388, Fevrier 1881.

Colombo, A., "Giuoco di Guerra Navale" (Naval War Game), *Rivista Marittima* (Roma), 24(12):347–363, Dec. 1891.

Connolly, Richard L., "The Principles of War," *United States Naval Institute Proceedings*, (whole no. 599):1–9, Jan. 1953.

Conrad, N. I., *Sun-Tzu, Treatise on Military Art*, translation and analysis, Moscow, Publishing House of the Academy of Science U.S.S.R., 1950.

Cooper, Margaret, *The Inventions of Leonardo Da Vinci*, New York, The Macmillan Company, 1965.

Coplon, William D., "Inter-Nation Simulation and Contemporary Theories of International Relations," *American Political Science Review*, 60(3):562–578, Sept. 1966.

Cormier, Everett L., and Walter N. Flournoy, "A Guerrilla War Game," *Army*, 18(2):46–53, Feb. 1968.

Cotten, Lt. Lyman A., USN, "The Naval Strategy of the Russo-Japanese War," *United States Naval Institute Proceedings*, (whole no. 133):41–60, March 1910.

Cowburn, Philip, *The Warship in History*, New York, The Macmillan Company, 1965.

Craven, Francis S., "The Painful Development of a Professional Navy," *United States Naval Institute Proceedings*, (whole no. 759):78–89, May 1966.

Creswell, John, *Naval Warfare—An Introductory Study*, London, Sampson, Low, Marston & Co., Ltd., 1942.

Cripwell, F. J., *The Concept of Computer-assisted Games*, Canada, Department of National Defence, Operational Research Division, ORD Informal Paper 66/P8, April 1966.

Cromer, Everard Baring, *Rules for the Conduct of the War Game*, London, Her Majesty's Stationery Office, 1872.

Croswell, Thomas L., "On the Analysis of Wars between Machines," *Proceedings for the Third Symposium of the East Coast War Games Council*, held Feb. 27–28, 1964, Miami Beach, Fla., Oswego, N.Y., IBM Space Guidance Center (IBM No. 64-825-1139).

Crowther, J. G., and R. Whiddington, *Science at War*, New York, Philosophical Library, Inc., 1948.

Cunningham, Andrew Browne, *A Sailor's Odyssey*, The Autobiography of Admiral of the Fleet Viscount Cunningham of Hyndhope, New York, E. P. Dutton & Co., Inc., 1951.

Cushen, W. Edward (ed.), *An Experiment in the Application of Scientific Methods to Foreign Policy Planning*, Cleveland, Case Institute of Technology, June 1962.

————, *Generalized Battle Games on a Digital Computer*, Chevy Chase, Md., The Johns Hopkins University, Operations Research Office, ORO-T-263 (AD 53283), Sept. 1954.

————, *Operational Gaming in Industry*, Chevy Chase, Md., The Johns Hopkins University, Operations Research Office, Informal Seminar Paper 18, March 23, 1955.

————, *The POLEX-DAIS Games: Game Analysis Techniques*,

Cambridge, Mass., Massachusetts Institute of Technology, Center for International Studies, 1966.

———, "War Games and Operations Research," *Philosophy of Science*, 22(4):309–320, Oct. 1955.

Custance, Adm. Sir Reginald, *A Study of War*, London, Constable & Co., Ltd., 1924.

Dalkey, Norman C., *Families of Models*, Santa Monica, Calif., The RAND Corporation, P-3198, Aug. 1965.

——— and Olaf Helmer, "An Experimental Application of the Delphi Method to the Use of Experts," *Management Science*, 9(3):458–467, April 1963.

Dannhauer, General der Infanterie z.D., "Das Reisswitzsche Kriegsspiel von seinen Beginn bis zum Tode des Erfinders 1827," *Militair-Wochenblatt*, 1874(56):527–532, July 11, 1874.

Dashiell, Fred K., and William W. Fain, "Solution of the Extended Lanchester Equations Used in a Tactical Warfare Simulation Programme," *Journal of the Canadian Operational Research Society*, 4(2):89–96, July 1966.

Date, Govind Tryambak, *The Art of War in Ancient India*, London, Oxford University Press, 1929.

Daveluy, R., "A Study of Naval Strategy," Fifth Part, Examples, translated by Philip R. Alger, *United States Naval Institute Proceedings*, (whole no. 134):391–428, June 1910.

———, "The Genius of Naval Warfare," translated by Philip R. Alger, *United States Naval Institute Proceedings*, (whole no. 135):753–804, Sept. 1910, and (whole no. 136):991–1042, Dec. 1910.

Davidson, Henry A., *A Short History of Chess*, New York, Greenberg: Publisher, Inc., 1949.

Davies, Max, and Michel Verhulst (eds.), *Operational Research in Practice*, Report of a NATO conference, New York, Pergamon Press, 1958.

———, R. T. Eddison, and Thornton Page (eds.), *Proceedings of the First International Conference on Operational Research*, Oxford, 1957, Baltimore, Operations Research Society of America, 1957.

Davis, Burke, *The Billy Mitchell Affair*, New York, Random House, Inc., 1967.

Davis, John B., Jr., and J. A. Tiedeman, "The Navy War Games Program," *United States Naval Institute Proceedings*, (whole no. 688):61–67, June 1960.

Davis, Lee J., "Map Maneuvers—Their Preparation and Conduct," *Military Review*, 31(8):16–24, Nov. 1951.

Dayan, Moshe, *Diary of the Sinai Campaign*, New York, Harper & Row, Publishers, Incorporated, 1966.

Deems, Paul S., "War Gaming and Exercises," *USAF Air University Quarterly Review*, 9(1):98–126, Winter 1956–1957.

DeQuoy, Alfred W., "Centaur Rides Again," *Army Information Digest*, 17(2):54–59, April 1962.

DeSelincourt, Aubrey, *The World of Herodotus*, London, Martin Secker & Warburg, Ltd., 1962.

Dewey, L. R., and the THEATERSPIEL Staff, "The THEATERSPIEL Game," unpublished paper, McLean, Va., Research Analysis Corporation, 1966.

Dikshitar, V. R. Ramachandra, *War in Ancient India*, Madras, India, Macmillan and Co., Inc., 1944.

Dodge, Theodore Ayrault, *Great Captains*, Boston, Houghton Mifflin Company, 1889.

Doenitz, Adm. Karl, *Memoirs, Ten Years and Twenty Days*, translated by R. H. Stevens in collaboration with David Woodward, London, Weidenfeld and Nicholson, 1959.

Donaldson, John L., Thomas R. Shaw, and Richard G. Williams, *A Semi-automatic War Gaming System*, Bethesda, Md., Research Analysis Corporation, RAC-ORO-TP-59 (AD 282262), June 1962.

Dondero, Lawrence J., R. C. Ling, P. F. Narten, N. W. Parsons, R. F. Patchett, L. S. Simcox, C. O. Smeak, and W. Whipple, *TACSPIEL War Game Procedures and Rules of Play for Guerrilla/Counterguerrilla Operations*, Addendum to Technical Paper RAC-TP-111, McLean, Va., Research Analysis Corporation, Oct. 1965; RAC-TP-223 (AD 811508), Aug. 1966.

Douhet, Giulio, *The Command of the Air*, translated by Gino Ferrari, New York, Coward-McCann, Inc., 1942.

Dowd, John T., "Airborne Operations," unpublished manuscript, HQ, USARPAC, Nov. 1966. (List of Airborne operations with dates, places, mission, and forces.) A chronological history of the employment of Airborne Forces (parachute and glider) from the first recorded operation in 1931 to date.

Dowdall, Lt. Harry G., and Joseph H. Gleason, *Sham-Battle*, New York, Alfred A. Knopf, Inc., 1921.

Dresch, Francis W., "The Construction of Economic Models with Application to Models for COSMAGON," *War Gaming—COSMAGON and ZIGSPIEL*, Chevy Chase, Md., The Johns Hopkins University, Operations Research Office, ORO-SP-12, June 3, 1955.

―――, *A Schematic Representation of a Highly Simplified Economic Model for the ZIGSPIEL Game*, Chevy Chase, Md., The Johns

Hopkins University, Operations Research Office, ORO-SP-12, June 3, 1955.
Dresher, Melvin, *Games of Strategy, Theory and Applications*, Englewood Cliffs, N.J., Prentice-Hall, Inc., 1961.
Dreyfus, Hubert L., *Alchemy and Artificial Intelligence*, Santa Monica, Calif., The RAND Corporation, P-3244, Dec. 1965.
Duckwall, R. L., "Tomorrow's Battles Today," *Army Information Digest*, 18(10):12–19, Oct. 1963.
Duke, Richard D., Gaming-Simulation in Urban Research, from "Gaming-Simulation Studies in Urban Land Use Allocation," Ph.D. dissertation, University of Michigan, Ann Arbor, 1964; Michigan State University, Institute for Community Development and Services.
―――――, "Gaming Urban Systems," Institute for Community Development, Reprint 30 from *Planning 1965*, Conference of the American Society of Planning Officials and the Community Planning Association of Canada, April, 25–29, 1965.
―――――, Paul H. Ray, and Allan G. Feldt, "Game-Simulations in the Teaching of Urban Sociology," unpublished paper, authors at Michigan State University, University of Michigan, and Cornell University, respectively.
Dull, Paul S., and Michael Takaaki Umemura, *The Tokyo Trials:* A Functional Index to the Proceedings of the International Military Tribunal for the Far East, Ann Arbor, The University of Michigan Press, 1957.
Duncan, Acheson, Jr., *Quality Control and Industrial Statistics*, rev. ed., Homewood, Ill., Richard D. Irwin, Inc., 1959.
Dunlap, J. W. (ed.), *Mathematical Models of Human Behavior*, Stamford, Conn., Dunlap and Associates, Inc., 1955.
Dunne, Frank, *The Draughts-Player's Guide and Companion*, A Guide to the Student, and a Companion for the Advanced Player, Warrington, England, Frank Dunne, 1890.
Durbin, Eugene P., *TARLOG: A Differential Ground Combat Model*, Santa Monica, Calif., The RAND Corporation, P-3301, Feb. 1966.
Dyer, George C., "Naval Amphibious Landmarks," *United States Institute Proceedings*, (whole no. 762):50–60, Aug. 1966.
Earle, Edward Mead (ed.), *Makers of Modern Strategy, Military Thought from Machiavelli to Hitler*, Princeton, N.J., Princeton University Press, 1952.
East Coast War Games Council, *Proceedings for the Third Symposium*, Miami Beach, Fla., Feb. 27–28, 1964, a collection of unpublished papers, printed in a limited edition for conference participants. (Editor and issuing source unidentified.)

Easton, David, *A Systems Analysis of Political Life*, New York, John Wiley & Sons, Inc., 1965.

Eddy, Alan G., and Paul C. Hewett, *Player Participation Gaming in Limited War Applications*, Bedford, Mass., Technical Operations, Inc., OMEGA Staff Memo 61-1, Feb. 16, 1961.

Edmonds, James E., *A Short History of World War I*, London, Oxford University Press, 1951.

Elliott, Charles F., "The Genesis of the Modern U.S. Navy," *United States Naval Institute Proceedings*, (whole no. 757):62–69, March 1966.

Ellis, J. W., Jr., and T. E. Greene, "The Contextual Study: A Structured Approach to the Study of Political and Military Aspects of Limited War," *Operations Research*, 8(5):539–651, Sept.–Oct. 1960.

Emme, Eugene M., *The Impact of Air Power—National Security and World Politics*, Princeton, N.J., D. Van Nostrand Company, Inc., 1959.

Engel, Joseph H., "Comments on a Paper by H. K. Weiss," *Operations Research*, 11(1):147–150, Jan.–Feb. 1963.

———, "A Verification of Lanchester's Law," *Journal of the Operations Research Society of America*, 2(2):163–171, May 1954.

Enos, John L., *An Analytic Model of Political Allegiance and Its Application to the Cuban Revolution*, Santa Monica, Calif., The RAND Corporation, P-3197, Aug. 1965.

Esposito, Vincent J. (ed.), *The West Point Atlas of American Wars*, New York, Frederick A. Praeger, Inc., 1959.

Evans, George W., II, Graham F. Wallace, and Georgia L. Sutherland, *Simulation Using Digital Computers*, Englewood Cliffs, N.J., Prentice-Hall, Inc., 1967.

Fall, Bernard B., *Hell in a Very Small Place; The Siege of Dien Bien Phu*, Philadelphia, J. B. Lippincott Company, 1967.

———, *Street Without Joy; Indochina at War 1946–1954*, Harrisburg, Pa., Stackpole Company, 1961.

———, *The Two Vietnams: A Political and Military Analysis*, 2d rev. ed., New York, Frederick A. Praeger, Inc., 1967.

———, *Viet-Nam Witness 1953–66*, New York, Frederick A. Praeger, Inc., 1966.

Falls, Cyril B., *The Art of War*, New York, Oxford University Press, 1961.

Featherstone, Donald F., *Advanced War Gaming*, London, Stanley Paul & Co., Ltd., 1969.

———, *Air War Games, Fighting Air Battles with Model Aircraft*, London, Stanley Paul & Co. Ltd., 1966.

———, *Naval War Games, Fighting Sea Battles with Model Ships*, London, Stanley Paul & Co. Ltd., 1965.
———, *War Games, Battles and Maneuvres with Model Soldiers*, London, Stanley Paul & Co. Ltd., 1962.
Feigenbaum, Edward, and Julian Feldman (eds.), *Computers and Thought*, New York, McGraw-Hill Book Company, 1963.
Fergusson, Charles M., Jr., "Strategic Thinking and Studies," *Military Review*, 44:9–24, April 1964.
Filipponi, Ernesto, "Le Scuole Navale Di Guerra," *Rivista Marittima*, 41(3):435–458, Marzo 1908.
———, "Les Écoles Navales de Guerre" (Naval War Schools), translated from "Le Scuole Navale di Guerra" in *Rivista Marittima*, translated into French by A. Fournier in *Revue Maritime* (Paris), Juillet-Août-Septembre, 1908, pp. 401–420.
Fishman, George S., *Problems in the Statistical Analysis of Simulation Experiments: The Comparison of Means and the Length of Sample Records*, Santa Monica, Calif., The RAND Corporation, RM-4880-PR, Feb. 1966.
Fiske, Bradley A., "American Naval Policy," *United States Naval Institute Proceedings*, (whole no. 113):1–80, March 1905.
Flagle, Charles D., William H. Huggins, and Robert H. Roy (eds.), *Operations Research and Systems Engineering*, Baltimore, The Johns Hopkins Press, 1960.
Flattops in the War Games, in "Evolution of Aircraft Carriers," *Naval Aviation News*, 43rd Year:22–27, Aug. 1961.
Foch, F., *De la Conduite de la Guerre, La Manoeuvre pour la Bataille*, Paris-Nancy, Berger-Levrault, Editeurs, 1915.
———, *The Principles of War*, translated by J. de Morinni, New York, The H. K. Fly Co., 1918.
Fogelsanger, D. K., "Minigaming," unpublished paper, presented to Fifth Symposium, East Coast War Games Council, Miami Beach, Fla., May 6, 1966, Fort Belvoir, Va., CORG, Technical Operations, Inc.
"Foreign War Games," *Revue Militaire de l'Étranger*, V, in *Selected Professional Papers*, No. XVIII, Government Printing Office, 1898.
Forrestel, E. P., *Admiral Raymond A. Spruance, USN, A Study in Command*, with Foreword by C. W. Nimitz, Government Printing Office, 1966.
Foster, Richard B., and Francis P. Hoeber, "Cost-Effectiveness Analysis for Strategic Decisions," *Journal of the Operations Research Society of America*, 3:482–493, Nov. 1955.
Franklin, William D., "Douhet Revisited," *Military Review*, 47(11):65–69, Nov. 1967.

Fremantle, Edmund R., "Naval Tactics," paper presented Feb. 19, 1886, *Journal of the Royal United Service Institution*, 30(133):199-231, 1886.

———, *The Navy As I Have Known It, 1849-1899*, London, Cassell & Co., Ltd., 1904.

Friedman, Saul, "The RAND Corporation and Our Policy Makers," *The Atlantic Monthly*, 212(3):61-68, Sept. 1963.

Froude, James Anthony, *The Spanish Story of the Armada and Other Essays*, New York, Charles Scribner's Sons, 1892.

Fuchida, Mitsuo, and Okumiya Masatake, in Clarke H. Kawakami and Roger Pineau (eds.), *Midway, The Battle That Doomed Japan*, Annapolis, U.S. Naval Institute, 1955.

Fulkerson, D. R., and S. M. Johnson, "A Tactical Air Game," *Operations Research*, 5(5):704-712, Oct. 1957.

Fuller, J. F. C., *The Conduct of War 1789-1961*, New Brunswick, N.J., Rutgers University Press, 1961.

———, *The Decisive Battles of the Western World and Their Influence Upon History*, London, Eyre & Spottiswoode (Publishers), Ltd., 1956.

———, *The Foundations of the Science of War*, London, Hutchinson & Co. (Publishers), Ltd., 1926.

Furer, J. A., "Research in the Navy," *Journal of Applied Physics*, 15(3):209-213, March 1944.

———, "Scientific Research and Modern Warfare," *United States Naval Institute Proceedings*, (whole no. 505):259-273, March 1945.

"Games Students Play," *Time*, 87(22):51, June 3, 1966.

Gamow, George A., *Certain Aspects of Battle Theory*, Chevy Chase, Md., The Johns Hopkins University, Operations Research Office, ORO-T-230, Aug. 1953.

———, "Monte Carlo Method in War-game Theory," in *Mathematical Models for Ground Combat*, Chevy Chase, Md., The Johns Hopkins University, Operations Research Office, ORO-SP-11 (AD 235891), April 1957.

——— and Richard E. Zimmerman, *Mathematical Models for Ground Combat*, Chevy Chase, Md., The Johns Hopkins University, Operations Research Office, ORO-SP-11 (AD 235891), April 1957.

Ganoe, William Addleman, *The History of the United States Army*, rev. ed., Ashton, Md., Eric Lundberg, 1964.

Garratt, John G., "Model Soldiers," *A Collector's Guide*, London, Seeley Service & Co., Ltd., 1959.

Germany, Reichsarchiv, *Der Weltkrieg, 1914 bis 1918*, Die Militarischen Operationen zum Lande. Zwerter Band, Die Befreiung Ostpreussens, Berlin, L. S. Mittler & Sohn, 1925.

Gibson, Ralph E., "The Recognition of Systems Engineering," chap. 4, in Charles D. Flagle et al. (eds.), *Operations Research and Systems Engineering*, Baltimore, The Johns Hopkins Press, 1960.

Giffin, Sidney F., *The Crisis Game. Simulating International Conflict*, Garden City, N.Y., Doubleday & Company, Inc., 1965.

Gilpin, Robert, and Christopher Wright (eds.), *Scientists and National Policy Making*, New York, Columbia University Press, 1964.

Girard, E. W., "The History of War Gaming," presented at the Second Symposium on War Gaming, Washington, D.C., Washington Operations Research Council, 1964.

―――, "The War Game as a Methodological Tool," unpublished paper, Bethesda, Md., Research Analysis Corporation, March 1962.

Goerlitz, Walter, *History of the German General Staff 1657–1945*, translated by Brian Battershaw, New York, Frederick A. Praeger, Inc., 1955.

―――, *Paulus and Stalingrad*, translated by R. H. Stevens, New York, The Citadel Press, 1963.

―――, *Paulus, Ich stehe hier auf Befehl*, Verlag fur Wehrwesen Bernard & Graefe, Frankfurt am Main, 1960; English translation by R. H. Stevens, New York, The Citadel Press, 1963.

Goldberg, Alfred (ed.), *A History of the United States Air Force, 1907–1957*, Princeton, N.J., D. Van Nostrand Company, Inc., 1957.

Goldhamer, Herbert, and Hans Speier, "Some Observations on Political Gaming," *World Politics*, 12(1):71–83, Oct. 1959.

―――― and ――――, "Some Observations on Political Gaming," in James A. Rosenau (ed.), *International Politics and Foreign Policy*, New York, The Free Press of Glencoe, Inc., 1961.

Goldman, Thomas, *Proceedings of Symposium on Military Logistics*, Sept. 1963, Washington, D.C., Washington Operations Research Council, 1963.

Golovine, Nicholas N., *The Russian Army in the World War*, New Haven, Conn., Yale University Press, 1931.

―――, *The Russian Campaign of 1914*, The Beginning of the War and Operations in East Prussia, translated by A. G. S. Muntz, Fort Leavenworth, Kans., The Command and General Staff School Press, 1933.

Gombrich, E. H., and E. Kris, *Caricature*, Harmondsworth, England, The King Penguin Books, Ltd., 1940.

Goode, H. H., and R. E. Machol, *System Engineering*, New York, McGraw-Hill Book Company, 1957.

Goodeve, Charles F., "The 'Scientific Method,'" *Proceedings of the First International Conference on Operational Research*, Baltimore, Operations Research Society of America, 1957.

Goodman, Nathan G. (ed.), *The Ingenious Dr. Franklin*, Philadelphia, University of Pennsylvania Press, 1931.

Gorden, Morton, *International Relations Theory in the TEMPER Simulation*, Cambridge, Mass., Abt Associates, Inc., 1965.

Great Britain War Department, *Rules for the Conduct of the War-Game*, London, printed under the Superintendence of Her Majesty's Stationery Office, 1884.

Great Britain War Department, *Rules for the Conduct of the War-Game on a Map*, London, Harrison and Sons, Ltd., 1899.

Great Britain War Office, Topographical and Statistical Department, "Regulation for the Training of Troops in the Field and for the Conduct of Peace Maneuvres," *Reforms in the French Army*, translated by Lt. E. Baring, London, Her Majesty's Stationery Office, 1871.

————, Intelligence Department, "An Outline of the 'Attack Formations' for Infantry in the Austrian, French, German and Italian Armies," *Reforms in the French Army*, London, Her Majesty's Stationery Office, March 1881.

————, Topographical and Statistical Department, "The Elementary Tactics of the Prussian Infantry," *Reforms in the French Army*, translated by Lt. E. Baring, London, Her Majesty's Stationery Office, 1872.

Greaves, Fielding L., "Peace in Our Time—Fact or Fable?", *Military Review*, 42(12):55–58, Dec. 1962.

Gretton, Peter, *Convoy Escort Commander*, London, Cassell & Co., Ltd., 1964.

Griffith, Samuel B. (translator), *Mao Tse-Tung on Guerrilla Warfare*, New York, Frederick A. Praeger, Inc., 1961.

Guderian, Heinz, *Panzer Leader*, translated from German by Constantine Fitzgibbon, London, Michael Joseph, Ltd., 1952.

Guetzkow, Harold, et al., *Simulation in International Relations: Developments for Research and Teaching*, Englewood Cliffs, N.J., Prentice-Hall, Inc., 1963.

————, "Simulations in International Relations," *Proceedings of the IBM Scientific Computing Symposium on Simulation Models and Gaming*, Yorktown Heights, N.Y., Thomas J. Watson Research Center, Dec. 1964.

————, *Simulation in Social Science: Readings*, Englewood Cliffs, N.J., Prentice-Hall, Inc., 1962.

————, "Some Uses of Mathematics in Simulation of International

Relations," in John M. Claunch (ed.), *Mathematical Applications in Political Science*, Dallas, Texas, Southern Methodist University, The Arnold Foundation, 1965.

―――, "A Use of Simulation in the Study of International Relations," *Behavioral Science*, 4 (3):183–191, July 1959.

Gurney, Gene, *A Pictorial History of the United States Army in War and Peace, from Colonial Times to Vietnam*, New York, Crown Publishers, Inc., 1966.

Halder, Franz, *War Diary*, vol. V, Oct. 31, 1940 to Feb. 20, 1941, undated translation of manuscript, New York, Foreign Policy Association.

Hall, Vernon W., "Coast Artillery War Game Material," *Journal of the United States Artillery*, (whole no. 133):343–345, May–June 1915.

Halsbury, Earl of, "From Plato to the Linear Program," *Journal of the Operations Research Society of America*, 3(3):239–254, Aug. 1955.

Hamburger, William, *Monopologs: An Inventory Management Game*, Santa Monica, Calif., The RAND Corporation, RM-1579, Jan. 3, 1956.

Hanes, R. M., and J. W. Gebhard, "The Computer's Role in Command Decision," *United States Naval Institute Proceedings*, (whole no. 763):60–68, Sept. 1966.

Hantzes, Harry N., et al., *Development of Intelligence Requirements Through Interrogation of War-game Players*, Bethesda, Md., The Johns Hopkins University, Operations Research Office, ORO-SP-112, Sept. 1969.

Harbottle, Thomas B., *Dictionary of Battles*, London, Swan Sonnenschein & Co., Ltd., 1904.

Hare, Robert R., Jr., "Models and System Effectiveness," *Military Review*, 45:26–30, Nov. 1965.

Harrison, Joseph O., Jr., *Computer-aided Information Systems for Gaming*, McLean, Va., Research Analysis Corporation, RAC-TP-133, Sept. 1964.

――― and M. F. Barrett, "Guide to the Literature," *Computer-aided Information Systems for Gaming*, McLean, Va., Research Analysis Corporation, RAC-TP-133, Sept. 1964.

――― and Edward M. Lee, *The Stratspiel Pilot Model*, Bethesda, Md., The Johns Hopkins University, Operations Research Office, ORO-TP-7, Aug. 1960.

Harsanyi, John C., "Mathematical Models for the Genesis of War," *World Politics*, 14(4):687–699, July 1962.

Hart, Joseph T., "Gaming as a Research Tool in International Relations," doctoral dissertation, American University, School of International Service, Washington, D.C., 1967.

———, "STRAT-X: A Strategic Exercise," unpublished paper, McLean, Va., Research Analysis Corporation, Sept. 1965.

Harte, Walter, *The History of Gustavus Adolphus*, King of Sweden surnamed The Great, 3d ed., vols. I and II, London, Luke Ansard & Sons, 1807.

Harvey, H. Paul Beck, *Description and Rules of the Game of Two Cities*, Philadelphia, George H. Buchanan and Co., 1903.

Haviland, W. A. de, *The ABC of GO, The National War-Game of Japan*, Yokohama, Japan, Kelly & Walsh, Ltd., 1910.

Hays, David G., *Simulation: An Introduction for Anthropologists*, Santa Monica, Calif., The RAND Corporation, P-2668, Nov. 1962.

Hays, James H., "Basic Concepts of Systems Analysis," *Military Review*, 45:4–13, April 1965.

Haywood, O. G., Jr., "Military Decision and Game Theory," *Journal of the Operations Research Society of America*, 2(4):365–385, Nov. 1954.

Hearle, Edward F. R., *How Useful Are "Scientific" Tools of Management*, Santa Monica, Calif., The RAND Corporation. P-2260, March 1961.

Hedrick, David I., "Research and Experimental Activities of the U.S. Naval Proving Ground," *Journal of Applied Physics*, 15(3):262–268, March 1944.

Heistand, H. O. S., "Foreign War Games," *Selected Professional Papers*, 18:233–289, Jan. 1, 1898 (translated from *Revue Militaire de l'Étranger*, Aug. and Oct. 1897), War Department, Government Printing Office.

Helmbold, R. L., *Historical Data and Lanchester's Theory of Combat*, Fort Monroe, Va., Technical Operations, Inc., Combat Operations Research Group, CORG-SP-128, 1961.

———, *Lanchester Parameters for Some Battles of the Last Two Hundred Years*, Fort Monroe, Va., Technical Operations, Inc., Combat Operations Research Group, CORG-SP-122, 1961.

Helmer, Olaf, *Strategic Gaming*, Santa Monica, Calif., The RAND Corporation, P-1902, Feb. 10, 1960.

———, "The Systematic Use of Expert Judgment in Operations Research," in G. Kreweras and G. Morlat (eds.), *Proceedings of the Third International Conference on Operational Research*, Oslo, 1963, Paris, Dunod, 1964.

——— and Nicholas Rescher, *On the Epistemology of the Inexact Sci-*

*ences*, Santa Monica, Calif., The RAND Corporation, R-353, Feb. 1960.

Hendrickson, Robert G., *Concept Paper: Computer-assisted War Games*, McLean, Va., Research Analysis Corporation, RAC-TP-192 (AD 477579), Nov. 1965.

―――, *Pros and Cons of War Gaming and Simulation*, McLean, Va., Research Analysis Corporation, RAC(ORO)-TP-49 (AD 480730), Oct. 1961.

Hermann, Charles F., *Crises in Foreign Policy Making: A Simulation in International Politics*, China Lake, Calif., Project Michelson, U.S. Naval Ordnance Test Station, 1965.

――― and Margaret G. Hermann, "An Attempt to Simulate the Outbreak of World War I," *American Political Science Review*, 61(3):400–416, June 1967.

Heymont, Irving, "Israeli Defense Forces," *Military Review*, 47(3):37–47, Feb. 1967.

Higgins, Gerald J., "Sky Lancers in Battle," in John T. Dowd (ed.), *Airborne Operations*, HQ, USARPAC, Nov. 1966.

Higgins, Trumbull, *Hitler and Russia*, The Third Reich in a Two-front War 1937–1943, New York, The Macmillan Company, 1966.

Higham, Robin, *The Military Intellectuals in Britain: 1918–1939*, New Brunswick, N.J., Rutgers University Press, 1966.

Hill, John G., "Notes on Political Decision Models," unpublished paper, McLean, Va., Research Analysis Corporation, Oct. 7, 1965.

―――, "The Theaterspiel Manual," vol. I, unpublished paper, McLean, Va., Research Analysis Corporation, RAC-TP-120, Dec. 1964.

Himes, Billy L., Sr., "The Theaterspiel Logistics Model," unpublished paper, McLean, Va., Research Analysis Corporation, Oct. 1961.

―――, Dino G. Pappas, and Horace H. Figuers, *An Experimental Cold War Model, THEATERSPIEL'S Fourth Research Game*, McLean, Va., Research Analysis Corporation, RAC-TP-120, Dec. 1964.

"Historical Trends Related to Weapon Lethality," Basic Historical Studies, Annex vol. I, Washington, D.C., Historical Evaluation and Research Organization, Oct. 15, 1964.

Hitch, Charles J., *Decision Making for Defense*, Berkeley, University of California Press, 1965.

―――, "An Appreciation of Systems Analysis," *Journal of the Operations Research Society of America*, 3(4):466–481, Nov. 1955.

Hitchman, Norman, and Paul F. Dunn, "A Design for a Hand-played Tactical War Game, Part I: The Structure of Rules and Mechanics of Play; Part II: The Combat Model," presented at the Sixth Annual Meeting of the Operations Research Society of America, May 16, 1958, Boston, Mass.; Abstract, *Operations Research* 6(4):626, July–Aug. 1958.

Hittle, J. D., *Jomini and His Summary of the Art of War*, Harrisburg, Pa., Military Service Publishing Co., 1947.

Hjalmarson, J. K., "The Development of War Games," *Canadian Army Journal*, 15(1):4–10, Winter 1961.

Ho, Kenmin, "Mao's 10 Principles of War," *Military Review*, 47(7):96–98, July 1967.

Hoffmann, Max von, *The War of Lost Opportunities*, New York, International Publishers Company, Inc., 1925.

———, *War Diaries and Other Papers*, 2 vols., translated from the German by Eric Sutton, London, M. Secker, 1929.

Hofmann, Rudolf, *War Games*, Draft translation by P. Luetzkendorf, Washington, D.C., U.S. Department of Army, Office of the Chief of Military History, MS P-094.

Holmen, Milton G., *Applications of Simulation in Command and Control Systems*, Santa Monica, Calif., Systems Development Corporation, SP-1455 (AD 427 817), Nov. 1963.

Holmes, W. J., "Naval Research," *United States Naval Institute Proceedings*, (whole no. 396):178–185, Feb. 1936.

Holmquist, C. O., and R. S. Greenbaum, "Navy's 'In-House' Research Laboratories," *United States Naval Institute Proceedings*, (whole no. 708):68–75, Feb. 1962.

Hornbaker, Glen E., and Thomas E. Goswick, *The Firing Assessment Model, Description and User's Guide*, Dahlgren, Va., U.S. Department of Navy, Naval Weapons Laboratory, TM-K-29/65, Feb. 1965.

Horowitz, Irving Louis, *The War Game—Studies of the New Civilian Militarists*, New York, Ballantine Books, Inc., 1963.

Hough, Richard, *The Fleet That Had to Die*, New York, The Viking Press, Inc., 1958.

Howard, Herbert S., "The David W. Taylor Model Basin," *Journal of Applied Physics*, 15(3):227–235, March 1944.

———, "The David W. Taylor Model Basin," *The Smithsonian Report for 1944*, Washington, D.C., Smithsonian Institution, 1945.

Humphrey, Hubert H., "The Technological Revolution and the World of the 1970's," *Department of State Bulletin*, 56(1440):164–168, Jan. 30, 1967.

Hutchinson, Elmer (ed.), "Mobilization of Scientific Resources—IV, The U.S. Navy," *Journal of Applied Physics*, 15(3):203, March 1944.
Hynden, C. M., Jr., *The Ship-to-Shore Model*, Dahlgren, Va., U.S. Department of the Navy, U.S. Naval Weapons Laboratory, TM-K-3/61, Feb. 1961.
Immanuel, Friedrich, *The Regimental War Game*, translated by Walter Krueger, Kansas City, Mo., Hudson Press, 1907.
Ingram, Henry A., "Research in the Bureau of Ships," *Journal of Applied Physics*, 15(3):215–220, March 1944.
Ivanoff, D. H., *International Environment Evaluation Systems*, Santa Monica, Calif., Douglas Aircraft Co., SM-51925, Jan. 1966.
―――, *Threat Analysis Briefing*, Santa Monica, Calif., Douglas Aircraft Co., Douglas Missile and Space Systems Division, June 1966.
――― and D. Harrison, "International Environment: 1965–1975," *Environment Data Handbook*, Santa Monica, Calif., Douglas Aircraft Co., SM-49226, Oct. 1965.
Jacobsen, H. A., and J. Rohwer (eds.), *Decisive Battles of World War II: The German View*, New York, G. P. Putnam's Sons, 1965.
Jacobson, Robert V., *DECAP, A Computerized War Game of Aircraft Penetration*, Bedford, Mass., Raytheon Company, BR-3079, July 1964.
Jane, Fred T., *How to Play the "Naval War Game,"* Official Rules Cancelling All Others, London, Sampson, Low, Marston & Co., Ltd., 1912.
―――, "The Naval War Game," *United States Naval Institute Proceedings*, (whole no. 107):595–660, Sept. 1903.
Johnson, Chalmers A., "Civilian Loyalties and Guerrilla Conflict," *World Politics*, 14(4):646–661, July 1960.
Johnson, Ellis A., *The Application of Operations Research to Industry*, Chevy Chase, Md., The Johns Hopkins University, Operations Research Office, 1953.
―――, *The Estimate of the Situation as a Checklist for Operations Research*, Chevy Chase, Md., The Johns Hopkins University, Operations Research Office, 1952.
―――, *The History and Future of War Gaming in Operations Research*, unpublished paper, Bethesda, Md., The Johns Hopkins University, Operations Research Office, 1960.
―――, "The Nature of War," unpublished paper, Chevy Chase, Md., The Johns Hopkins University, Operations Research Office, Nov. 1955.

———, "The Scope of Operations Research," presented at Informal Seminar in Operations Research, Chevy Chase, Md., The Johns Hopkins University, Operations Research Office, Seminar Paper 1, 1952–1953, Oct. 1952.

———, *The Technological Conflict*, Bethesda, Md., The Johns Hopkins University, Operations Research Office, 1960.

Johnson, Max S., "War Gaming and Management," *Proceedings of the Third Annual Conference on Operations Research*, New York, Society for the Advancement of Management, Feb. 1958.

Johnstone, H. M., *The Foundations of Strategy*, London, George Allen & Unwin, Ltd., 1914.

Jomini, Le Baron Henri de, *The Art of War*, translated by G. H. Mendell and W. P. Craighill, Philadelphia, J. B. Lippincott Company, 1892.

———, *Précis de l'art de la Guerre*, Paris, Anselin, Libraire pour l' Art Militaire, les Science et les Arts, 1838.

Kahn, Herman, *Applications of Monte Carlo*, Santa Monica, Calif., The RAND Corporation, RM-1237, April 19, 1954.

——— and I. Mann, *Monte Carlo*, Santa Monica, Calif., The RAND Corporation, P-1165, July 30, 1957.

——— and ———, *Techniques of Systems Analysis*, Santa Monica, Calif., The RAND Corporation, RM-1829-1, Dec. 3, 1956.

Kane, Francis X., "Security Is Too Important to be Left to Computers," *Fortune*, 69(4):146–147 ff., April 1964.

Kao, John H. K., *Statistical Confidence Intervals. Their Uses and Misuses in Reliability Engineering*, New York, New York University, July 1965.

Kaplan, Morton A., "Balance of Power, Bipolarity and Other Models of International Systems," *American Political Science Review*, 51(3):684–695, Sept. 1957.

———, *System and Process in International Politics*, New York, John Wiley & Sons, Inc., 1957.

Keenan, E. A., "The CAORE War Game—Background and Operation," *Canadian Army Journal*, 15(2):15–18, Spring 1961.

Keener, Bruce, III, "The Principles of War: A Thesis for Change," *United States Naval Institute Proceedings*, (whole no. 777):26–36, Nov. 1967.

Kennedy, Ludovic, *Nelson's Captains*, New York, W. W. Norton & Company, Inc., 1951.

Kennedy, Robert F., *Thirteen Days. A Memoir of the Cuban Missile Crisis*, New York, W. W. Norton & Company, Inc., 1969.

Kibbee, Joel M., Clifford J. Craft, and Burt Nanus, *Management*

*Games—A New Technique for Executive Development,* New York, Reinhold Publishing Corporation, 1961.
Kimber, Thomas, *Vauban's First System of Fortification,* 3d ed., London, Longmans, Green, Longmans, and Roberts, 1861.
Kingsford, P. W., *F. W. Lanchester, A Life of an Engineer,* London, Edward Arnold (Publishers), Ltd., 1960.
Kinnard, Harry W. O., "Activation to Combat—in 90 Days. The Story of the 1st Cavalry Division (Airmobile)," *Army Information Digest,* 21(4):24–31, April 1966.
———, "A Victory in the Ia Drang: The Triumph of a Concept. The 1st Air Cavalry Division in Battle," *Army,* 17(9):71–91, Sept. 1967.
Kisi, Takasi, and Tadasi Hirose, "Winning Probability in an Ambush Engagement," *Operations Research,* 14(6):1137–1138, Nov.–Dec. 1966.
Kitchell, T. J., R. A. Wulf, and O. R. Swigart, Jr., *Script of Amphibious War Game Program Briefing,* Dahlgren, Va., U.S. Naval Weapons Laboratory, 1966.
Kitchener, F. W., *Rules for War Games on Maps and on Tactical Models,* Simla, India, Government Central Printing Office, 1895.
Knight, Kenneth E., "Changes in Computer Performance," *Datamation,* 12(9):40–46, Sept. 1966.
Koopman, Bernard O., "Analytical Treatment of a War Game," in G. Kreweras and G. Morlat (eds.), *Proceedings of the Third International Conference on Operational Research,* Oslo, 1963, Paris, Dunod, 1964.
Kormendi, Ferenc, "The Emperor's War Games," *The Reporter,* 35(8):50–56, Nov. 17, 1966.
Korothev, I., "Development of Soviet Military Theory in the Post War Years," *Voyeuno-istoricheskiy Zhurnal* (Moscow) (4):39–50, April 1964. Published in the series "Soviet Military Translations No. 144 (Military Doctrine) JPRS 24,667 (May 19, 1964), Washington, D.C., U.S. Department of Commerce, Office of Technical Services.
Kreidler, Robert N., "The President's Science Advisors and National Science Policy," in Robert Gilpin and Christopher Wright (eds.), *Scientists and National Policy Making,* New York, Columbia University Press, 1964.
Kreweras, G., and G. Morlat (eds.), *Proceedings of the Third International Conference on Operational Research,* Oslo, 1963, Paris, Dunod, 1964.
*Das Kriegsspiel,* Unregungen, Erfahrungen und Beispiele, Berlin, Verlag von R. Eisenschmidt, 1903.

Lachoque, Henry, *The Anatomy of Glory*, Napoleon and His Guard, adapted from the French by Anne S. K. Brown, Providence, R.I., Brown University Press, 1962.

Lanchester, F. W., *Aircraft in Warfare, The Dawn of the Fourth Arm*, London, Constable & Co., Ltd., 1916.

Landstrom, Björn, *The Ship*, Garden City, N.Y., Doubleday & Company, Inc., 1961.

Laubot, Chasseloup, *Reforms in the French Army*, Part I: "The Law of Recruiting," translated by Captain Home; Part II: "General Organization," translated by C. B. Brackenbury, London, Her Majesty's Stationery Office, 1872–1873.

Laulicht, Jerome, "The Vietnam War Game," *New Society*, Jan. 27, 1966. (Typescript copy forwarded by author.)

———— and John Martin, "Vietnam at Leeds," unpublished paper, March 1966.

————, Paul Smoker, Robin Jenkins, and John MacRae, "Participant's Manual," incomplete, unpublished manuscript, Peace Research Centre, Lancaster, England, and Peace Research Institute, Clarkson, Ontario.

Lea, Homer, *The Day of the Saxon*, New York, Harper & Brothers, 1912.

————, *The Valor of Ignorance*, New York, Harper & Brothers, 1942.

Lee, Edward M., "A Model of Combat with Both Space and Time Variables," unpublished paper, Chevy Chase, Md., The Johns Hopkins University, Operations Research Office, Nov. 9, 1954.

Lewis, Michael, *Armada Guns*, A Comparative Study of English and Spanish Armaments, London, George Allen & Unwin, Ltd., 1961.

————, *The Spanish Armada*, New York, The Macmillan Company, 1960.

Liddell Hart, Basil H., *Great Captains Unveiled*, Edinburgh, William Blackwood & Sons, Ltd., 1927.

————, *The Liddell Hart Memoirs—1895–1938*, vol. I, New York, G. P. Putnam's Sons, 1965.

————, *The Liddell Hart Memoirs—The Later Years*, vol. II, New York, G. P. Putnam's Sons, 1966.

———— (ed.), *The Rommel Papers*, translated by Paul Findlay, New York, Harcourt, Brace and Company, Inc., 1953.

————, *Through the Fog of War*, London, Faber & Faber, Ltd., 1938.

————, *Why Don't We Learn from History?*, London, George Allen & Unwin, Ltd., 1944.

Little, W. McCarty, "The Strategic Naval War Game or Chart Maneuver," *United States Naval Institute Proceedings*, (whole no. 144):1213–1233, Dec. 1912.

Livermore, W. R., *The American Kriegspiel*, A Game for Practicing the Art of War upon a Topographical Map, Boston, Houghton Mifflin Company, 1882.

———, *The American Kriegspiel*, A Game for Practicing the Art of War upon a Topographical Map, Tables and Plates, new and rev. ed., Boston, W. B. Clarke Co., 1898.

Lloyd, E. W., and A. G. Hadcock, *Artillery: Its Progress and Present Position*, Portsmouth, England, J. Griffin & Co., 1893.

Lockley, Lawrence C., *Operations Research*, New York, National Industrial Conference Board, 1957.

Loeb, Leonard B., "Naval Research in Peace and War," *United States Naval Institute Proceedings*, (whole no. 512):1169–1191, Oct. 1945.

"LOGSIM-W, A Logistics Simulation of the Wholesale Army Supply System," for the United States Army Logistics Management Center and Logistics Gaming Group, by The Johns Hopkins University, Operations Research Office, Bethesda, Md., ORO-SP-106, May 1959.

"LOGSIM-W, A Logistics Simulation of the Wholesale Army Supply System, Instructor's Manual," for the United States Army Logistics Management Center and Logistics Gaming Group, by The Johns Hopkins University, Operations Research Office, Bethesda, Md., ORO-SP-107, May 1959.

Lopez, Antonio, *Explanation of the Plates to the Theory of the Infantry Movements*, London, William Clowes & Sons, 1846.

Luce, R. Duncan, and Howard Raiffa, *Games and Decisions, Introduction and Critical Survey*, New York, John Wiley & Sons, Inc., 1957.

Luce, Stephen B., "The U.S. Naval War College," *United States Naval Institute Proceedings*, (whole no. 134):559–586, June 1910.

———, "The U.S. Naval War College" (concluded), *United States Naval Institute Proceedings*, (whole no. 135):683–696, Sept. 1910.

Lutman, R. E., and R. G. Hinkle, *The Embarkation Planning Support Model*, Dahlgren, Va., U.S. Department of Navy, U.S. Naval Weapons Laboratory, TM-K-85/65, Sept. 1965.

Luvaas, Jay, *The Education of an Army*, Chicago, The University of Chicago Press, 1964.

Lycan, Daniel L., "Space: A Challenge for the Army," *Military Review*, 46(12):82–87, Dec. 1966.

Lytel, Allan, *ABC's of Computers*, Indianapolis, The Bobbs-Merrill Company, Inc., 1966.

MacArthur, Douglas, *Reminiscences*, New York, McGraw-Hill Book Company, 1964.

McCartney, Edward E., and Adam Ensch, "Plastic Paper Pulp and Land Features of the War Game," *Journal of the United States Artillery*, (whole no. 127):315–316, May–June 1964.

McCloskey, Joseph F., "The Characteristics of Operations Research," presented at Informal Seminar in Operations Research, 1954–1955, Chevy Chase, Md., The Johns Hopkins University, Operations Research Office, Seminar Paper 24, May 1955.

―――, "Of Horseless Carriages, Flying Machines, and Operations Research," *Operations Research*, 4(2):141–147, April 1956.

――― and Florence N. Trefethen (eds.), *Operations Research for Management*, Baltimore, The Johns Hopkins Press, 1954.

MacCurdy, Edward (ed.), *The Notebooks of Leonardo da Vinci*, vols. I and II, New York, George Braziller, Inc., 1958.

MacDonald, Scot, The Japanese Developments, in "Evolution of Aircraft Carriers," *Naval Aviation News*, 43rd Year:23–27, Oct. 1962.

―――, Last of the Fleet Problems, in "Evolution of Aircraft Carriers," *Naval Aviation News*, 43rd Year:22–26, Sept. 1962.

McDonald, Thomas T., "JCS Politico-Military Desk Games," in *Second War Gaming Symposium Proceedings*, March 16, 17, 1964, pp. 63–74, Washington, D.C., Washington Operations Research Council, June 1964.

MacGowan, Roger A., "Modular Programming," *Data Processing*, 6(10):49–53, Oct. 1964.

McHugh, Francis J., *Fundamentals of War Gaming*, 2d ed., Newport, R.I., The United States Naval War College, Nov. 1961.

―――, *Fundamentals of War Gaming*, 3d ed., Newport, R.I., The United States Naval War College, March 1966.

McKinsey, J. C. C., *Introduction to the Theory of Games*, Santa Monica, Calif., The RAND Corporation, R-228, July 1952.

McLeary, O. S., *The Principles of War*, privately printed, 1930.

McNamara, Robert S., Address before American Society of Newspaper Editors, Montreal, Canada; news release, Washington, D.C., Department of Defense, Office of Assistant Secretary of Defense for Public Affairs, May 18, 1966.

MacRae, John, and Paul Smoker, "A Vietnam Simulation: A Report on the Canadian/English Joint Project," *Journal of Peace Research* (Oslo, Norway), no. 1, 1967.

McRae, V. V., "Gaming as a Military Research Procedure," unpublished paper, Bethesda, Md., Research Analysis Corporation, 1962.

Mahan, A T., *From Sail to Steam, Recollections of Naval Life,* New York, Harper & Brothers, 1907.

———, *The Influence of Sea Power Upon History 1660–1783,* London, Sampson, Low, Marston & Co., Ltd., 1890.

———, *The Life of Nelson,* The Embodiment of the Sea Power of Great Britain, Boston, Little, Brown and Company, 1943.

Maloney, E. S., "Modern War Gaming: State of the Art," *Marine Corps Gazette,* 44(11):MCA 10–12, Nov. 1960.

Manstein, Erich von, *Lost Victories,* Chicago, Henry Regnery Company, 1958.

Mao Tse-Tung, *On Guerrilla Warfare,* translated by Samuel B. Griffith, New York, Frederick A. Praeger, Inc., 1961.

Margenau, Henry, "The Competence and Limitations of Scientific Method," *Journal of the Operations Research Society of America,* 3(2):135–146, May 1955.

Markel, Gene A., *Towards a General Methodology for Systems Evaluation,* Science Park, State College, Pa., HRB-Singer, Inc., July 1965.

Martin, Francis F., *Computer Modeling and Simulations,* New York, John Wiley & Sons, Inc., 1968.

Massachusetts Institute of Technology, Center for International Studies, *The Detex-Exdet Political-Military Exercises on Naval Weapons Systems During Crises,* Final Report, Cambridge, Mass., Massachusetts Institute of Technology, D/65-5, June 1965.

Maxim, Hudson, *The Game of War,* Invention of Hudson Maxim, Landing, N.J., privately printed, 1910.

Mayfield, D. W., "Simulation of Total Atomic Global Exchange (STAGE)," *Proceedings for the Third Symposium of the East Coast War Games Council,* Miami Beach, Fla., Feb. 27–28, 1964.

Meals, Donald W., "Trends in Military Operations Research," *Operations Research,* 9(2):252–257, March–April 1961.

Meier, Richard L., and Richard D. Duke, *Gaming Simulation for Urban Planning,* East Lansing, Mich., Michigan State University, Institute for Community Development and Services, Reprint 31, 1965–1966.

Meier, Robert F., *PAM-80, SIMULATION DIGEST I,* Silver Spring, Md., The Johns Hopkins University, Applied Physics Laboratory, Aug. 23, 1964.

Mellenthin, F. W. von, *Panzer Battles,* 1939–1945, translated by H. Betzler, edited by L. C. F. Turner, London, Cassell & Co., Ltd., 1955.

Merritt, Wesley, *The Armies of Today,* New York, Harper & Brothers, 1892.

———, "The Army of the United States," *The Armies of Today*, New York, Harper & Brothers, 1893.

Metaxas, Alexandre, "Blunders in the Kremlin," *The Sunday Times* (London), Dec. 9, 1956.

Metcalfe, Howard H., and Lin Conger, *Final Report on the Computer-assisted Rapid Air and Missile Battle Analysis Game (CARAMBA), Mark III Version*, Los Angeles, Calif., Planning Research Corporation, R-710, May 30, 1956.

*M.E.T.R.O. Report on Phase I*, Lansing, Mich., Tri-county Regional Planning Commission, M.E.T.R.O., Project Technical Report 5, Jan. 1966.

Mihori, Fukumensi, *Japanese Game of "GO,"* translated by Z. T. Iwado, Japanese Government Railways, Board of Tourist Industry, 1939.

Milkman, Raymond H., "Operations Research in World War II," *United States Naval Institute Proceedings*, (whole no. 783):78–83, May 1968.

Miller, David, "The Government-Management Game," unpublished manuscript, Washington, U.S. Department of Commerce, National Bureau of Standards, 1966.

Milling, J. McM., "The War Game," *Journal of the United Service Institution of India* (Simla, India), 63(n.p.), Oct. 1933.

Minor, W. Thane, "Political-Military Gaming," unpublished paper, script for video briefing, Washington, D.C., The Pentagon, The Joint War Games Agency, Cold War Division, 1966.

Montgomery, Bernard L., *El Alamein to the River Sangro*, New York, E. P. Dutton & Co., Inc., 1949.

Montgomery, Field Marshal Viscount of Alamein, *A History of Warfare*, Cleveland, The World Publishing Company, 1968.

Mood, Alexander M., *War Gaming as a Technique of Analysis*, Santa Monica, Calif., The RAND Corporation, RAND Paper P-899, Sept. 1954.

——— and F. A. Graybill, *Introduction to the Theory of Statistics*, New York, McGraw-Hill Book Company, 1963.

——— and R. D. Specht, *Gaming as a Technique of Analysis*, Santa Monica, Calif., The RAND Corporation, P-579, Oct. 1954.

Mooney, Craig M., "Operational Research—A Deciding Military Science," *Canadian Business*, 27(7):19–21 ff., July 1954.

Moore, John V., "Scenarios—A Basis for Gaming," *Proceedings for the Third Symposium of the East Coast War Games Council*, Miami Beach, Fla., Feb. 27–28, 1964.

Moorehead, Alan, *Montgomery—A Biography*, New York, Coward-McCann, Inc., 1946.

Mordal, Jacques (Pseudonym), Herve Cras, *Twenty-five Centuries of Sea Warfare*, translated by Len Ortzen, New York, Clarkson N. Potter, Inc., 1959.

Morison, Samuel Eliot, "Coral Sea, Midway and Submarine Actions May 1942–August 1942," *History of the United States Naval Operations in World War II*, vol. IV, Boston, Little, Brown and Company, 1949.

———, "The Rising Sun in the Pacific 1939–April 1942," *History of United States Naval Operations in World War II*, vol. III, Boston, Little, Brown and Company, 1948.

———, "The Struggle for Guadalcanal, August 1942–February 1943," *History of United States Naval Operations in World War II*, vol. V, Boston, Little, Brown and Company, 1949.

Moro-Lin, F., "Duello Navale," *Rivista Marittima* (Roma), 24 (Terzo Trimester):123–148, 1891.

Morse, Philip M., and George E. Kimball, *Methods of Operations Research*, New York, The Technology Press of the Massachusetts Institute of Technology and John Wiley & Sons, Inc., 1951.

Morton, Louis, "Crisis in the Pacific," *Military Review*, 46(4):12–21, April 1966.

Morton, N. W., "A Brief History of the Development of Canadian Military Operations Research," *Operations Research*, 4(2):187–192, April 1956.

Moses, W. M., "Research in the Bureau of Ordnance," *Journal of Applied Physics*, 15(3):249–254, March 1944.

Moss, James A., *Applied Minor Tactics*, Washington, D.C., National Capital Press, Inc., March 1912.

Mouillard, L. P., "The Empire of the Air: An Ornithological Essay on the Flight of Birds," *Annual Report of the Board of Regents of the Smithsonian Institution*, Government Printing Office, 1893.

Murphy, Charles J. V., "Defense: The Revolution Gets Revolutionary," *Fortune*, 53:100–103, May 1956.

Murray, H. J. R., *A History of Board Games Other than Chess*, Oxford, Clarendon Press, 1952.

———, *A History of Chess*, Oxford, Clarendon Press, 1913.

Musgrove, Edgar F., "No Game," *Marine Corps Gazette*, 49(8):53–56, Aug. 1965.

Nachin, Lucien, *Sun Tse et Les Anciens Chinois, Ou Tse et Se Ma Fa*, Paris, Berger-Levrault & Cie, 1948.

*The Navy War Games Manual*, Washington, D.C., U.S. Department of Navy, Chief of Naval Operations, Office of the Assistant for War Gaming Matters, June 1, 1965.

Newcomb, Paul R., "The Analysis of War Gaming, unpublished

paper, Chevy Chase, Md., The Johns Hopkins University, Operations Research Office, May 24, 1956.

Newell, Allen, and Herbert A. Simon, "Problem Solving," in Edward Feigenbaum and Julian Feldman (eds.), *Computers and Thought*, New York, McGraw-Hill Book Company, 1963.

_____ and _____, "Problem-solving Machines," *International Science and Technology*, no. 36, pp. 48–62, Dec. 1964.

_____, J. C. Shaw, H. A. Simon, and A. L. Samuel, "Machines That Play Games," in Edward Feigenbaum and Julian Feldman (eds.), *Computers and Thought*, New York, McGraw-Hill Book Company, 1963.

Niblack, A. P., "The Jane Naval War Game in the Scientific American," *United States Naval Institute Proceedings*, (whole no. 107):581–593, Sept. 1903.

Nims, Charles F., *Thebes of the Pharaohs*, New York, Stein and Day Incorporated, 1965.

Ninth United States Army Staff, *Conquer, The Story of the Ninth Army*, prepared under the direction of Maj. Gen. James A. Moore, assisted by Cols. Theodore W. Parker and William S. Thompson, Washington, D.C., The Infantry Journal Press, 1947.

Niu Sien-chong, "Two Forgotten American Strategists," *Military Review*, 46(10):53–59, Nov. 1966.

Norman, Lloyd, "Games of Peace," *Newsweek*, 63:56–57, Jan. 27, 1964.

_____, "War Without Gadgets," *Army*, 16(12):53–59, Dec. 1966.

"Notes by the Editor," *The New Philosophy*, quarterly magazine of Swedenborg Scientific Association, 40(4):353, Oct. 1937.

Oman, Carola, *Lord Nelson*, London, William Collins Sons & Co., Ltd., 1954.

Oman, Charles W. C., *The Art of War in the Middle Ages A.D. 378–1515*, Ithaca, N.Y., Cornell University Press, 1953.

_____, *A History of the Art of War*, The Middle Ages from the Fourth to the Fourteenth Century, London, Methuen & Co., Ltd., 1898.

O'Neill, Larry J., and E. E. Steck, "Preparation and Conduct of Field Exercises," *Military Review*, 30(6):57–62, Sept. 1950.

Optner, Stanford L., *Systems Analysis for Business and Industrial Problem Solving*, Englewood Cliffs, N.J., Prentice-Hall, Inc., 1965.

Padelford, Norman J., and George A. Lincoln, *The Dynamics of International Politics*, New York, The Macmillan Company, 1962.

Page, Thornton L., "A Survey of Operations Research Tools and Techniques," in Charles D. Flagle et al. (eds.), *Operations Research*

*and Systems Engineering*, Baltimore, The Johns Hopkins Press, 1960.
Papoulis, Athanasios, *Probability, Random Variables, and Stochastic Processes*, New York, McGraw-Hill Book Company, 1965.
Pappas, Dino G., "STRAT-X, A Gaming Concept for Regional Crises," unpublished manuscript, McLean, Va., Research Analysis Corporation, Jan. 1966.
Parson, Nels A., "The Impact of Guided Missiles on Ground Warfare," *Military Review*, 32(5):16–22, Aug. 1952.
Parsons, William Barclay, *Engineers and Engineering in the Renaissance*, Baltimore, The Williams & Wilkins Company, 1939.
Paxson, E. W., *War Gaming*, Santa Monica, Calif., The RAND Corporation, RM-3489-PR, Feb. 1963.
Pearson, Augustus W., *Rules for Playing Defence*, Salem, Mass., Press of Barry & Lufkin, 1895.
Pennington, Arthur W., "A Description of the STAGE Global War Simulation System and Its Uses," presented at the 23d National Meeting of the Operations Research Society of America, Cleveland, May 27–28, 1963. Abstract in *Operations Research*, 11(Suppl. 1):B-48, Spring 1963 Bulletin.
Pfeiffer, John, Machines That Man Can Talk With, in "The Boundless Age of the Computer, Part III," *Fortune*, 70:153–156 ff., May 1964.
Phillips, Thomas R. (ed.), *Roots of Strategy*, A Collection of Classics Containing the Art of War, by Sun Tzu, 500 B.C.; The Military Institutions of the Romans by Vegetius, A.D. 390; My Reveries on the Art of War by Marshal Maurice de Saxe, 1932; The Instructions of Frederick the Great for His Generals, 1747; The Military Maxims of Napoleon, Harrisburg, Pa., The Military Service Publishing Co., 1940.
Pickett, George B., Jr., "The Army's Tactical Mobility Concept," *U.S. Army Aviation Digest*, 10(11):1–5, Nov. 1964.
Playfair, John, "Memoir Relating to the Naval Tactics of the Late John Clerk, Esq. of Eldin; being a Fragment of an Intended Account of His Life," *Transactions of the Royal Society of Edinburgh*, 9:113–137, 1823.
Pogue, Forrest C., *George C. Marshall: Education of a General*, New York, The Viking Press, Inc., 1963.
―――, *George C. Marshall: Ordeal and Hope*, New York, The Viking Press, Inc., 1966.
Pool, Ithiel de Sola, and Robert Abelson, "The Simulmatics Project," *Public Opinion Quarterly*, 25(2):167–183, Summer 1961.
Potter, E. B., "Chester William Nimitz, 1885–1966," *United States*

*Naval Institute Proceedings*, (whole no. 761):30–55, July 1966.
———— (ed.), *Sea Power, A Naval History*, Englewood Cliffs, N.J., Prentice-Hall, Inc., 1960.
Potter, John Dean, *Yamamoto, The Man Who Menaced America*, New York, The Viking Press, Inc., 1965.
Price, Don K., *Government and Science*, New York, New York University Press, 1954.
————, "The Scientific Establishment," in Robert Gilpin and Christopher Wright (eds.), *Scientists and National Policy Making*, New York, Columbia University Press, 1964.
————, *The Scientific Estate*, Cambridge, Mass., The Belknap Press, Harvard University Press, 1965.
Price, Maurice T., "Wargaming the Cold War," *United States Naval Institute Proceedings*, (whole no. 673):44–47, March 1959.
"The Proper Study of Mankind . . . ," *Newsweek*, 68:80–82, Aug. 13, 1966.
Puleston, W. D., *Mahan, The Life and Work of Captain Alfred Thayer Mahan, USN*, New Haven, Conn., Yale University Press, 1939.
Quade, E. S. (ed.), *Analysis for Military Decisions*, Chicago, Rand McNally & Company, 1964.
————, and W. I. Boucher, *Systems Analysis and Policy Planning. Applications in Defense*, New York, American Elsevier Publishing Company, Inc., 1968.
Quandt, Richard E., "On the Use of Game Models in Theories of International Relations," *World Politics*, 14(1):69–76, Oct. 1961.
The RAND Corporation, Social Science Division, *Experimental Research on Political Gaming*, Santa Monica, Calif., The RAND Corporation, P-1540-RC, Nov. 10, 1958.
Rapoport, Anatol, *Fights, Games and Debates*, Ann Arbor, The University of Michigan Press, 1960.
————, *The Role of Game Theory in Uncovering Non-strategic Principles of Decision*, Ann Arbor, Mich., University of Michigan, Mental Health Research, July 2, 1964.
————, "The Sources of Anguish," *Bulletin of the Atomic Scientists*, 21(10):31–36, Dec. 1956.
————, *Strategy and Conscience*, New York, Harper & Row, Publishers, Incorporated, 1964.
————, "Three Modes of Conflict," *Management Science*, 7(3):210–218, April 1961.
————, "The Use and Misuse of Game Theory," *Scientific American*, 207(6):108–118, Dec. 1962.
———— and Carol Orwant, "Experimental Games: A Review," *Behavorial Science*, 7(1):1–37, Jan. 1962.

Rapp, William T., "What Price Sea Power?" *United States Naval Institute Proceedings*, (whole no. 761):56–61, July 1966.

Raser, John R., and Wayman J. Crow, *WINSAFE II, An Inter-Nation Simulation Study of Deterrence Postures Embodying Capacity to Delay Response*, La Jolla, Calif., Western Behavioral Sciences Institute, July 1965.

Rauner, R. M., and W. A. Steger, *Game-Simulation and Long-Range Planning*, Santa Monica, Calif., The RAND Corporation, P-2355, June 1961.

Raymond, Charles Walker, "Kriegsspiel," a paper read before the United States Military Service Institute at West Point, N.Y., Feb. 17, 1881, printed at the United States Artillery School, Fort Monroe, Va., 1881.

Raymond, R. A., "More on War Games," *The Reserve Officer* (n.p.), Nov. 1938.

———— and Harry W. Baer, Jr., "A History of War Games," *The Reserve Officer*, Oct. 1938, pp. 19–20.

Rayner, A. C., and G. W. Simmons, *Summary of Capabilities*, Washington, U.S. Department of the Army, Corps of Engineers, Coastal Engineering Research Center, MP 3-64, April 1964.

Raytheon Company, *TEMPER*, vol. I, *Orientation Manual*, prepared under the direction of the National Military Command System Support Center for the Joint War Games Agency (AD 470069L), Washington, D.C., July 9, 1965.

Rehkop, A. M., *Experience with the Management Decision Simulation Game, Monopologs*, Santa Monica, Calif., The RAND Corporation, RM-1917 and P-1131, July 17, 1957.

Reinhardt, George C., "The Doctrinal Gap," *United States Naval Institute Proceedings*, (whole no. 672):61–69, Aug. 1966.

————, "War Gaming as a Research Tool at RAND," unpublished manuscript, presented as a talk at LRL, Berkeley, Calif., Santa Monica, Calif., The RAND Corporation, Sept. 24, 1962.

Reisswitz, Lt. G. H. R. J. von, "Anzeige" (Notice), *Militair-Wochenblatt* (Berlin), no. 42, pp. 2973–2974, Marz 6, 1824.

Rhyne, Russell F., "Operations Research and Counterinsurgency," *Military Review*, 44:26–34, June 1964.

Richardson, Charles, *War-Chess, or the Game of Battle*, New York, privately printed by C. B. Richardson, 1866.

Richardson, Lewis F., *Arms and Insecurity, A Mathematical Study of the Causes and Origins of War*, Pittsburgh, The Boxwood Press, 1960.

————, Mathematical Psychology of War, Oxford, Wm. Hunt, Feb. 1919; a limited edition, typescript manuscript, privately circulated by Richardson in 1719 and obtainable at that time from Wm. Hunt.

———, *Statistics of Deadly Quarrels*, Pittsburgh, The Boxwood Press, 1960.

Riddleberger, Peter B., *A Preliminary Bibliography on Studies of the Roles of Military Establishments in Developing Nations*, Washington, D.C., The American University, Special Operations Research Office, July 1963. SORO became AIR/CRESS.[1]

Ridgway, James, "Games in Fabuland," *The New Republic*, 155:17–18, Aug. 27, 1966.

Rigg, Robert B., "Orwellian World: Computerized Statistics Sitting in Judgment of Command Performance," *Army*, 17(3):48–53, March 1967.

Roberts, Charles R., *Development of CENTAUR, A Computerized War Game*, Part I: "General Considerations," Bethesda, Md., U.S. Department of Army, Strategy and Tactics Analysis Group, Dec. 1962.

Robertson, Dalton S., "Operational Research in Industry," *Canadian Business*, 27(7):22–23 ff., July 1954.

Robinson, James A., *Crisis Decision Making: An Inventory and Appraisal of Concepts, Theories, Hypotheses and Techniques of Analysis*, Columbus, Ohio, Ohio State University, The Research Foundation, Project Michelson, Phase II and Final Report, Aug. 31, 1965.

———, Lee F. Anderson, Margaret G. Hermann, and Richard C. Snyder, "Teaching with Inter-Nation Simulation and Case Studies," *American Political Science Review*, 60(1):53–65, March 1966.

Robinson, Patrick J., "Operational Research in Canada," *Proceedings of the First International Conference on Operational Research*, Baltimore, Operations Research Society of America, 1957.

Robison, S. S., and Mary L. Robison, *A History of Naval Tactics from 1530 to 1930*, Annapolis, Md., United States Naval Institute, 1942.

Rodgers, W. L., "The Naval War College Course," *United States Naval Institute Proceedings*, (whole no. 144):1235–1240, Dec. 1912.

———, "The Relations of the War College to the Navy Department," *United States Naval Institute Proceedings*, (whole no. 143):835–850, Sept. 1912.

---

[1] AIR/CRESS, American Institutes for Research/Center for Research in Social Systems, Kensington, Maryland.

Roelants, J. N., and L. DeVlaming, *Handleiding voor de Tactische Oefeningen op De Kaart*, Te Breda, The Netherlands, ter Drukkerij van Broese & Co., Koninklijke Militaire Academie, 1886.

Rogers, Gordon B., "Battles Without Bloodshed," *Army Information Digest*, 15(12):33–39, Dec. 1960.

Rohne, H., "The Artillery-fire Game," translated by John P. Wisser, *Journal of the United States Artillery*, (whole no. 6):122–140 ff., Jan. 1893.

―――, "The Artillery-fire Game Continued," translated by John P. Wisser, *Journal of the United States Artillery*, (whole no. 7):260–288, April 1893.

―――, "The Artillery-fire Game Continued," translated by John P. Wisser, *Journal of the United States Artillery*, (whole no. 8):383–407, July 1893.

―――, "The Artillery-fire Game Concluded," translated by John P. Wisser, *Journal of the United States Artillery*, (whole no. 9):608–639, Oct. 1893.

Rollins, A. P., Jr., "The Waterways Experiment Station," *The Military Engineer*, 50(337):353–355, Sept.–Oct. 1958.

Rosenau, James A. (ed.), *International Politics and Foreign Policy; A Reader in Research and Theory*, New York, The Free Press of Glencoe, Inc., 1961.

Rowny, Edward L., "After the Air Mobile Tests," *Army*, 15(10):36–39, May 1965.

Ruger, Bruno, *Das Go-Spiel*, Leipzig, Julius Klinkhardt/Verlagsbuchlandlung, 1941.

Russell, John H., "A Fragment of Naval War College History," *United States Naval Institute Proceedings*, (whole no. 354):1164–1165, Aug. 1932.

"Russia Bets Its Future on Computer Knowhow," *Business Week*, no. 1928, p. 92, Aug. 13, 1966.

Samuel, A. L., "Some Studies in Machine Learning Using the Game of Checkers," in Edward Feigenbaum and Julian Feldman (eds.), *Computers and Thought*, New York, McGraw-Hill Book Company, 1963.

Sas, Anthony, "Military Campaigns—Strategy in the Mediterranean," *Military Review*, 46(10):3–7, Oct. 1966.

Sayre, Farrand, *Map Maneuvers and Tactical Rides*, 3d ed., Fort Leavenworth, Kans., The Army Service Schools Press, 1908–1910.

―――, *Map Maneuvers and Tactical Rides*, 5th ed., adopted for use in the Army Service Schools, Springfield, Mass., Springfield Printing and Binding Co., 1912.

Sayre, Wallace, "Scientists and American Science Policy," in Robert Gilpin and Christopher Wright (eds.), *Scientists and National Policy Making*, New York, Columbia University Press, 1964.

Schelling, T. C., "Experimental Games and Bargaining Theory," *World Politics*, 14(1):47–68, Oct. 1961.

———, *The Strategy of Conflict*, Cambridge, Mass., Harvard University Press, 1960.

———, "War Without Pain and Other Models," *World Politics*, 15(3):465–487, April 1963, a review of *Conflict and Defense: A General Theory* by Kenneth E. Boulding, New York, Harper & Brothers, 1962.

Schilling, Warner R., "Scientists, Foreign Policy and Politics," in Robert Gilpin and Christopher Wright (eds.), *Scientists and National Policy Making*, New York, Columbia University Press, 1964.

Schindler, W. G., "Research Activities of the Naval Ordnance Laboratory," *Journal of Applied Physics*, 15(3):255–261, March 1944.

Schwartz, David C., "Problems in Political Gaming," *Orbis*, 9(3):677–693, Fall 1965.

Schwarz, Urs, *American Strategy: A New Perspective*, Garden City, N.Y., Doubleday & Company, Inc., 1966.

Scott, Andrew M., William A. Lucas, and Trudi M. Lucas, *Simulation and National Development*, New York, John Wiley & Sons, Inc., 1966.

Scruby, Jack, *All About War Games*, Visalia, Calif., Jack Scruby Military Miniatures, 1957.

Sebastian, R. A., R. G. Kahr, and D. H. Schiller, *A War Game for the Evaluation of Field Army Air Defense*, for the Ballistic Research Laboratories by Caywood-Schiller Associates, Chicago, Ill. (AD 265631), May 1961.

Segoe, Kensaku, *Go Proverbs Illustrated*, translated by John Bauer, vol. 1 of *GO Library in English*, Tokyo, The Japanese GO Association, 1960.

Selfridge, Oliver G., and Ulric Neisser, "Pattern Recognition," in Edward Feigenbaum and Julian Feldman (eds.), *Computers and Thought*, New York, McGraw-Hill Book Company, 1963.

Shepard, R. W., *Some Reasons for Present Limitations in the Application of the Theory of Games to Army Problems*, Great Britain, Army Operational Research Establishment, Sept. 1965.

———, "War Gaming as a Technique in the Study of Operational Research Problems," *Operational Research Quarterly*, 14(2):119–130, June 1963.

————, "War Gaming as a Tool for the Optimisation of Resource Development," presented at NATO Conference on the Application of Operations Research to Military Resource Allocation and Planning, West Byfleet, England, Defence Operational Analysis Establishment, United Kingdom, Aug. 1965.

Shirer, William L., *The Rise and Fall of the Third Reich*, A History of Nazi Germany, New York, Simon & Schuster, Inc., 1960.

Shortley, George, "Operations Research in Wartime Naval Mining," *Operations Research*, 15(1):1–10, Jan.–Feb. 1967.

Shubik, Martin (ed.), *Game Theory and Related Approaches to Social Behavior*, New York, John Wiley & Sons, Inc., 1964.

————, "Some Reflections on the Design of Game Theoretic Models for the Study of Negotiation and Threats," *Journal of Conflict Resolution*, 7(1):1–12, March 1963.

————, "Towards a Theory of Threats," paper presented at Conference on the Theory of Games, Toulon, France, June 29–July 3, 1964, Yorktown Heights, N.Y., Thomas J. Watson Research Center, June–July 1964.

Shure, Gerald H., and Robert J. Meeker, *Bargaining and Negotiating Behavior*, Quarterly Technical Progress Report, Santa Monica, Calif., Systems Development Corporation, TM-2304/100/00, Feb. 7, 1967.

————, ————, and Earle A. Hansford, "The Effectiveness of Pacifist Strategies in Bargaining Games," *Journal of Conflict Resolution*, 9(1):106–117, March 1965.

Siguad, Louis A., *Douhet and Aerial Warfare*, New York, G. P. Putnam's Sons, 1941.

Silkwood, Darold W., "A New War Gaming Methodology for Military Planners," presented at 23d National Meeting of the Operations Research Society of America, Cleveland, May 27–28, 1963. Abstract in *Operations Research*, 11(Suppl. 1):49, Spring 1963 Bulletin.

Silverstone, Paul H., *U.S. Warships of World War II*, Garden City, N.Y., Doubleday & Company, Inc., 1966.

Simcox, L. S., "Lessons Learned in Guerrilla War Gaming," unpublished paper, McLean, Va., Research Analysis Corporation, May 18, 1966.

————, "The TACSPIEL War Game," unpublished paper, McLean, Va., Research Analysis Corporation, July 1, 1967.

Simes, Thomas, *The Military Medley*, 2d ed., London, 1768.

Sims, William S., "The Fleet Gridiron," *The World's Work*, 46(5):505–509, Sept. 1923.

Singer, Charles, E. J. Holmyard, and A. R. Hall (eds.), *A History of Technology*, vol. I, New York, Oxford University Press, 1956.

Slate, John, "The Warning Impulse Timing and Computing Haversack (TWITCH)," *Fortune*, 70:157–158 ff., May 1964.

Smith, Arthur, *The Game of GO*, The National Game of Japan, New York, Moffat, Yard & Co., 1908.

Smith, Bruce L. R., *The RAND Corporation*, Case Study of a Nonprofit Advisory Corporation, Cambridge, Mass., Harvard University Press, 1966.

Smith, John Maynard, and Donald Michie, "Machines That Play Games," *New Scientist*, no. 206, pp., 367–369, Nov. 9, 1961.

Smith, Nicholas M., "A Calculus of Ethics: A Theory of the Structure of Value," Part I, *Behavorial Science*, 1(2):111–142, April 1956.

―――, "A Calculus of Ethics: A Theory of the Structure of Value," Part II, *Behavorial Science*, 1(3):186–211, July 1956.

―――, "A Rationale for Operational Gaming," a paper presented to the Eighth National Meeting of the Operations Research Society of America, Ottawa, Canada, Jan. 9–10, 1956. Abstract in *Operations Research*, 4(1):128, Feb. 1956.

Smith, Stanley E. (ed.), *The United States Navy in World War II*, New York, William Morrow & Company, Inc., 1966.

Smyth, Albert Henry (ed.), *The Writings of Benjamin Franklin*, vol. I, New York, The Macmillan Company, 1905.

――― (ed.), *The Writings of Benjamin Franklin*, vol. V, New York, The Macmillan Company, 1905.

Smythe, John, in J. R. Hale (ed.), *Certain Discourses Military*, for the Folger Shakespeare Library, Ithaca, N.Y., Cornell University Press, 1946.

Snow, R. N., *Contributions to Lanchester Attrition Theory*, Santa Monica, Calif., The RAND Corporation, RA-15078, April 5, 1948.

Snyder, Richard C., "Game Theory and Analysis of Political Behavior," in James A. Rosenau (ed.), *International Politics and Foreign Policy*, New York, The Free Press of Glencoe, Inc., 1961.

―――, "Toward Greater Order in the Study of International Politics," in James A. Rosenau (ed.), *International Politics and Foreign Policy*, New York, The Free Press of Glencoe, Inc., 1961.

―――, H. W. Bruck, and Burton Sapin, "The Decision-making Approach to the Study of International Politics," in James A. Rosenau (ed.), *International Politics and Foreign Policy*, New York, The Free Press of Glencoe, Inc., 1961.

―――, ―――, and ―――, (eds.), *Foreign Policy Decision-making*, New York, The Free Press of Glencoe, Inc., 1962.

Solandt, Omond, "Observation, Experiment and Measurement in Operations Research," presented at Informal Seminar in Operations Research, Chevy Chase, Md., The Johns Hopkins University, Operations Research Office, April 6, 1954.

———, "Observation, Experiment and Measurement in Operations Research," *Journal of the Operations Research Society of America*, 3(1):1-14, Feb. 1955.

Sonderegger, Emil, *Anlage und Leitung von Kriegsspiel-Uebungen*, Frauenfeld, Switzerland, 1897.

———, *Infanterieangriff und Strategische Operation*, Ausblicke und Vorschlage, Frauenfeld, Switzerland, Verlag von Huber & Co., Aktiengesellschaft, 1929.

Sorenson, Theodore C., *Decision-making in the White House*, New York, Columbia University Press, 1963.

Specht, Robert D., "Gaming as a Technique of Analysis," paper presented to the Fifth National Meeting of the Operations Research Society of America, Washington, D.C., Nov. 19-20, 1954. Abstract in *Journal of the Operations Research Society of America*, 3(1):120, Feb. 1955.

———, "RAND—A Personal View of Its History," *Operations Research*, 8(6):825-839, Nov.-Dec. 1960.

———, "The Why and How of Model Building," in E. S. Quade (ed.), *Analysis for Military Decisions*, Chicago, Rand McNally & Company, 1964.

Spinks, J. W. T., "Wartime Operational Research," *Chemistry in Canada* (Ottawa), 7(n.p.), June 1955.

Stapler, John, "The Naval War College, a Brief History," *United States Naval Institute Proceedings*, (whole no. 354):1157-1163, Aug. 1932.

Steger, W. A., *The Use of Gaming and Simulation Devices in Business*, Santa Monica, Calif., The RAND Corporation, P-1219, Nov. 20, 1957.

Steinberg, Jonathan, *Yesterday's Deterrent, Tirpitz and the Birth of the German Battle Fleet*, London, MacDonald & Co., Publishers, Ltd., 1965.

Sterne, Theodore E., "An Introduction to War Games," unpublished notes, McLean, Va., Research Analysis Corporation, Jan. 12, 1965.

———, "War Games: I. What They Are and How They Evolved," *Army*, 16(3):40-46, March 1966.

———, "War Games: II. Validity and Interpretation," *Army*, 16(4):64-68, April 1966.

———, "War Games: III. Play, Characteristics and Suitability," *Army*, 16(5):72-77, May 1966.

Stewart, Irvin, *Organizing Scientific Research for War*, Administrative History of the Office of Scientific Research and Development, Boston, Little, Brown and Company, 1948.

Stolzenbach, C. Darwin, and John H. Roseboom, Jr., "A Mobilization Model for Strategic Gaming: A Computer Application," paper presented to the Tenth National Meeting of the Operations Research Society of America, San Francisco, Calif., Nov. 15–16, 1956. Abstract in *Operations Research*, 5(1):138, Feb. 1957.

Sturges, Joshua, *Sturges' Guide to the Game of Draughts*, New York, Prall Lewis & Co., 1850.

Sun Tzu, *The Art of War*, Military Manual written circa 510 B.C., translated by Cheng Lin, with the original Chinese text appended, Chungking, China, World Encyclopedia Institute, 1945.

———, *The Principles of War*, translated by E. Machell-Cox, Ceylon, Royal Air Force, 1943.

Ssun-DS' (Sun Tzu), *Traktat Uber Die Kriegskunst (The Struggle at the Bridge)*, translated into Russian by J. I. Sidorenko, and from the Russian into German by Ina Balcerowiak, Berlin, Ministeriums fur Nationale Verteidigung, 1957.

Supreme Commander for the Allied Forces, Charles A. Willoughby (editor-in-chief), *Reports of General MacArthur, Japanese Operations in the Southwest Pacific Area*, vol. II, pt. 1, Washington, D.C., Government Printing Office, 1966.

Sutherland, William H., *Some Basic Problems in War Gaming of Ground Combat*, McLean, Va., Research Analysis Corporation, RAC-TP-153 (AD 470102), March 1965.

Swedenborg, Emanuel, *The Mechanical Inventions of Emanuel Swedenborg*, translated and edited by Alfred Acton, Philadelphia, Swedenborg Scientific Association, 1939.

Tada, Kazuo, and Macon Fry, "Military Operations Research in Japan," *Proceedings of the First International Conference on Operations Research*, Baltimore, Operations Research Society of America, 1957.

Takagawa, Kaku, *How to Play GO*, Japan, 1956.

Tankins, Edwin S., "War Gaming," *Military Review*, 47(12):88–96, Dec. 1967.

Taylor, Frederick Lewis, *The Art of War in Italy, 1494–1529*, New York, Cambridge University Press, 1921.

Taylor, Irving W., "The Coast Artillery War Game of the State of New York," *Journal of the United States Artillery*, (whole no. 138):212–216, March–April 1916.

———, "Service of the Piece Game," *Coast Artillery Bulletin 43, N.*

G., New York, in *Journal of the United States Artillery*, (whole no. 139):319–322, May–June 1916.

Technical Operations, Inc., *Joint War Gaming Feasibility Study*, prepared for the Defense Atomic Support Agency, Washington, D.C., Technical Operations, Inc., TO-B61-23, April 1961.

Theil, Henri, John C. G. Boot, and Teun Kloek, *Operations Research and Quantitative Economics*, translated from the Dutch *(Voorspellen en Beslissen)*, New York, McGraw-Hill Book Company, 1965.

Thomas, Clayton J., "The Genesis and Practices of Operational Gaming," *Proceedings of the First International Conference on Operational Research*, Baltimore, Operations Research Society of America, 1957.

———, "Military Gaming," in Russell L. Ackoff (ed.), *Progress in Operations Research*, New York, John Wiley & Sons, Inc., 1961.

——— and Walter L. Deemer, Jr., "The Role of Operational Gaming in Operations Research," *Operations Research*, 5(1):1–27, Feb. 1957.

Thomas, Shirley, *Computers, Their History, Present Applications and Future*, New York, Holt, Rinehart and Winston, Inc., 1965.

Thompson, Frederick B., "War Gaming," paper presented to the Tenth National Meeting of the Operations Research Society of America, San Francisco, Nov. 15–16, 1956. Abstract in *Operations Research*, 5(1):141, Feb. 1957.

Throop, T. A., "Thoughts on the Development of Computer Learning Programs," unpublished paper, McLean, Va., Research Analysis Corporation, Oct. 1965.

Tinbergen, Jan, H. C. Bos, et al., *Econometric Models of Education, Some Applications*, Paris, Organization for Economic Cooperation and Development, 1965.

Todd, F. H., "The Fundamentals of Ship Model Testing," paper presented to New England Section, The Society of Naval Architects and Marine Engineers (SNAME), Weymouth, Mass., April 1951.

Tomlinson, William H., "The Father of Airpower Doctrine," *Military Review*, 46(9):27–31, Sept. 1966.

Tonge, Fred M., "The Use of Heuristic Programming in Management Science," *Management Science*, 7(3):231–237, April 1961.

Totten, Charles A., *STRATEGOS: A Series of American Games of War Based upon Military Principles*, vol. 1, New York, D. Appleton & Company, 1880.

Toynbee, Arnold J., *War and Civilization*, New York, Oxford University Press, 1950.

Tron', A. P., *Certain Problems in the Application of the Theory of Games to Naval Studies—USSR*, Washington, U.S. Department of Commerce, Office of Technical Services, Joint Publications Research Service, JPRS: 19374, May 23, 1963.

Trotha, Thilo von, *Kaukasische Kasaken-Brigade im Balkan-Feldzug 1877-78*, Berlin, Ernst Siegfried Mittler & Sohn, 1894.

Tuchman, Barbara W., *The Guns of August*, New York, The Macmillan Company, 1962.

Turing, A. M., "Can a Machine Think?" in Edward Feigenbaum and Julian Feldman (eds.), *Computers and Thought*, New York, McGraw-Hill Book Company, 1963.

Tyushkevich, S. A., *Neobkhodimost' i Sluchaynost' v Voyne*, Moscow, Ministry of Defense, U.S.S.R., 1962. (Access through an unpublished draft translation.)

U.S. Air Force, Headquarters, Air Battle Analysis Center, *The Annotated Bibliography of Computer Programs*, Washington, D.C., U.S. Air Force, Headquarters, Jan. 15, 1965.

U.S. Army Electronic Proving Ground, *Investigation of Model Techniques*, Fort Huachuca, Ariz., Report USAEPG-SIG 940-43R1, July 1961.

U.S. Army War College, *War Gaming: Selected Readings*, Course 4, Carlisle Barracks, Pa., U.S. Army War College, Nov. 10, 1960.

U.S. Department of Army, *A Historical and Pictorial Review of the Parachute Battalions*, Fort Benning, Ga., 1942, Baton Rouge, La., The Army and Navy Publishing Co., Inc., 1942.

———, Office of Chief of Military History, *Command Decisions*, New York, Harcourt, Brace and Company, Inc., 1959.

U.S. Department of Navy, Office, Chief of Naval Operations, *Naval Warfare, Appendix C, Principles of War*, Washington, D.C., U.S. Department of Navy, NWP-10, May 1954.

———, Chief of Naval Operations, Planning Analysis Group, *Fleet Anti-Air Warfare Game (FAAWG-MOD 1), General Description*, Washington, July 8, 1964.

U.S. Naval Research Laboratory, *Naval Technology in Today's War*, Washington, U.S. Naval Research Laboratory, Anacostia Station, Sept. 1942.

U.S. Naval War College, Members of the Staff, "Notes on the Applicatory System of Solving War Problems, with Examples Showing the Adaptation of the System to Naval Problems," *United States Naval Institute Proceedings*, (whole no. 143):1011-1036, Sept. 1912.

———, School of Naval Warfare, *Curriculum 1966-67*, Newport, R.I., U.S. Naval War College, July 1966.

U.S. War Department, Adjutant General's Office, Military Information Division, *Sources of Information on Military Professional Subjects*, Document 55, Government Printing Office, 1898.

Valentine, A. W., *More Sand Table Exercises*, 8th ed., London, Aldershot, Gale & Polden, Ltd., 1941.

Van Doren, Carl, *Benjamin Franklin*, New York, The Viking Press, Inc., 1938.

Van Keuren, A. H., "The U.S. Naval Research Laboratory," *Journal of Applied Physics*, 15(3):221–226, March 1944.

Verba, Sidney, "Simulation, Reality, and Theory in International Relations," *World Politics*, 16(3):490–519, April 1964.

Verdy du Vernois, Julius von, *Beitrag zum Kriegsspiel*, Berlin, Ernst Siegfried Mittler & Sohn, 1876.

———, *Essai de Simplification du Jeu de Guerre*, Bruxelles, Librairie Militaire C. Muquardt, 1877.

———, *A Simplified War Game*, translated from the French by Eben Swift, Kansas City, Mo., Hudson-Kimberly Publishing Co., 1897.

Virden, Frank, "The Naval War College Today," *United States Naval Institute Proceedings*, (whole no. 602):365–371, April 1953.

Von Neumann, John, *The Computer and the Brain*, New Haven, Conn., Yale University Press, 1958.

——— and Oskar Morgenstern, *Theory of Games and Economic Behavior*, 3d ed., Princeton, N.J., Princeton University Press, 1953.

Waddell, M. C., *Air Battle Analyzer Handbook*, Silver Spring, Md., The Johns Hopkins University, Applied Physics Laboratory, TG-421, Oct. 1964.

Wade, Leroy E., "War Gaming Civil Affairs Operations," *Proceedings for the Third Symposium of the East Coast War Games Council*, Miami Beach, Fla., Feb. 27–28, 1964.

Walker, Benjamin S., "Push-button War," *John Hopkins Magazine*, pp. 10–11 ff. March 1946.

Wallace, G. W., *The Military Game of "Grant's Line,"* Rules, Remarks and Illustrations for Playing the Game, Cincinnati, Ohio, H. C. Cady, 1868.

Wang, Hao, "Games, Logic and Computers," *Scientific American*, 213:98–104 ff., Nov. 1965.

Wang, Wei, "Young 'Meijin' Keeps China's GO Tradition Alive," *Free China Weekly*, 5:2, Aug. 21, 1966.

War Control Planners, Inc., *War Safety Control Report*, A Conceptual Study of All-nation Defense System, written by a Citizens Task Force, colligated by Howard G. Kurtz and Harriet B. Kurtz, Chappaqua, N.Y., The Planners, 1963.

"The War Game and How It Is Played, Mimic Warfare for the Schooling of Army Officers," *Scientific American*, 113:470–471, Dec. 5, 1915.

"War Games," *Military Review*, 41(6):68–77, June 1961.

Warner, Oliver, *The Battle of the Nile*, New York, The Macmillan Company, 1960.

———, *Great Sea Battles*, New York, The Macmillan Company, 1963.

Washington Operations Research Council, John L. Overhold (ed.), *First War Gaming Symposium Proceedings*, Washington, D.C., Washington Operations Research Council, Nov. 1961.

———, Murray Greyson (ed.), *Second War Gaming Symposium Proceedings*, Washington, D.C., Washington Operations Research Council, March 1964.

Weiner, M. G., "Gaming Methods and Applications," in E. S. Quade (ed.), *Analysis for Military Decisions*, Chicago, Rand McNally & Company, 1964.

———, *An Introduction to War Games*, Santa Monica, Calif., The RAND Corporation, P-1773, Aug. 17, 1959.

———, *War Gaming Methodology*, Santa Monica, Calif., The RAND Corporation, RM-2413, July 10, 1959.

Weiss, Herbert K., "Combat Models and Historical Data: The U.S. Civil War," *Operations Research*, 14(5):759–790, Sept.–Oct. 1966.

———, "The Fiske Model of Warfare," *Operations Research*, 10(4):569–571, July–Aug. 1962.

———, "Lanchester-type Models of Warfare," *Proceedings of First International Conference on Operational Research*, Baltimore, Operations Research Society of America, 1957.

Wells, H. G., *Floor Games*, Boston, Small, Maynard and Company, 1912.

———, *Little Wars*, A Game for Boys, Boston, Small, Maynard and Company, 1913.

Westcott, Allan F. (ed.), *American Sea Power Since 1755*, Philadelphia, J. B. Lippincott Company, 1947.

White, Charles, "European EDP: A View from England," *Datamation*, 12(9):22–24, Sept. 1966.

Whitson, William L., "The Growth of the Operations Research Office in the U.S. Army," *Operations Research*, 8(6):808–824, Nov.–Dec. 1960.

———, "The History of Operations Research (I)," paper given at Informal Seminar in Operations Research, Chevy Chase, Md., The Johns Hopkins University, Operations Research Office, Seminar Paper 2, Oct. 15, 1952.

Selected Bibliography 373

―――, "The History of Operations Research (II)," paper given at Informal Seminar in Operations Research, Chevy Chase, Md., The Johns Hopkins University, Operations Research Office, Seminar Paper 3, Oct. 22, 1952.

―――, "War Gaming at Combat Operations Research Group," paper presented to the Fifth National Meeting of the Operations Research Society of America, Washington, D.C., Nov. 19–20, 1954. Abstract in *Journal of the Operations Research Society*, 3(1):120, Feb. 1955.

Wiener, Norbert, *The Human Use of Human Beings, Cybernetics and Society*, 2d ed., Garden City, N.Y., Doubleday & Company, Inc., 1954.

Wilhelm, John C., *Militaire, or Republican Chess*, Philadelphia, privately published, 1877.

Wilkinson, J. G., *Manners and Customs of the Ancient Egyptians*, vol. I, London, John Murray (Publishers), Inc., 1837.

―――, *Manners and Customs of the Ancient Egyptians*, London, John Murray (Publishers), Inc., 1878.

―――, *A Popular Account of the Ancient Egyptians*, vol. I, London, John Murray (Publishers), Inc., 1854.

Willard, Daniel, *Lanchester as Force in History: An Analysis of Land Battles of the Years 1618–1905*, Bethesda, Md., Research Analysis Corporation, RAC-TP-74, Nov. 1962.

Williams, J. D., *The Compleat Strategyst, Being a Primer on the Theory of Gaming*, New York, McGraw-Hill Book Company, 1954.

Willow Run Laboratories, *Proceedings, Third War Games Symposium*, Oct. 1960, Ann Arbor, Mich., Willow Run Laboratories, Nov. 1960.

Wilson, Andrew, *The Bomb and the Computer*, London, Barrie and Rockliff, The Cresset Press, 1968.

Winthrop, Beckman, "Be Ready for War," *United States Naval Institute Proceedings*, (whole no. 143):829–834, Sept. 1912.

Wintringham, Tom, *The Story of Weapons and Tactics from Troy to Stalingrad*, Boston, Houghton Mifflin Company, 1943.

Wisser, John P., *Practical Field Exercises in Tactics and Strategy*, Kansas City, Mo., Hudson-Kimberly Publishing Co., 1903.

―――, "The Uses of the Artillery-fire Game," *Journal of the United States Artillery*, (whole no. 15):255–264, April 1895.

Wohlstetter, Albert, "Strategy and the National Scientists," in Robert Gilpin and Christopher Wright (eds.), *Scientists and National Policy Making*, New York, Columbia University Press, 1964.

Wohlstetter, Roberta, *Cuba and Pearl Harbor: Hindsight and Foresight*, Santa Monica, Calif., The RAND Corporation, RM-4328-ISA (AD 461774), April 1965.

_____, *Pearl Harbor, Warning and Decision*, Stanford, Calif., Stanford University Press, 1962.

Wood, David, "The Middle East and the Arab World: The Military Context," *Adelphi Papers* (London), no. 20, July 1965.

Wood, Marshall K., "Remarks on Anatol Rapoport's Paper," *Management Science*, 7(3):224–230, April 1961.

Wood, Robert C., "Scientists and Politics: The Rise of an Apolitical Elite," in Robert Gilpin and Christopher Wright (eds.), *Scientists and National Policy Making*, New York, Columbia University Press, 1964.

Wright, Christopher, "Scientists and the Establishment of Science Affairs," in Robert Gilpin and Christopher Wright (eds.), *Scientists and National Policy Making*, New York, Columbia University Press, 1964.

Wright, Quincy A., *A Study of War*, 2 vols., Chicago, The University of Chicago Press, 1942.

_____, *A Study of War*, abridged by Louise Leonard Wright, Chicago, The University of Chicago Press, 1965.

Wriston, Henry M., *Strategy of Peace*, Boston, World Peace Foundation, 1944.

Yadin, Yigael, *The Art of Warfare in Biblical Lands*, New York, McGraw-Hill Book Company, 1963.

Young, John P., "A Brief History of War Gaming," unpublished paper, Chevy Chase, Md., The Johns Hopkins University, Operations Research Office, Oct. 1956.

_____, *History and Bibliography of War Gaming*, Chevy Chase, Md., The Johns Hopkins University, Operations Research Office, ORO-SP-13 (AD 235893), April 1957.

_____, *A Survey of Historical Developments in War Games*, Bethesda, Md., The Johns Hopkins University, Operations Research Office, ORO-SP-98 (AD 210865), March 1959.

_____, Andrew J. Eckles, Norman W. Parsons, and L. F. Koehler, *Project Pinpoint—Disclosure of Antitank Weapons to Overwatching Tanks*, Bethesda, Md., The Johns Hopkins University, Operations Research Office, ORO-T-362, Jan. 1958.

Zacharias, Ellis M., *Secret Missions—The Story of an Intelligence Officer*, New York, G. P. Putnam's Sons, 1946.

Zimmerman, R. E., "The Application of Electronic Computers to Monte Carlo War Game Problems," *Mathematical Models for*

*Ground Combat*, Chevy Chase, Md., The Johns Hopkins University, Operations Research Office, ORO-SP-11 (AD 235891), April 1957.

―――, *CARMONETTE, A Concept of Tactical War Games*, based on work done by H. E. Adams, W. E. Cushen, J. F. Kraft, and R. E. Zimmerman, Chevy Chase, Md., The Johns Hopkins University, Operations Research Office, ORO-SP-33 (AD 204089), Nov. 1957.

―――, "Maximum Complexity Computer Battle," in *Mathematical Models for Ground Combat*, Chevy Chase, Md., The Johns Hopkins University, Operations Research Office, ORO-SP-11 (AD 235891), April 1957.

―――, *Monte Carlo Computer War Gaming, A Feasibility Study*, Chevy Chase, Md., The Johns Hopkins University, Operations Research Office, ORO-T-325 (AD 94459), March 1956.

―――, "Simulation of Tactical War Games," in Charles D. Flagle et al. (eds.), *Operations Research and Systems Engineering*, Baltimore, The Johns Hopkins Press, 1960.

―――, "Some Thoughts on the Future of War Gaming Simulation," unpublished paper, McLean, Va., Research Analysis Corporation, Sept. 1965.

―――, H. E. Adams, R. E. Forrester, J. F. Kraft, and B. A. Oosterhout, "CARMONETTE: A Computer Combat Simulation," presented at Fifth SHAPE Operations Research/Scientific Advisory Conference, Jan. 3, 1960, Bethesda, Md., The Johns Hopkins University, Operations Research Office.

Zook, David H., Jr., and Robin Higham, *A Short History of Warfare*, New York, Twayne Publishers, Inc., 1966.

Zuckerman, Sir Solly, "Judgment and Control in Modern Warfare," *Foreign Affairs*, 40(2):196–212, Jan. 1962.

"Zur Bergeschichte des v. Reisswitz'schen Kriegspiels," *Militair-Wochenblatt*, (Nr. 73):693–694, Sept. 9, 1874.

# Subject Index

Afrika Korps, Rommel's, 32-33
Allied operations:
 OVERLORD, 33-34
 planning and testing of, 34-35
American Association of Collegiate Schools of Business, 198-199
American Management Association:
 Academy for Advanced Management, 195
 business games and management simulations, 194-196, 198, 199, 204-207, 210, 220, 328
American Political Science Association, 230
American Society of Mechanical Engineers (ASME), 189
American Society of Training Directors, 197, 209, 215n., 327
Analysis:
 problem, 300
 side, 296-297
 terrain, 115-117, 297-298
Analytical gaming, 61-81, 108, 115, 152, 294, 297, 301
 (See also War gaming, analytical)
Applied Physics Laboratory, The Johns Hopkins University (APL/JHU), games and simulations, 149, 192n., 326
Arab-Israeli Wars, 35-37
Arab-Israeli (six-day) War (1967), 35, 47
 Palestinian War (1948-1949), 35
 Sinai War (1956), 35
Arctic studies, problems and maneuvers of Canadian and U.S. Armies, 289
Arctic Institute, 289
Arms control, 2, 303

Army of the Republic of Vietnam (ARVN), 261
Army Operational Research Establishment (AORE), U.K., 71-76
 formerly Army Operational Research Group (AORG), 71
Army War College, 144, 290
 war games manual, 144, 145n.
Assumptions:
 avoid fruitless efforts, 281
 avoid unnecessary calculations, 282
 examples of, 280
 exclude unmanageable factors, 280
 exclude unneeded aspects, 280
 in games, 226
 and objectives, 279-280, 283
 of other nations' capabilities, 283
 of parity, 281
 reduce number of variables, 281
 significant, 282
 simplifying, 282
 statistical, 283-284
 where stated, 279
Australian Army, 143
Automatic game play:
 in simulations, 298
 by submodels, 298-300
Automation, 295, 298

Battles, great (see Historical examples of great battles)
Booz, Allen Applied Research, Inc. (BAARINC), 144n.
British Army war gamers, 71, 73, 79
British Army war games, 32-35, 71-79, 143, 151

*377*

378   Subject Index

British business games, 199–200
  six management simulations, 199
Brookings Institution, The, 230
Budgets, military, 2
  (See also Costs; Maneuvers, cost of)
Business education, 189–191, 197–202, 221
Business games and simulations, 80
  in agricultural business, 202
  companies conducting, 190–196, 198, 210–211, 213–214
  computer supported, 195, 199, 200, 204–207
  equipment for, 84
  extent of, 197–198
  facilitate learning, 218–222
  "feedback" in, 190, 191, 197
  functional type (selected departments of a business), 199, 214–217
  general (overall) management type, 199, 210–214
  how conducted, 207–208
  how constructed, 209–210, 220–221
  in "in-service" training, 221–222
  management types, 80, 188–202, 210–214, 221–222
  most frequently used, 200
  named, 192–194, 196, 198–200, 202, 204, 210–217, 220
  noncomputerized, 195–196, 199–200
  origin of, 188–189, 194–196
  in other organizations, 191–200
  proliferation of, 197–199
  similarities to war games, 202–204
  for small businesses, 213–214
  spread to other countries, 202
  types of, 191–200, 202, 210–217, 221–222
  in universities, 191–195, 198–200, 202, 210–213, 215–217
  uses of, 198
    in business education, 190–202, 218–219
  what happens in, 204–207
  (See also specific games under their given names)
Business Games Handbook of American Management Association, lists of business games, 198
Business management exercises (see British business games; Business games and simulations, management types)
Business schools, graduate level, in universities, 190–192

Business simulations (see Business games and simulations)

Canadian Army:
  arctic maneuvers, 289
  war games (see CAORE war game in Index of Games and Simulations, Part 1)
Canadian Army Operational Research Establishment (CAORE), 78
  (See also Canadian Army; CAORE war game in Sec. III, Part 1)
Canadian Peace Research Center, 232
Case study method, 190, 191, 270
Center for Advanced Study in the Behavioral Sciences, Palo Alto, California, 230
Center for International Studies (CENIS) at MIT, 233n., 266, 273
Chance, element of, 221, 291, 292
Chance events:
  examples of, 285
  and simulations, 221, 278
  (See also Probability)
Checkers (see Draughts)
Chess:
  origin of, 3
  resembles a single battle, 3
  War Chess, 5
  (See also Games of war)
China, Red, 10, 302
Chinese medicine, 305
Chinese wise man, legendary, 307–308
Civil affairs inputs in military games (see FAME game in Index of Games and Simulations, Part I)
Civil authority, supremacy of, 223
Civil-military relationships (see Political-military relationships)
Cold war:
  gamers, 250, 253n.
  gaming of, 265
  THEATERSPIEL model, 243n., 253–261
Colleges active in gaming, 230, 303
  (See also Universities active in gaming)
Combat Developments Experimentation Center (CDEC), U.S. Army, Fort Ord, 94–95
Combat operations, modeling of, 100
  (See also Military operations and war gaming)
Combat Operations Research Group (CORG), 69–72, 77n., 139, 145–146, 149, 151

Subject Index    379

Command Post Exercise (CPX), 93, 94, 193n.
   (See also LOGEX, a military logistics CPX in Index of Games and Simulations, Part 1)
Commanders, role of, in war games, 12–13, 125–130, 175, 181, 184, 286, 299–300
Commander's decisions:
   in games, 298
   overriding automatic processing, 299
Companies active in gaming and simulations:
   Allied Chemical Corp., 216
   Allstate Insurance Companies, 202
   American Telephone and Telegraph Company, 198
   Boeing Airplane Co., 198, 202, 215, 217, 271
   Burroughs Corp., 214
   Dayco Corp., 213
   Douglas Airplane Co. (now McDonald Douglas Corp.), 234–242, 271
   General Electric Co. (DISPATCH-O, TEMPO, etc.), 200, 210–211, 215–216, 271, 324
   Humble Oil and Refining Co., 211, 214
   International Business Machines Corporation (IBM), 194–195, 199, 210
   Kroger Company, 214, 220
   McKinsey and Company, 210
   Minneapolis-Honeywell Corp., 198
   Pillsbury Co., 198, 211
   Raytheon (TEMPER simulation), 250, 266, 271
   Remington-Rand UNIVAC Corp., 197–198
   Science Research Associates, 230
   Trans-Canada Airlines, 217
   Westinghouse Electric Co. (MELPAR), 271
Comparative gaming, 78
Computer models:
   for automatic processing, 297
   improved, 296–297
   independent, 301
   programming of, 294
   programming languages, 300
   subroutines, 300–301
Computers:
   in gaming support, 44, 83–85, 88, 290, 296–299
   remote control, 296
   storage capacity, 296
Concepts, new:
   air-assault, 47–48

Concepts, new (Cont.):
   airmobile, 47–49
   "lightning war," 35–37
   (See also Guerrilla war games; Nuclear weapons)
Condottiere mercenaries, 277
Constraints:
   examples of, 280–281
   imposed on commanders, 280–281
   not publicized, 281
   where stated, 280
Control Center (see Control room)
Control room, use of, by control staff, 12–14, 84, 141, 150, 176–185, 196, 208, 245
Controllers or umpires, role of, in war games, 12–13, 125–130, 172, 175–180, 183–186, 286, 299–300
Cost-effectiveness studies, 159, 160
Costs:
   of CPX's, 93–94
   of field testing, 94–96
   of games, 91–94
   of maneuvers, 91–96
   of wars, 91–92
Council for Economic and Industrial Research (CEIR), 149n.
Crises:
   Bay of Pigs, 224n.
   Berlin, 232
   Cuban missile, 224n., 264, 302
   international, chart of, 237–239
   residual sovereignty of Ryukyu Islands, 224n.
   Suez (1956), 224n.
   U-2 overflights of the U.S.S.R., 224n.
Crisis decision-making, 272
Crisis games, 242–243, 264, 265, 304
Cuban revolution, 272

Data:
   aggregated, 220, 296
   analysis of, 283–284
   availability of, 102
   generation of, 42, 63, 71–72, 296
   input, 87, 97, 103–104, 110, 291
      (See also Input data)
   needed, 101–105
   nonfunctional, 119
   on opponent, 282–283
   processing of, 297, 299–300
      automatic, 299–300
   qualitative, 104–105
   quantitative, 105–111
   scaling of, 104
   synthesis of, 301

Data (Cont.):
    tables of, 112
    unavailable, 102–103
    valid, 282–283
    (See also Information and intelligence)
Data banks, 295–296
Decision making:
    and expertise, 286–287
    and gaming, 275, 295, 298, 303
    and submodels, 298–299
    (See also Commander's decisions)
Defense Communications Agency, 273
Department of the Army (U.S. Department of Defense), 159
Desk games, 265
    (See also JCS political-military games in Index of Games and Simulations, Part 3)
Developments, unforeseen, in games and in war, 278
Dien Bien Phu, Battle of, 244
Diplomacy (see Political affairs; Political gaming; Political-military games)
Directorate of Land/Air Operational Research (Canada), 138
Disarmament (see Arms control)
Documentation, facilitation of, 300
Draughts:
    early name of Checkers, 3
    origin of, 3
    played in ancient Egypt, 3–4
    (See also Games of war)

Educational games, 2, 190–202, 218–219, 302–304
Educational gaming (see Business schools; School systems active in gaming; Universities active in gaming)
Eighth (U.S.) Army Staff Game, 157–159
Estimate of the situation, a military procedure, 34
Evaluation and assessment:
    of courses of action, 301
    of games and simulations, 291–293
Executive development through management simulations, 195–197, 200–207

"Feedback," 190, 191, 197
Field exercises, 63
Field experiments, 63, 71–72
    to develop data needed in games, 71–72
    to try out new developments, 94–95
    (See also Field testing and field tests; Combat Developments Experimentation Center)

Field testing and field tests, 103, 109
Firepower:
    determination of, 102–103
    scores, 102
1st Cavalry Division (Airmobile), 48–49, 351, 363
Fiske, Rear Adm. Bradley A., USN (1905), 108–109, 341
Fiske's model, 109, 120–121

Game board:
    as used in: Chess and Go, 3
    early games of war, 4–5
    later war games, 13–15
    map maneuvers, 125
    (See also Terrain model)
Game directives, elements included, preparation of, general and special situations, scenarios, 83, 85, 88
    (See also Game preparation)
Game directors, role of, in war games, 12–13, 124–130, 133, 173, 187
Game equipment needed, 83–85, 87–88, 182
Game facilities, 12–15, 75, 77, 83–85, 125, 133, 139, 141, 150, 175–185, 245
Game organization, 89, 142, 174, 186
Game participants, personnel (other personnel, controller, umpire, "Red" and "Blue" Commanders, game director), 12, 44, 124–130, 133, 299
Game periods, 86–91
    playing, 88–90
    post-play, 90–91
    pre-play, 86–88
Game preparation, 86–87, 296
    activities involved in, 87
Game rooms:
    three-room setup, 14
    (See also Control room; Player rooms)
Game rules, 82–83, 85
    extent of, 112
    manual of, 112
Game scenario, 85
Game starting conditions, 85
Games:
    analysis of, 90–91
    for analytical studies, 61–68, 132–151
    assessment of, 90–91
    characteristics of, 125–151
    closed, 124–126
    computer assisted, 127, 199
    custom crafted, 295
    for data generation, 9
    differing requirements of, 295
    equipment used in, 127

Games (Cont.):
  free, 124, 264–266
  hobby type, 139
  limitations of, 220–221, 290–291
  manual type, 126–127
  map type, 127–128
  micro-type, 139
  for occupational choice, 303
  one-sided, 127–128
  open type, 124–125, 128
  organization for, 83–91, 142, 174, 186
  personnel requirements for, 12, 44, 154, 286–287
  predictive value of, 121
  preparation for, 83–91
  requirements for, 142
  for research, 130, 221–222
  rigid type, 123–124, 266
  rules for, 85, 130, 153
  short-cut, 290
    (See also Quick gaming)
  standardization of, 295
  to test plans, 130
  time required to play, 142, 160, 169
  two-sided, 128
  types of, 125–151
  for urban planning, 302–303
  for use in schools, 302–304
  for vocational guidance, 303
    (See also Business games and simulations; Educational games; Political games; War games; and specific games under their given names)
Games of war (forerunner of war games):
  Checkers, Chess, Draughts, and Go, 3–4
  in 18th and 19th centuries, 5–8
  in Renaissance period, 4
    (See also War games)
Gaming:
  alternative courses of action, 2
  analytical, 61–81, 108, 115, 152, 290, 294, 297
  to answer questions, 158
  and creativeness, 10–11
  crises, 83, 242, 304
  data for, 85
  difficulties, unsolved, 275
  essentials for, 82–91
  to establish doctrine, 45–49
  fields for (see Gaming applications)
  future of, 294–308
  limitations of, 224n., 268–269, 288–291
  of new concepts, 45–47
  new fields for, 275, 302–304
  obstacles encountered in, 275
  procedures, 85

Gaming (Cont.):
  state of the art of, 274, 291
  strategic, 242
  validity in, 291–293
    (See also Business games and simulations; Educational games; Political games; War games; War gaming)
Gaming applications, recent and new fields for:
  armistice and truce policing, 303
  arms control and disarmament, 303, 305
  crime control, 304
  diplomacy (see Crises; Peace gaming; Political gaming; Political-military games)
  economic unions, 303
  gambling strategies, 304
  international disputes, 303
  international relations and ethics, 304, 306
  international trade agreements, 304
  labor disputes, 304
  land use, 303
  law enforcement, 304
  metropolitan area projects, 303–304
  municipal government, 302–304
  peace, 305–308
  political science, 302–308
  protective occupations (military), 303
  regional development programs, 304
  riot control, 304
  social sciences, 302–303
  sports, 304
  strikes, 304
  traffic management, 304
  transportation problems, 304
  treaty negotiations, 303
  urban planning, 302–303
  urban problems, 303–304
  vocational guidance, 303
  world law, 305
    (See also Business games and simulations; Educational games; Political gaming; War games; War gaming)
Gaming organizations, 79–81, 302–304
  (See also Companies active in gaming and simulations; Independent research organizations active in gaming; Nations reported as active in military gaming and simulations; Universities active in gaming)
Gaming personnel, military, scientific, and technical support for, 44
Gaming process as related to models, computers, and simulation, 2

382  Subject Index

Gaming purposes:
  for analytical studies, 158, 275
  to answer questions, 158
  to develop plans, 157–159
  to generate needed data, 9
  for problem solving, 39–44
  for research, 152
  to test plans, 157–159
  for training, 157–159
Gantt charts, 84, 189, 210
General Research Office (later Operations Research Office), 64
Geographic areas represented in business and/or military games (Europe, Far East, Middle East, North Africa, North America, Northeast Asia, Pacific Area, South Asia, Southeast Asia)(see Nations reported as active in military gaming and simulation; and specific subject)
German war gaming, 5–8, 22–30, 55–56
Go, the national game of Japan:
  origin of, and resemblance to Chess, 3
  represents a military campaign, 3
  (See also Games of war)
Guerrilla war games, 244–264

High schools active in gaming, 230, 303
Historical examples of great battles:
  Ardennes, Battle of the Bulge, 28–29
  Coral Sea, 31
  Dien Bien Phu, 244
  Midway, 31
  Pearl Harbor, 30–31, 49–53, 95–96
  Tannenberg, 22–25
  Thermopylae, 276
  Trafalgar, 106–108
  Waterloo, 108, 110
Historical records:
  examples of, used in training, 21
  of formulas and models, validity of, 101
Historical roots of war games:
  in China, Egypt, India, and Japan, 3–4
  in Europe, 3–9
Hobby war games (floor or parlor games):
  Featherstone's games, 139n.
    books on (see Selected Bibliography)
  Well's games, 139
    books on (see Selected Bibliography)
Hudson Institute, 233n.
Human decisions in games (see Decision making)

Independent research organizations active in gaming:
  Abt Associates, Inc., 328
  Booz, Allen & Hamilton, 194n.
  Booz, Allen Applied Research, Inc. (BAARINC), 144n.
  Council for Economic and Industrial Research (CEIR), 149n.
  Hudson Institute, 233n.
  McKinsey and Company, 196
  Operations Research Office (see Operations Research Office)
  RAND Corporation (see RAND Corporation)
  Research Analysis Corporation (see Research Analysis Corporation)
  Stanford Research Institute, 149, 251–253
  Systems Development Corporation, 348
  Technical Operations, Inc. (see Technical Operations, Inc.)
  Western Behavioral Sciences Institute, 233n., 303, 361
  (See also quasi-independent organizations, e.g., American Management Association, Academy for Advanced Management; Applied Physics Laboratory; Center for International Studies)
Index of combat effectiveness (ICE), 153
Indo-China (see Vietnam)
Industrial College of the Armed Forces, 268, 271
Industrial engineering (see Management science)
Information and intelligence:
  accessibility of, 296
  display of, 297
  inadequate, 278–279
  on intangibles, 104, 276
  required in games, 296
  retrieval, 296
  sources of, 296
  storage in computer, 296
  types of, in games, 296
  unknown, 277, 281, 291–292
  (See also Data; Data banks)
"In-house" gaming, 226
Innovation, 10–11, 304
Innovative weapons, finding ways to employ, 69
Input data, 87, 97, 103–104, 110, 291
  simplification of, 298–300
  standardization of, 295
  time saving methods for, 296

Institute for Defense Analyses, 149n., 247
Institute of Management Sciences, The (TIMS), 188n.
Institute of Special Studies (USACDC), 151
Insurgency operations, gaming of, 244–264
Intangibles, dealing with, 104, 276
Intelligence (see Information and intelligence)
Intercontinental Ballistic Missiles (ICBMs), 283
International politics, 2, 301
  power factors in, 301
  (See also Political gaming)
International relations (see International politics)
International tensions, 241–242
  Bay of Pigs invasion of Cuba, 224n.
  cold war, U.S.-U.S.S.R., 242
  Cuba missile crisis, 224n.
Israeli war gaming, 37
Israeli wars, 35–37
  (See also Arab-Israeli Wars)

Japanese Army War College, 30
Japanese military actions and operations:
  attack on Pearl Harbor, 30–31, 49–53, 56–57, 95–96
  Kamikaze attacks, 32
  military-naval maneuvers, 94–96
  World War II campaigns, 30
Japanese Naval General Staff, 31
Japanese Naval War College, 30
Japanese Total War Research Institute, 30
Japanese war gaming, 22, 30–32, 43–44
Johns Hopkins University, The:
  Applied Physics Laboratory (APL/JHU) (see Applied Physics Laboratory)
  Operations Research Office (ORO) (see Operations Research Office)
Joint Chiefs of Staff (JCS), 226, 251, 266, 273
  games, 300
Joint War Games Agency (JWGA) of Joint Chiefs of Staff, 80, 151, 266, 273

Kamikaze attacks, 32

League of Nations, The, 305
"Lightning wars":
  Arab-Israeli War, 35–37
  German blitzkrieg, 36

"Lightning wars" (Cont.):
  Sinai War, 35
Limitation of Armaments Conference (1921–1922), 305n.
Logic of a battle, 102, 118, 277–278
Logistics:
  exercises, games and simulations (see LOGEX, LOGSIM, SIGMALOG in Index of Games and Simulations, Part 1)
  model, 113, 114
  simulations, 159, 160, 162–165

Management games (see Business games and simulations)
Management science (scientific management):
  evolution of, 188–190
  leaders in, 189
Management simulations (see Business games and simulations)
Management techniques:
  emphasis in the military, 2
  represented in games (see Business games and simulations)
Maneuvers:
  to acquire data, 288–289
  arctic, 289
  cost of, 91–96
  map type, 127–130, 308
  in relation to exercises, field tests, and games, 62–63, 148, 288
  troop: Desert Strike, 93
    Polar Siege, 289
    Swift Strike, 148
Mao Tse-tung, Chairman, Peoples Republic of China, principles of guerrilla war of, 58
Mass production, 294–295
Military doctrine:
  and advancing technology, 279
  development of, 40–48
  as established practice, 18n., 100, 279, 298
  gaming of, 45–49
  and obsolescence, 279
  as standing operating procedures, 100, 279, 298
  strategic, 45–47
  tactical, 45
  testing of, 45–49
Military education:
  historical examples, shortcomings of, 21, 292
  map problems in, 128–130
Military games (see War games)

Military gaming:
    of new concepts, 44-48
    purposes of, 18
    for research, 37-44
    to test plans, 22-37
    to train officers, 19-22
    (*See also* War gaming)
Military history, 292
    (*See also* Historical examples of great battles)
Military judgments, 290-292
Military maps from aerial photography, 297
    (*See also* Photogrammetry; Universal Automatic Compilation System)
Military operations and war gaming:
    Ardennes offensive, 27-29
    Atlantic convoys, 29
    attack on Pearl Harbor, 30-31, 49-53, 55-58
    BARBAROSSA, 26-27
    Coral Sea, 31
    German Navy, "wolf-pack" tactics of, 29
    Guadalcanal, 31
    Midway, 31
    mine warfare against Japan, 39-44
    OVERLORD, 33-34
    SEA LION (SEELOEWE), 26-27
    TORCH landings, North Africa, 33
    U-boat warfare, 29
Military personnel for gaming:
    level of experience required, 286-287
    roles of, 12-13, 44, 83-84, 286-287
    (*See also* Commanders; Controllers or umpires; Game directors)
Military weapons and weapons systems:
    developed by enemy powers in World War II (magnetic sea mines, toxic gases, V-1 missiles, V-2 rockets), 69
    developed by allied powers (atomic, proximity fuse, radar), 69
Mitre Corporation (Massachusetts Institute of Technology), 144*n*.
Models, 98-122
    analogical, 112
    of battle, 99-102, 117-119
    of combat, 99-100
    deterministic, 105-109, 292-293
    development of, 117-119
    to facilitate analysis, 296
    macro- and micro-, 99-100
    master, 112
    mathematical, 64, 67, 101-102, 114
    nature of, 98-119
    nonfunctioning data in, 119
    normally incomplete, 98, 100-105
    probabilistic, 109-111, 293
    purpose of, 98, 100-101

Models (*Cont.*):
    represent functions, 118
    representative of real life, 118-119
    as simplified real life, 98, 100-119, 278, 281
    and submodels, 112-113
    as a substitute for real life, 98
    (*See also* Games; Simulations; War games)
Monte Carlo methods, 64


National Military Command System Support Center, 149*n*.
National policy, 223-225
National power:
    factors, 305
    indicators, 240
    ranking of nations, 240-241
National Security Council (U.S.), 265
National War College (U.S.), 290
Nations reported as active in business games and simulation, 202
Nations reported as active in military gaming and simulation:
    Australia, 9, 143
    Austria, 5
    Belgium, 132
    Canada, 9, 138, 143-145
    China, 58, 307-308
    Denmark, 132
    France, 15, 132
    Germany, 5-9, 23-30, 132, 156, 167
    Hungary (Kormendi), 351
    India, 19, 54
    Israel, 37
    Italy, 5, 132
    Japan, 30-32, 49-51
    Korea, 157-159, 167
    Netherlands, 132
    Norway, 132
    Turkey, 132
    U.S.S.R., 6, 22-25
    United Kingdom, 9, 32-34, 71-79, 132, 138-139, 143-145
    United States of America, 9, 30-32, 34, 39-43, 45-48, 52-53, 132, 145-187, 253-264
Naval Ordnance Laboratory (U.S.), 39-44, 63
    gamers, 41
    games, 40-45
    Navy's Mine Warfare Operations Research Group (NMWORG), 42-43
    Operations Research Group, 39-44
Naval Research Laboratory (U.S.), 39*n*.

Naval War College (U.S.), 38n., 125, 134, 136, 150, 194, 297
  games, 125, 134, 136, 194, 225n., 290
Nelson, Adm. Horatio (Lord Nelson), 106-109
  and Trafalgar, 106-108
North Atlantic Treaty Organization (NATO) games, 131-132
Nuclear weapons:
  gaming of, 45-47, 157, 170
  threat of, 223, 226

Obsolescence, 279, 287
Office of Naval Research (U.S.), 271, 289
Operations research (systems analysis):
  and analytical gaming, 61-81, 284
  case study, mine warfare, 40-44
  complete cycle, 40-44
  and national security, 63
  as a new profession in World War II, 8
  origin of, in analysis of military operations, 8
  and relation to war gaming, 8, 10, 38-44, 61-81
  (See also Management science)
Operations Research Office (ORO) of The Johns Hopkins University:
  activities of, 64-71, 132-136, 144-146, 148, 192-193, 226-227, 244, 266
  gamers, 64, 70, 133, 135-136, 166, 174-186, 271-273
Outcomes or results of games:
  prediction of, 291
  variability of, 222, 291

Peace:
  by deterrence, 306
  by direct approach, 305
  maintenance of, 306, 307
Peace gaming, 2, 305-308
  model construction, 306
  similar to war gaming, 305
Peace Research Institutes (in Canada, England, Scotland, Sweden, and United States), 307
Pearl Harbor, 302, 306
  Japanese attack on, 49-53, 95-96
  Japanese gaming of attack, 30-32, 306
  training for attack on, 94-97
Photogrammetry:
  and computer processing, 297, 298
  and terrain data, 297
Player rooms as operations or war room for one participating team in a game, 12-14, 175, 180, 183-184, 245

Polaris missiles and submarines, 283
Political affairs, 224-244
Political games, 229-234
Political gaming, 2, 223-274, 301
  arms control and disarmament, 2, 305
  crisis games, 306
  international, 301-302
  international trade agreements, 304
  intranational, 301
  labor disputes and strikes, 304
  monitoring and policing truces, 303
  regional development programs, 304
  in treaty negotiations, 303
  (See also Gaming applications; Peace gaming)
Political-military gamers, 249-251, 253n., 266-267n., 272
Political-military games, 242-243, 265, 273
  JCS type, desk games, 265
Political-military relationships, 223
Principles:
  of war, 18n., 37, 58
  of guerrilla war, 58
Probability:
  in games, 111
  in models, 109-111
  random numbers, 110, 111, 121, 300
Profession of arms:
  ineffectiveness, penalty for, 19-20
  vicarious practice, need for, 20-21
Programming:
  for extraction of data, 300-301
  improvements in, 300-301
  modular, 300

Qualitative Materiel Development Objective(s) (QMDO), 292
Qualitative Materiel Requirements (QMR), 292
Quick gaming, 153, 155-157, 160-163, 169, 303

Rameses II of Egypt playing Draughts (circa 1250 B.C.), illustrated, 4
RAND Corporation, 77n., 144, 150, 151n., 192, 193, 194n., 226-229, 243, 252, 366
  gamers, 248, 286
Raytheon Company, 144n., 250, 266, 273
  (See also TEMPER in Index of Games and Simulations, Part 1)
Reliability:
  and "fluke" events, 285-286
  of game results, 285
  and repetition of runs, 286-287

Reliability (*Cont.*):
  suited to simulations, 286
  unsuited to games, 286–287
Remagen (Germany), capture of bridge at, in World War II, 299
Republic of Korea Army, 156–159
Research Analysis Corporation (RAC), 78, 134, 151n., 155, 157, 159, 167–187, 243, 272, 286, 290, 297–298
  gamers, 134–136, 148, 159, 174, 186, 250, 253n., 271–273, 290
Research organizations (*see* Independent research organizations active in gaming)
Review Board:
  as example of, and members of RAC's, 287–288
  as expert judges, evaluators, and as check on validity, 174, 186, 287
Role playing, 191
Royal Armament Research and Development Establishment (RARDE), U.K., 75–79
Russian war gaming, 22–23
Russo-Finnish War (1939), 288

School systems active in gaming:
  Baltimore and Seattle, 230
  Los Angeles, San Diego, and Westchester County, New York, 303
Scientific management (*see* Management science)
Security (secrecy), 283
Short-cut games (*see* Quick gaming)
Simulation techniques related to gaming, 2, 44
Simulations:
  computer, 44, 278
  (*See also* Business games and simulations; Games; Gaming)
Social Science Research Council, 230
Specified and unified commands, 167
Standardization of games, input data, machine parts, and models, 294–295
Stanford Research Institute (SRI), 251–253
  war games, 149
State of the art of gaming, 295
State Department, U.S., 229, 240
Statistical matters, 283–286
Strategic gamers, 249–251, 266, 267n.
Strategic games, 223–274
Strategic hamlets in Vietnam, 253–264, 270
Strategic policy (*see* National policy)
Strategic war games, 226–229

Strategic warfare, 225
Strategy and Tactics Analysis Group of U.S. Army (STAG), 146, 148–150, 167
  war games developed at, 147–149, 300
Supreme Headquarters Allied Expeditionary Force (SHAEF), 34
  General Eisenhower, Supreme Commander, 33
  war gaming at, 33–34
Surveillance, aerial, 302, 306
Synthesis of complex actions, 301
Systems analysis (*see* Operations research)
Systems Development Corporation (originally RAND's Systems Research Laboratory), 348
Systems Research Laboratory, RAND, 227–228

Tannenberg, Battle of:
  gamed by the Germans, 23–25
  gamed by the Russians, 22–25
Technical Operations, Inc. (TOI), 69–71, 148, 151, 369
Technology, advancing:
  and game inputs, 279
  and game models, 279
  impact of, 302
  lag in application of, 304
  new fields of, 301–302
Terrain in military operations, features of, grid system, 115–117
Terrain analysis, elements in, 115–117, 297–298
Terrain model as chart, map, sand table, three-dimension representation, and playing surface for military games, 5, 11–13, 15, 93, 95, 125, 133, 141, 176–181
Theater staff games, 165–172
  (*See also* Theater Staff Games *in Index of Games and Simulations, Part 1*)
"Think tank" organizations, 40, 62, 191
Threat analysis, 234, 271
  Douglas Aircraft Company model of, 234–242, 264
Tokyo Center for Economic Research, 211
Total War Research Institute (Japan), 30
Trafalgar, Battle of, 106–108, 358

U.S.S.R., 68, 288–289
United Kingdom (*see* British Army war gamers; British Army war games; British business games)

## Subject Index 387

United Nations, 306, 308
United Nations Command in Korea, gaming in, 157–159
United States Army:
  Army Map Service, 297
  Army Tactical Mobility Requirements Board (Howze Board), 47
  Combat Developments Command (USACDC or CDC), 139
  Combat Developments Experimentation Center (CDEC), 94–95
  Combat Operations Research Group (CORG) at CONARC, 69–73, 145–146
  Continental Army Command (CONARC), Fort Monroe, Virginia, 69–73, 145–146
  Deputy Chief of Staff for Operations, 169
  (U.S.) Eighth Army (in Korea), 157–159
  1st Cavalry Division (Airmobile), 48–49, 60
  Force Planning Analysis Directorate, 165
  Logistics Center, Fort Lee, Virginia, 193
  Management School, Fort Belvoir, Virginia, 193, 198
  maneuvers of, 289
  9th Armored Division (in World War II), 299$n$.
  (U.S.) Ninth Army (in World War II), 34
  (U.S.) Seventh Army (in Germany), 157
  Strategy and Tactics Analysis Group (STAG), 80, 150$n$.
U.S. Geological Survey, 297
United States Marine Corps, Landing Force Development Center, Quantico, Virginia, 146
  (*See also* Marine Corps Landing Force Game in *Index of Games and Simulations, Part 1*)
United States Military Academy, 20
United States Navy:
  (U.S.) Naval Academy, 57, 360
  Naval Ordnance Laboratory (NOL), 39
    gamers, 41
  Operations Research Group, 39
  Naval War College (NWC), 32, 125, 150
    gamers, 136
  (*See also* Applied Physics Laboratory of The Johns Hopkins University; Navy Electronic Warfare Simulator in *Index of Games and Simulations, Part 1*; Pearl Harbor)
Universal Automatic Compilation System (UNAMACE) for mapping, 297

Universities active in gaming:
  Arizona State University, 200, 324
  Birmingham College of Advanced Technology (U.K.), 199
  British Columbia, University of, 230
  California, University of: at Berkeley, 196$n$.
    at Irvine, 230
    at Los Angeles (UCLA), 200, 211, 215–216
  Carnegie Institute of Technology (CIT), 191, 200, 211–212
  Chicago, University of, 196$n$., 212–213
  Clarkson College of Technology, 200, 324
  Columbia University, 232
  Cornell University, 213, 303
  Dartmouth University, 232
  Duke University, 232
  English Universities (U.K.), 199$n$., 210
  Harvard University, 211, 232
  Indiana University, 211
  Iowa, University of, 191$n$.
  Johns Hopkins University, 193, 230, 303
  Kansas State Teachers College at Emporia, 230
  Kansas University, 191$n$., 211$n$.
  Lancaster University (U.K.), 230
  Leeds University (U.K.), 230, 307
  Maryland, University of, 230
  Massachusetts Institute of Technology, (MIT), 230, 232, 243, 266, 273, 306
  Michigan, University of, 230, 302
  Michigan State University, 211, 233, 302, 303
  North Carolina, University of, 202$n$., 230
  Northwestern University, 230, 249
  Ohio State University, 230, 272
  Oklahoma, University of, 231
  Oregon, University of, 211
  Pennsylvania State University, 200, 324
  Princeton University, 217, 273
  Purdue University, 202
  San Francisco State College, 230
  Smith College, 230
  Southern California, University of (USC), 303
  Stanford University, 230
  Texas, University of, 198$n$.
  Tokyo University, 211, 230
  Toronto, University of, 230
  Wayne State University, 214, 230
  Wisconsin, University of, 230
  Yale University, 232

Urban problems, simulation and gaming of, 304

Validity:
of assumptions, 282–283
of data, 282–283
of findings, 283–285
of game results, 285, 301
Viet Cong, 261
Viet-Minh, 244
Vietnam, 58, 232
and airmobile tactics, 47–49
as game settings, 253–264, 270

War:
alternative to, 1, 305–308
artificial, 38
complexity of, 98, 100, 276
deterrent to, 1
"fog of," 278–279
folly of, 307, 308
genesis of, 307
imitation, 38
learnings in, carryovers to other pursuits, 2, 3
outlaw of, 305
simulated, 38
study of, 1–10
War Control Planners, 306
War games:
capabilities of, 8–11, 15–16
chance events in, 111
characteristics of, 82–83
common features of, 11–15
compensating variables, 111
coordinated program of, 159–165
costs of, 11–12, 160
defined, 8–10
described, 2, 8–15
economy of, 9, 15–16, 21, 91–96
effects of, 22
equipment for, 84–85
examples of, 149–150
extreme results in, 111
facilities for, 14, 84–85
historical roots of (see Historical roots of war games)
hobby type, 139
(See also Hobby war games)
and military successes, 22–54, 306
number of, 150–151
organization for, 142, 174, 186
origin of, 3
related to arms control and disarmament, 2

War games (Cont.):
requirements for, 83–85
rules for, 130
scope of, 11–12
strategic, 226–229
at tactical levels, 132–148, 157–158, 172–178, 251–264
to test military plans, 18, 22–37
at theater levels, 147–151, 159–172, 179–187
for training purposes, 19–22, 128, 130–132
types of, 8, 19, 22, 63, 127–128
umpired, 130
uncertainty of results, 110–111
uses for: general, 9–16, 18
specific: capability requirements of equipment, 15–16
concept development, 15–16
data generation, 9
as learning activities, 15
for planning and plans testing, 9, 15–16, 19, 22
pretesting weapons and weapons systems in simulated warfare, 22
requirements, determination of military, 15–16
for research, 9, 13, 15–16, 19, 22, 37
for testing new concepts, doctrine, 22, 38
for training of military personnel, 9, 13, 15, 19
utility of, 10
validity of, 276
(See also Games; Games of war; Maneuvers; Military gaming)
War gaming:
advantages and disadvantages of, 21–22
analytical, 61–81, 108, 115
for combat readiness, 157–159
extent of, 150–151, 226
as a laboratory for military studies, 10
limitations of, 276–278, 288–293
in the military school system, 21
personnel for, military officer, scientific, technical, support, 83, 154
results of, 42–43
as simulated warfare, 9, 12
techniques: for problem solving, 2
related to simulations, 2
uses of, 2, 18, 302–304
as an underwriter of peace, 305–308
values of, 10–11, 14–15
(See also Gaming; Military gaming)
War room (see Control room; Player rooms)

Warfare:
  artificial, 130
  countless variables in, 276
  incomplete understanding of, 276–277
  new developments in, 44–48, 304–308
  study of, 20–21
Wars:
  costs of, 91–92
  statistics of, 20, 49
Washington Operations Research
    Council (WORC), 273
Waterloo, Battle of, 108, 110

Weapons and weapons systems:
  characteristics, specifications, and
    gaming of, 292
  nuclear, 45–47, 131, 302
Weapons Systems Evaluation Group
    (WSEG) and Institute for Defense
    Analyses (WSEG/IDA), 149n.
Western Behavioral Sciences Institute,
    La Jolla, California, 233n., 303, 361
White House, the, interest in, 229, 265
  Camp David, games at, 265
  staff, 265

# Index of Persons Active in Simulation and Gaming*

Abt, Dr. Clark C. (Raytheon Co. and Abt Associates, Inc.), 233, 250, 266, 273, 274, 327, 328
Adams, Col. C. Y. (JCS and SRI), 250–252, 271
Adams, Dr. H. E. (ORO/RAC), 134, 135, 186, 328, 375
Adams, Gen. Paul D., 93
Alcock, N. (Canadian Peace Research Institute), 232, 233
Alger, C. F. (INS), 231, 233
Andlinger, G. R. (business gaming), 195–196, 199, 322
Andrews, Marshall (ORO/RAC), 13, 328
Archer, Dr. William L. (CORG/TOI), 70, 329

Baker, Lt. Col. I. H. (ORO), 175
Bardeen, Dr. John (NOL), 41
Baring, E. (U.K., 1872), 73
Benn, Edward (U.K.), 71, 73, 74, 79
Bennett, Dr. Ralph (NOL), 41
Benson, Prof. Oliver (INS), 231, 233, 307, 330
Bentz, R. M. (RAC), 179
Beswick, Lt. Col. J. E. (U.K.), 79
Bloomfield, Dr. Lincoln (CENIS, MIT), 233, 249, 270, 273, 306, 307, 330–331

Blumstein, Dr. Alfred (IDA), 247, 251, 271, 331
Boatwright, Lt. Col. J. R. (ORO), 175
Bonawitz, Col. N. C., USAF (Ret.) (ORO/RAC), 184
Bonnett, Gabriel (French) (guerrilla warfare model), 246
Boocock, Sarane S. (JHU), 233, 234, 331
Bradley, Gen. Omar N., 28, 33, 55, 299n., 331
Brody, Dr. R. A. (INS), 231, 233, 249, 332
Brooks, Dr. Franklin (ORO/TOI at CORG), 70
Brossman, M. W. (ORO/RAC), 134, 135, 157n., 166, 332
Bruner, Dr. Joseph A. (ORO/RAC), 71, 333
Busacker, Robert (ORO/RAC), 166
Butchers, Maj. Gen. R. J. (RAC), 183

Calhoon, Brig. Gen. W. R. (JWGA), 273
Carlson, Elliott (educational games), 212n., 323
Chamberlaine, William (1912), 334
Christiansen, Maj. Gen. J. G. (ORO/RAC), 134, 135, 173, 299n.
Churchill, Prime Minister Winston (U.K.), 33, 334

---

* Includes persons whose work is cited in this book as related to simulation and gaming. All officers are of U.S. Army unless otherwise noted.

Clark, Dr. Dorothy K. (ORO/RAC), 176, 334
Clausewitz, Karl von (German), 223, 334–335
Clerk, John, of Scotland (1779):
 analysis of naval battles by, 108
 essay on naval tactics by, 108, 335
 tactics of, tested, 108
Coffin, S. (ORO), 166
Cohen, Kalman J. (CIT), 195$n$., 212$n$., 323
Colby, Archie (ORO at CORG), 70
Cole, Dr. Hugh M. (ORO/RAC), 28$n$., 55, 335
Coleman, Dr. J. S. (JHU), 233, 234, 303, 335
Conant, Dr. James B. (Harvard), 39, 58
Cooper, Rear Adm. J., USN (Ret.) (RAC), 186
Corbett, Col. James B. (ORO/RAC), 13
Cormier, Col. Everett L. (ORO/RAC), 336
Crow, Dr. Wayman J. (INS at WBSI), 233, 249, 274, 361
Cushen, Dr. W. Edward (ORO/RAC), 336–337, 375

Dale, Alfred G. (University of Texas), 198$n$., 323
Dalkey, N. C. (RAND), 248, 337
Dewey, Maj. Gen. L. R. (ORO/RAC), 186, 338
Dill, W. R. (CIT), 191, 323
Doenitz, Grand Adm. Karl (German), 29, 55, 338
Dondero, L. J. (ORO/RAC), 173, 250, 272, 338
Dorsett, Capt. J. D. F., USN (Ret.) (ORO/RAC), 179
Dresch, Francis W. (ORO), 338–339
Dresher, M. (RAND), 248, 339
Duke, Dr. R. D. (MSU), 233, 303, 339
Dunford, Maj. Gen. D. (RAC), 184
Dunn, Paul F. (ORO), 271

Echardt, Walter (ORO/RAC), 134, 135
Edison, Thomas A., 39$n$.
Eisenhower, Gen. Dwight D., 33, 55, 299$n$.
Ellis, Dr. J. W. J. (RAND), 234, 270, 340
Elmore, Brig. Gen. J. A. (RAC), 185, 250

Fairhead, J. N. (U.K.), 199–201, 201$n$., 210, 220, 323

Fall, Dr. B. B. (French), 244–246, 270, 271, 340
Farrell, Col. Norman (Institute of Special Studies, U.S. Army), 151
Featherstone, Donald F. (U.K.), 139$n$., 340–341
Field, Lt. Col. G. W. H. (British Army), 73, 79
Figuers, Col. H. H., USMC (Ret.) (ORO/RAC), 184, 250, 253$n$.
Flannagan, John C. (TOI/CORG), 70
Flournoy, Col. W. N., USMC (Ret.) (RAC), 174, 177, 336
Fogelsanger, D. K. (TOI/CORG), 151, 341
Fountain, Lt. Col. R. (RAC), 174
Fuller, Gen. J. F. C. (British), 25$n$., 342

Gadsby, G. Neville (U.K.), 71, 79
Gamow, Dr. George A. (ORO), 64–68, 342
Gantt, Henry L., 189, 210, 323
Giffin, Brig. Gen. S. F., USAF (Ret.) (IDA), 264, 265$n$., 272–273, 343
Gilbreth, Frank, 189, 327
Gilbreth, Lillian, 189, 327
Girard, E. W. (ORO/RAC), 133–135, 173–174, 343
Gleditsch, N. P. (INS), 249
Goldhamer, Dr. H. (RAND), 229, 230, 233, 274, 343
Goodley, Col. J. T. (ORO/RAC), 186
Gorden, Morton (Raytheon Co. and Abt Associates, Inc.), 273, 344
Gowen, P. (ORO), 166
Graham, Robert G. (AMA), 198$n$., 202$n$., 209$n$., 211$n$., 214$n$., 217$n$., 324
Grasse, Adm. F. J. P. de (French), 108
Gray, Clifford F. (AMA), 198$n$., 202$n$., 209$n$., 211$n$., 214$n$., 217$n$., 324
Greene, Jay R. (General Electric Co.), 196, 324
Greene, T. E. (RAND), 234, 270, 340
Greenlaw, Paul S. (business games), 198, 202$n$., 211$n$., 213–217$n$., 324
Guetzkow, Dr. Harold (INS), 229–231, 233, 249, 274, 307, 344–345

Hafner, Ralph (ORO/RAC), 166
Hamburger, William (RAND), 193, 324, 345
Handy, Gen. Thomas T. (ORO/RAC), 174, 186, 288
Hantzes, Dr. Harry N. (ORO/RAC), 133, 345
Harper, Col. J. E. (RAC), 184

Harrison, Dr. J. O., Jr. (ORO), 68, 134, 135, 345
Harsanyi, J. C., 306n., 307, 345
Hart, Dr. Joseph T. (ORO/RAC), 269n., 273, 346
Heistand, Maj. H. O. (1898), 5n., 6n., 8n., 17, 346
Helmer, Olaf (RAND), 248, 346–347
Herder, Dr. John D. (Southern New England Telephone Co.), 209–210, 324
Hermann, Charles F. (INS), 274, 347
Hermann, Margaret G. (INS), 274, 347
Herron, Lowell W. (Dean of Clarkson College of Technology), 200, 324
Hill, Brig. Gen. J. G. (ORO/RAC), 134, 135, 179, 186, 187, 271, 347
Himes, Billy H., Sr. (ORO/RAC), 134, 135, 179, 253n., 347
Ho Chi Minh (Vietnamese), 244, 245
Hoffman, Gen. Max von (German), 25n., 348
Hoffman, S. (ORO/RAC), 166
Hofmann, Gen. Rudolph (German), 27, 55, 348
Hoge, Brig. Gen. W. M., 299n.
Hoisington, Dr. Lee H. (NOL), 41
Howze, Maj. Gen. Hamilton (the "Howze Board"), 47
Hulse, Brig. Gen. Allen (ORO/RAC/TOI-CORG), 70

Inglis, Lt. Col. E. (ORO/RAC), 174
Ivanoff, Dimitri N. (Douglas Aircraft Co.), 233–242, 349

Jackson, James R. (UCLA), 196n., 211n., 324
Jackson, Col. W. (ORO/RAC), 186
Johnson, Dr. Ellis A. (ORO), 39–45, 59, 63–64, 146n., 349–350
Johnson, J. W. (ORO/RAC), 134, 135, 179

Kadel, Lt. Col. R. B. (ORO), 13
Kashiwai, Col. S. (Japanese), Secretary of the General Staff, Ground Staff, Japanese Defense Agency, 43n.
Keating, Brig. Gen. J. W. (RAC), 186
Ker, Maj. Gen. Howard (ORO/RAC), 186, 187
Kerlin, E. D. (ORO/RAC), 185
Kibbee, J. M. (AMA), 195n., 197, 198, 202n., 209n., 211n., 214n., 215n., 325, 326, 350–351

Kinnard, Maj. Gen. Harry W. O., 48, 59–60, 351
Klasson, Charles R. (University of Texas), 198n., 323
Komlosy, Graham F. (U.K.), 71, 79, 151
Kurtz, Harriet, 306
Kurtz, Howard, 306

Lanchester, F. W. (U.K.), 105–109, 120–122, 247, 352, 372–373
equations, 105–109, 120–122, 148
linear law, 120
$N^2$ law, 106–109
Laulicht, Dr. J. A. (INS), 232–233, 270, 307, 352
Law, C. E. (Canada), 145
Lee, E. M. (ORO/RAC), 186, 345, 352
Leonard, Maj. Gen. J. W., 299n.
Lind, J. R. (RAND), 248
Ling, Dr. R. C. (ORO/RAC), 174, 177
Lipscomb, Maj. Gen. T. H., 139
Little, Rear Adm. M. N., USN (Ret.), (ORO/RAC), 179
Love, J. D. (ORO), 271
Lynch, C. J. (RAC), 177

McDonald, Col. T. J. (JWGA), 254, 273
McHugh, Francis J. (NWC), 38n., 99, 136, 290n., 354
McKenney, James L. (Harvard), 211, 325
McNamara, Robert, S., Secretary of Defense, 226, 354
McRae, John (INS), 274, 354
Magruder, Gen. Carter B., 157, 174, 186, 288
Manstein, Field Marshall Eric von (German), 55, 355
Manteuffel, Gen. Hasso von (German), 29
Mao Tse-tung, 155n., 245, 251, 355
Martin, E. W., Jr. (business games), 325
Mead, Maj. Gen. A. D. (ORO/RAC), 186, 187
Meals, Dr. Donald W. (ORO, CORG, TOI), 70, 71, 355
Michels, Dr. Walter C. (NOL), 41
Minor, Col. W. T., USAF (JWGA), 265n., 266n., 273, 356
Model, Field Marshall Walter (German), 27
Montgomery, Lt. Gen. Bernard L. (later Field Marshall Viscount of Alamein) (British), 25n., 32–34, 356
Montgomery, Lt. Col. R. A. (ORO/RAC), 13

## Index of Persons Active in Simulation and Gaming 393

Moore, Gen. J. E. (ORO/RAC), 174, 186
Morison, Dr. Samuel Eliot, Rear Adm. (USNR), 30, 56–58, 95, 357
Müffling, Lt. Gen. von, Chief of the German General Staff (1824), 6, 17
Murphy, Lt. Col. C. B. (ORO/RAC), 174, 177, 182
Murphy, E. J. (NWC), 136

Narden, T. (INS), 249
Narten, P. F. (ORO/RAC), 174, 177, 250
Neeley, Brig. Gen. R. B. (ORO/RAC), 174
Nelson, Adm. Horatio (later Lord Nelson), (British), 106, 108, 358, 372
Nelson, Maj. Gen. R. T. (ORO/RAC), 186
Ngo Dinh Diem (President of Republic of Vietnam), 247n.
Ngo Dinh Nhu (Republic of Vietnam), 247, 251
 guerrilla warfare model of, 247–251
Nimitz, Adm. Chester W., USN, 32

Opp, Col. R. D. (RAC), 174, 178

Padelford, Prof. N. J. (MIT), 249, 270, 358
Page, Dr. Thornton L. (NOL, ORO), 41, 100, 105, 120, 358–359
Palmer, Gen. C. D. (ORO/RAC), 174, 186, 288
Pappas, Dino (ORO/RAC), 186, 250, 253n., 271, 359
Parker, Col. D. (RAC), 250
Parrott, Dr. Lyman G. (NOL), 41
Parsons, Lt. Col. Norman (CORG, ORO/RAC), 70, 174, 177, 250
Patchett, R. N. (RAC), 174, 177
Patton, Gen. George S., Jr., 28
Paxson, E. W. (RAND), 150, 248, 305, 306, 359
Peaslee, Lt. Col. Jessee C., USAF (NATO), 132
Pedersen, Lt. Col. E. K., USMC (Ret.) (RAC, CORG), 174
Philipps, Brig. Gen. P. D. (RAC), 159n.
Pierce, W. L. (ORO/RAC), 134, 135, 174, 176
Pineau, Roger, Lt., USNR, 30–32, 57
Porterfield, Wanda (ORO), 166
Potter, Prof. John Deane (USNA), 57, 360

Quade, Dr. E. S. (RAND), 63, 360

Rae, E. W. (Canada), 145
Raring, Capt. G. L., USN (Ret.) (RAC), 186
Raser, Dr. J. R. (INS at WBSI), 233, 249, 274, 361
Rawden, Richard H. (AMA and Clarkson College of Technology), 194n., 198, 324, 326
Reese, Dr. Howard (RAC), 246
Reinhardt, Col. G. C. (RAND), 248, 361
Reisswitz, Baron von (German) (originator of modern war game, KRIEGSSPIEL, 1811), 5
Reisswitz, Lt. George Heinrich Rudolph Johann von (German) (developer of modern war game, KRIEGSSPIEL, circa 1816), 5–8, 361
Rennenkampf, Gen. Paul (Russian), 23
Rhenman, Eric (CIT and Stockholm School of Economics, Sweden), 195n., 212n.
Ricciardi, F. M. (AMA), 194n., 195n., 204n., 326
Rich, R. P. (APL/JHU), 192n., 326
Richardson, Lewis F. (U.K.), 122, 306n., 307, 361–362
Robinson, Dr. J. A. (INS), 230, 233, 249, 270, 272, 274, 362
Rodney, Adm. (Sir George) (British), 108
Rommel, Gen. Erwin W. (German), 32–34
Ross, A. W. (U.K.), 71, 79
Rumbaugh, Dr. Lynn H. (NOL, ORO), 41
Rundstedt, Field Marshall Gerd von (German), 28, 55
Ryan, R. B. (ORO), 134, 135

Salligar, Dr. F. M. (RAND), 229, 234
Samsonov, Gen. Alexander, (Russian), 23
Sayre, Farrand (1908), 5n., 128, 363
Schellenberger, Robert E. (University of North Carolina), 202n., 326
Schelling, T. C. (Harvard), 249, 364
Schlieffen, Gen. Alfred (German), 23
Schwartz, David C., 224n., 269n., 274, 364
Scott, Dr. Andrew (University of North Carolina), 231, 270, 274, 307, 364
Seely, J. S. (RAC), 300n.
Seibert, Col. L. R., USMC (Ret.), (RAC), 185
Shafroth, Vice Adm. John F., USN, 49
Shepherd, Ronald W. (U.K.), 71, 73, 75–77, 79, 364–365
Shortley, Dr. George (NOL, ORO, BAARINC), 41, 365
Shubik, Marten (General Electric Co.), 222n., 326, 365

Shure, G. (INS), 249, 365
Sibley, Brig. Gen. C. C. (ORO/RAC), 181
Simcox, Col. L. S. (ORO/RAC), 133, 136, 173–175, 177, 250, 272, 365
Sisson, Roger L. (business games), 200, 324
Smeak, C. O. (ORO/RAC), 174, 176, 177
Smoker, Paul (U.K.) (INS) (at Peace Research Centre, Lancaster, England), 232–233, 249, 274, 352
Snyder, Dr. Richard C. (INS), 231, 233, 366
Speier, Dr. Hans (RAND), 229, 230, 233, 274
Srivastava, Dr. S. S. (India), 19, 54
Stafford, Maj. J. (Canada), 145
Stephens, Brig. Gen. J. D. (RAC), 186
Stewart, Lois (AMA), 195n., 198
Sundt, Col. H. S. (ORO/RAC), 174
Sutherland, R. J. (Canada), 145
Sutherland, W. H. (ORO/RAC), 179, 271, 368

Taylor, Frederich W. (1911), 189, 327
Taylor, Irving W. (1916), 368–369
Tholen, A. D. (ORO/RAC), 136, 159n.
Thorelli, Hans B. (INTOP), 196n., 212n., 213n., 327
Timberman, Maj. Gen. T. S. (ORO/RAC), 181
Totten, Charles A. (1880), 7, 369
Towne, Henry R. (1886), 189
Treadwell, E. (U.K.), 71, 79
Troxel, Maj. Gen. O. C. (RAC), 185, 186
Tuchman, Barbara W., 23, 25, 370

Van Arsdall, Rear Adm. Clyde, USN (JWGA), 272, 273
Vance, Stanley C. (management simulations), 191n., 327
VanderHeide, Maj. Gen. H. J. (RAC), 184, 185
Vinci, Leonardo da (Italy) (military inventions), 4–5

Visco, E. P. (ORO/RAC), 136, 246, 247
Voigt, Ruth F. (ORO/RAC), 183

Wehle, Maj. Gen. P. C. (RAC), 183, 184
Weikhamann, Christopher (the King's game, 1664), 5
Weiner, M. G. (RAND), 248, 372
Weldon, Foster (NOL, ORO), 41
Wells, H. G. (U.K.), 139, 372
Wenger, L. H. (RAND), 248
Whaley, Dr. Barton (MIT), 233, 270, 273
Wheeler, Gen. E. G., USA, Chairman of the Joint Chiefs of Staff, 226
Whipple, Brig. Gen. S. (RAC), 177
Whitney, Eli (1798), 294
Whitson, Dr. William L. (ORO, CORG, President of Clarkson College of Technology), 70, 372–373
Williams, D. (U.K.), 79
Williams, Edward R. (TOI/CORG), 70, 151
Williams, Lt. Col. L. W. (RAC), 175
Williams, R. G. (ORO/RAC), 136, 298n., 300n.
Wohlstetter, Roberta, 30, 56, 57, 374
Worthington, Brig. Gen. J. M. (RAC), 183
Wright, Quincy, 122, 306, 307, 374

Yamamoto, Adm. Isoroku (Japanese), 56, 306
Young, Dr. John P. (ORO), 5n., 26n.–28n., 56, 374

Zacharias, Capt. Ellis M., USN, 56, 94–97, 374
Zimmerman, Richard A. (ORO/RAC), 64, 66n., 68, 118, 167n., 187, 271, 342, 374–375
Zumwalt, Adm. E. R., USN, 167n.
Zwicker, Maj. Gen. R. W. (ORO/RAC), 184, 185

# Index of Games and Simulations

**Part 1. Military Simulations and War Games**

Aldershot training games (1872 in U.K.), 73
AORE war games (U.K.), 74–76, 78, 143, 151
Atomic Air-Ground War Game (NATO), 131
AUTOTAG war game (CORG), 71, 72, 73$n$.
CAORE war game (Canada), 78, 143, 145
CARMONETTE war game or simulation (ORO/RAC), 77$n$., 78, 86, 117, 153, 375
  CARMONETTE III, 111, 137–138, 148
Coast Artillery War Game (New York State Militia, 1912), 334
Comprehensive Blast and Radiation Assessment System (COBRA), a simulation by CEIR for NMCSSC, 149
COSMAGON war game (ORO) (1955), 334, 338
Counter-Insurgency War Games (ORO/RAC and SRI):
  SRI, 251–253
  TACSPIEL, 250, 261–264, 336
  THEATERSPIEL, 253–261
FAME (Future Army Missions and Equipment), military game with political inputs (ORO), 243
FOE, a war game model representing company actions (ORO), 334
FORT ARMS, FORT IRWIN, FORT ROOT, and FORT SIMULATION, military management simulations developed at the U.S. Army Management School, 193–194, 198

Forward Air Strike Evaluation (FAST-VAL), a simulation by RAND for USAF, 149
Helwig's Game, or War Chess (circa 1780 in Germany), 5
INDIGO war game (ORO), 77$n$., 132–134, 136, 145, 148
KRIEGSSPIEL war game (Germany), 5–8, 67, 73, 85–86, 123, 126, 143, 148, 337, 371, 375
  first true war game (1811–1824), 5–8, 337
  spread throughout Europe, 6–8
LEGION war game (STAG), 300
LOGEX, a military logistics CPX, developed at U.S. Army Logistics Center, Fort Lee, Virginia, 93–94, 148, 193$n$.
LOGSIM, a military logistics simulation (ORO and U.S. Army), 165, 166
  (See also LOGSIM in Part 2)
Marine Corps Landing Force Game (USMC), 145–147
MAXIMUM COMPLEXITY COMPUTER BATTLE war game (ORO), 67–68, 375
MINIGAME, a tactical level war game for study of small unit operations, developed by CORG and Special Studies Group of the U.S. Combat Developments Command, Fort Belvoir, Virginia, 138–142, 148, 151, 341
Mixed Air Battle Simulation (MABS), by SRI for USAF, 149
NATO war game, 131

Navy Electronic Warfare Simulator (NEWS), at Naval War College, 86, 149, 150, 290, 297
(*See also* McHugh, Francis J., *in Index of Persons Active in Simulation and Gaming*)
Nuclear Assessment Routine [NAR(M-2)], by STAG for U.S. Army, 149
Nuclear Cost Assessment Technique (NUCAT), by CORG for USACDC, 149
QUICK GAME, a short-cut war game (RAC), 153–155, 160–163, 169
computerized, 155–157
RARDE war games (U.K.), 75–78
SCHNELLSPIEL, a theater staff game, by ORO/RAC for U.S. Army, 157, 167
Sea Warfare Integrated Model (SWIM), by APL/JHU for U.S. Navy, 149
SIERRA war game (RAND), 77n.
SIGMALOG, a military logistics simulation, by ORO/RAC for U.S. Army, 160, 162–165
STRATSPIEL, an early strategic simulation model (ORO), 345
Super-Quick Game, a short-cut war game by RAC for U.S. Army, 155–157
SYNTAC war game, by ORO/TOI at CORG for U.S. Army, 70, 72, 73n., 86, 145–146
TABWAG war game, by ORO/TOI at CORG for U.S. Army, 71
TACSPIEL Counter-guerrilla Model war game (RAC), 250, 336
TACSPIEL war game, by ORO/RAC for U.S. Army, 86, 111, 113, 134, 145, 148, 151, 169, 172–178, 263, 287, 298, 365
Tactical Penetration Model (PENTAC), by Boeing Airplane Company for USAF, 149
TARTARUS war game, by STAG for U.S. Army, 147–148
TEMPER, a political-military-economics game and simulation complex, by Raytheon Co., and for JWGA, 86, 250, 266–269, 273–274, 300
complexity of, 266–268
constraints on, 268–269, 274
Theater Battle Model war games:
TBM-63, by STAG for U.S. Army, 147, 167
TBM-68, by RAC for U.S. Army, 168–172
Theater Staff Games:
at Seventh Army, 157, 167
at Eighth Army, 157–159
(*See also* Theater Battle Model war games)
Theater war games, 86, 111, 113, 139n., 147–149, 157–159, 167–172, 179–187
THEATERSPIEL war game, by ORO/RAC for U.S. Army, 86, 111, 113, 139n., 148, 168, 179–187, 286, 287, 298, 300, 347
Cold War Model, 243n., 253–261, 347
SEA I, 243n.
TIN SOLDIER war game, at ORO, 64–67, 117
VALOR war game, by STAG for U.S. Army, 86, 148
War-at-sea, by WSEG/IDA, for U.S. Dept. of Defense, 149
ZIGSPIEL, a simulation of air defense of the U.S. by ORO/RAC for U.S. Army, 77n., 149, 192, 227, 229, 334, 338

**Part 2. Business Games and Management Simulations**

Airline Sales Game (Trans-Canada Air Lines), 215, 217
Automobile Dealer Simulation (Wayne State University), 214
BUSINESS MANAGEMENT GAME (also called ANDLINGER or McKINSEY & CO. GAME), 198, 210
Business Strategy Simulation (General Electric Co.), 211
Carnegie Institute of Technology (CIT) Management Game (also called the Carnegie Tech Management Game), 200, 211–212
Clarkson College of Technology Executive Action Simulation, 211
Cornell Land Use Game, for urban planning, 303
Cornell University Management Decision Game—Small Business, 213
Dayton Tire Simulation, a small-business game, by Dayco Corp., 213
DISPATCH-O, a production scheduling game using Gantt charts, by General Electric Co., 215–216
Dynamic Management Decision games, a group of seven games representing all levels of management, developed by General Electric Company and Arizona State University, 200, 324
Esso Service Station Operation Game

(Humble Oil and Refining Co.), 214
Executive Action Simulation (Clarkson College of Technology), a management type business game, 200
(*See also* Herron, Lowell W., *in Index of Persons Active in Simulation and Gaming*)
Executive Management Exercise, a management decision game (U.K.), 210
(*See also* Fairhead, J. N., *in Index of Persons Active in Simulation and Gaming*)
FORT ARMS, FORT IRWIN, FORT ROOT, and FORT SIMULATION, parallel military/business/management simulations developed at U.S. Army Management School, 193–194, 198
General Management Simulation (Tokyo Center for Economic Research), 211
Harvard Business School Game (HARBUS), 211
In Basket Exercise:
adaptable to desk jobs, 215
one-person business game, 193, 214–215
Indiana University Executive Decision game, 211
INTOP, the International Operations Simulation, developed at the University of Chicago, 196n., 212–213
INVENTROL (General Electric Co.), 215, 216
LOGSIM, as a business game, 193, 194, 198
Maintenance Management Game (Allied Chemical Corp.), 216
Management Decision Exercise (Pillsbury Co.), 211
Management Decision Simulation (Indiana University), 211

Management Decision Simulation (University of Oregon), 211
Management Decision-making Laboratory, a management game (IBM), 210
Managerial Game for Insurance Companies (Allstate Insurance Co.), 202n.
Manufacturing Management Simulation (AMA), 215
Materials Management Simulation (AMA), 215
Michigan State University Business Policy Game, 211
MIT Marketing Game, 215–217
MONOPOLOGS, an inventory management game (RAND):
as a business game, 192–194
a one-player game, 192
Multipurpose Retail Management Game (University of North Carolina), 202n.
Operation Federal Reserve Game (Boeing Airplane Co.), 202n., 215, 217
Physical Distribution Simulation (AMA), 215
SIGMALOG as a business game or management simulation, 193, 194
SOBIG, a functional security business game, developed at Princeton University, 215, 217
Supermarket Battle Maneuvers Game (Burroughs Corp.), 214
Supermarket Decision Simulation (Kroger Co.), 214, 220
TOP BRASS business game (Minneapolis-Honeywell Regulator Company), 198
Top Management Decision Simulation (AMA), 195, 204–207, 210
UCLA Executive Decision Game, 200, 211
UCLA Inventory Game, 215–216
University of Washington Top Management Decision Game, 211

## Part 3. Political and Multi-purpose Games and Simulations *

DETEX-EXDET political-military games, at MIT, 273
Inter-Nation Simulation (INS), 86, 230–234, 243, 270, 307
JCS Political-Military Game(s), 265
Life Career Game, an educational game, 303

METRO, an urban planning game, 302
Metropolis, an urban planning game, 303
(*See also* METRO *above*)
POLEX I and POLEX II, Political Game (MIT), 86, 232, 243, 266
gamers, 249

---

* See also Part 1. Military Simulations and War Games (military games with political inputs): Counter-Insurgency War Games; FAME; TACSPIEL Counter-guerrilla Model war game; THEATERSPIEL war game, Cold War Model. And, in Subject Index: Crisis games; Desk games; Games, free; Games, rigid type.

POMEX, a Political-Military experimental game (RAC), 243–247

STRAT-X and STRAT-XI, a regional-crisis, strategic-political game at RAC, 243, 271, 359

Sumerian game, for teaching economics in schools, by JHU, 303

TEMPER simulation (Raytheon Co.), 266–269, 273–274

Vietnam War Game, an educational game for use in schools (used in Los Angeles high schools with cooperation of University of Southern California), 303

WINSAFE II, an Inter-Nation Simulation (WBSI), 361

355.48
H376

88447

**DATE DUE**

| AP 30 '85 | MAY 1 '86 | | |